SOLUTIONS TO RED EXERCISES

SELECTED SOLUTIONS
ROXY WILSON

SEVENTH EDITION

CHEMISTRY
THE CENTRAL SCIENCE

BROWN LEMAY BURSTEN

PRENTICE HALL Upper Saddle River NJ 07458

D1160947

Production Editor: *Kimberly Dellas*
Production Coordinator: *Ben Smith*
Acquisitions Editor: *Ben Roberts*
Associate Editor: *Mary Hornby*
Special Projects Manager: *Barbara A. Murray*
Cover Designer: *PM Workshop Inc.*
Supplement Cover Manager: *Paul Gourhan*

© 1997 by **PRENTICE-HALL, INC.**
Simon & Schuster/A Viacom Company
Upper Saddle River, NJ 07458

All rights reserved. No part of this book may be repro-
duced, in any form or by any means,
without permission in writing from the publisher.

Printed in the United States of America

10 9 8 7 6 5 4 3 2

ISBN 0-13-578329-1

Prentice-Hall International (UK) Limited, *London*
Prentice-Hall of Australia Pty. Limited, *Sydney*
Prentice-Hall Canada, Inc., *Toronto*
Prentice-Hall Hispanoamericana, S.A., *Mexico*
Prentice-Hall of India Private Limited, *New Delhi*
Prentice-Hall of Japan, Inc., *Tokyo*
Simon & Schuster Asia Pte. Ltd., *Singapore*
Editora Prentice-Hall do Brasil, Ltda., *Rio de Janeiro*

Contents

Introduction

Chemistry: The Central Science, 7th edition, contains nearly 2200 end-of-chapter exercises. Considerable attention has been given to these exercises because one of the best ways for students to master chemistry is by solving problems. Grouping the exercises according to subject matter is intended to aid the student in selecting and recognizing particular types of problems. Within each subject matter group, similar problems are arranged in pairs. This provides the student with an opportunity to reinforce a particular kind of problem. There are also a substantial number of general exercises in each chapter to supplement those grouped by topic. Integrative exercises, which require students to integrate concepts from several chapters, are a new feature of the 7th edition. Answers to the odd numbered topical exercises plus selected general and integrative exercises, about 900 in all, are provided in the text. These appendix answers help to make the text a useful self-contained vehicle for learning.

This manual, **Solutions to Exercises in Chemistry: The Central Science, 7th edition**, was written to enhance the end-of-chapter exercises by providing documented solutions for those problems answered in the appendix of the text. The manual assists the instructor by saving time spent generating solutions for assigned problem sets and aids the student by offering a convenient independent source to check their understanding of the material. Most solutions have been worked in the same detail as the in-chapter sample exercises to help guide students in their studies.

When using this manual, keep in mind that the numerical result of any calculation is influenced by the precision of the numbers used in the calculation. In this manual, for example, atomic masses and physical constants are typically expressed to four significant figures, or at least as precisely as the data given in the problem. If students use slightly different values to solve problems, their answers will differ slightly from those listed in the appendix of the text or this manual. This is a normal and a common occurrence when comparing results from different calculations or experiments.

Rounding methods are another source of differences between calculated values. In this manual, when a solution is given in steps, intermediate results will be rounded to the correct number of significant figures; however, unrounded numbers will be used in subsequent calculations. By following this scheme, calculators need not be cleared to re-enter rounded intermediate results in the middle of a calculation sequence. The final answer will appear with the correct number of significant figures. This may result in a small discrepancy in the last significant digit between student-calculated answers and those given in this manual. Variations due to rounding can occur in any analysis of numerical data.

The first step in checking your solution and resolving differences between your answer and the listed value is to look for similarities and differences in problem-solving methods. Ultimately, resolving the small numerical differences described above is less important than understanding the general method for solving a problem. The goal of this manual is to provide a reference for sound and consistent problem-solving methods in addition to accurate answers to text exercises.

Extraordinary efforts have been made to keep this manual as error-free as possible. All exercises were worked by at least two chemists and proof-read by two others to ensure clarity in methods and accuracy in mathematics. The work of Julie Grundman, Leslie Kinsland, Jennifer Ridlen and Scott Wilson has been invaluable to this project. However, in a written work as technically challenging as this manual, typos and errors inevitably creep in. Please help us find and eliminate them. We hope that both instructors and students will find this manual accurate, helpful and instructive.

Roxy B. Wilson
University of Illinois
School of Chemical Sciences
601 S. Mathews Ave.
Urbana, IL 61801

1 Introduction: Matter and Measurement

Classification and Properties of Matter

1.1   (a) gas   (b) solid   (c) liquid   (d) gas

1.3   (a) heterogeneous mixture   (b) homogeneous mixture   (If there are undissolved particles, such as sand or decaying plants, the mixture is heterogeneous.)   (c) pure substance   (d) homogeneous mixture

1.5   Pure water is a pure substance, while a solution of salt in water is a mixture. We should be able to separate the components of the mixture by a physical process such as evaporation. Take a small quantity of the liquid and allow it to evaporate. If the liquid is salt water, there will be a solid white residue (salt). If the liquid is water, there will be no residue.

1.7   (a) Al   (b) Na   (c) Fe   (d) K   (e) P   (f) Br   (g) N   (h) Hg

1.9   (a) hydrogen   (b) magnesium   (c) lead   (d) silicon   (e) fluorine   (f) tin   (g) copper   (h) calcium

1.11 $A(s) \xrightarrow{\text{heat}} B(s) + C(g)$

When carbon(s) is burned in excess oxygen the two elements combine to form a gaseous compound, carbon dioxide. Clearly substance C is this compound.

Since C is produced when A is heated in the absence of oxygen (from air), both the carbon and oxygen in C must have been present in A originally. A is, therefore, a compound composed of two or more elements chemically combined. Without more information on the chemical or physical properties of B, we cannot determine absolutely whether it is an element or a compound. However, few if any elements exist as white solids, so B is probably also a compound.

1.13   Physical properties: silvery white (color); lustrous; melting point = 649°C; boiling point = 1105°C; density at 20°C = 1.738 g/mL; pounded into sheets (malleable); drawn into wires (ductile); good conductor. Chemical properties: burns in air to give intense white light; reacts with Cl_2 to produce brittle white solid.

1.15   (a) chemical   (b) physical   (c) physical   (d) chemical   (e) chemical

Units and Measurement

1.17   (a) 1×10^{-1} (b) 1×10^{-2} (c) 1×10^{-15} (d) 1×10^{-6} (e) 1×10^{6} (f) 1×10^{3}
       (g) 1×10^{-9} (h) 1×10^{-3} (i) 1×10^{-12}

1.19   (a) $454 \text{ mg} \times \dfrac{1 \times 10^{-3} \text{ g}}{1 \text{ mg}} = 0.454 \text{ g}$

       (b) $5.0 \times 10^{-9} \text{ m} \times \dfrac{1 \text{ pm}}{1 \times 10^{-12} \text{ m}} = 5.0 \times 10^{3} \text{ pm}$

       (c) $3.5 \times 10^{-2} \text{ mm} \times \dfrac{1 \times 10^{-3} \text{ m}}{1 \text{ mm}} \times \dfrac{1 \text{ } \mu m}{1 \times 10^{-6} \text{ m}} = 35 \text{ } \mu m$

1.21   (a) time   (b) density   (c) length   (d) area   (e) temperature
       (f) volume   (g) temperature

1.23   (a) density $= \dfrac{\text{mass}}{\text{volume}} = \dfrac{39.75 \text{ g}}{25.0 \text{ mL}} \times \dfrac{1 \text{ mL}}{1 \text{ cm}^3} = 1.59 \text{ g/cm}^3$

The units cm^3 and mL will be used interchangeably in this manual.

       (b) $75.0 \text{ cm}^3 \times 23.4 \dfrac{\text{g}}{\text{cm}^3} = 1.76 \times 10^{3} \text{ g}$ (1.76 kg)

       (c) $275 \text{ g} \times \dfrac{1 \text{ cm}^3}{1.74 \text{ g}} = 158 \text{ cm}^3$ (158 mL)

1.25   (a)   density $= \dfrac{38.5 \text{ g}}{45 \text{ mL}} = 0.86 \text{ g/mL}$

           The substance is probably toluene, density $= 0.866$ g/mL.

       (b) $45.0 \text{ g} \times \dfrac{1 \text{ mL}}{1.114 \text{ g}} = 40.4 \text{ mL ethylene glycol}$

       (c) $(5.00)^3 \text{ cm}^3 \times \dfrac{8.90 \text{ gm}}{1 \text{ cm}^3} = 1.11 \times 10^{3} \text{ g}$ (1.11 kg) nickel

1.27   Calculate the volume of the aluminum foil:

$$5.175 \text{ g} \times \dfrac{1 \text{ cm}^3}{2.70 \text{ g}} = 1.92 \text{ cm}^3$$

Divide volume by area to get thickness

$$1.92 \text{ cm}^3 \times \dfrac{1}{12.0 \text{ in}} \times \dfrac{1}{15.5 \text{ in}} \times \dfrac{1 \text{ in}^2}{(2.54)^2 \text{ cm}^2} \times \dfrac{10 \text{ mm}}{1 \text{ cm}} = 1.60 \times 10^{-2} \text{ mm}$$

1.29 (a) $°C = 5/9 (°F - 32°)$; $5/9 (62 - 32) = 17°C$

(b) $°F = 9/5 (°C) + 32°$; $9/5 (-16.7) + 32 = 1.9°F$

(c) $K = °C + 273.15$; $-33°C + 273.15 = 240 K$

(d) $315 K - 273 = 42°C$; $9/5 (42°C) + 32 = 108°F$

(e) $°C = 5/9 (°F - 32°)$; $5/9 (2500 - 32) = 1371°C$; $1371°C + 273 = 1644 K$

(assuming 2500 °C has 4 sig figs)

Uncertainty In Measurement

1.31 Exact: (c), (d), (e), (f)

1.33 (a) 4   (b) 3   (c) 4   (d) 3   (e) 5

1.35 (a) 3.002×10^2 (b) 4.565×10^5 (c) 6.543×10^{-3}
(d) 9.578×10^{-4} (e) 5.078×10^4 (f) -3.500×10^{-2}

1.37 (a) 77.04   (b) -51   (c) 9.995×10^4 (d) 3.13×10^4

Dimensional Analysis

1.39 Arrange conversion factors so that the starting units cancel and the new units remain in the appropriate place, either numerator or denominator.

1.41 (a) $0.076 \text{ L} \times \dfrac{1000 \text{ mL}}{1 \text{ L}} = 76 \text{ mL}$

(b) $5.0 \times 10^{-8} \text{ m} \times \dfrac{1 \text{ nm}}{1 \times 10^{-9} \text{ m}} = 50. \text{ nm}$

(c) $6.88 \times 10^5 \text{ ns} \times \dfrac{1 \times 10^{-9} \text{ s}}{1 \text{ ns}} = 6.88 \times 10^{-4} \text{ s}$

(d) $1.55 \dfrac{\text{kg}}{\text{m}^3} \times \dfrac{1000 \text{ g}}{1 \text{ kg}} \times \dfrac{1 \text{ m}^3}{(10)^3 \text{ dm}^3} \times \dfrac{1 \text{ dm}^3}{1 \text{ L}} = 1.55 \text{ g/L}$

1.43 (a) $8.60 \text{ mi} \times \dfrac{1.609 \text{ km}}{1 \text{ mi}} \times \dfrac{1000 \text{ m}}{1 \text{ km}} = 1.38 \times 10^4 \text{ m}$

(b) $3.00 \text{ days} \times \dfrac{24 \text{ hr}}{1 \text{ day}} \times \dfrac{60 \text{ min}}{1 \text{ hr}} \times \dfrac{60 \text{ s}}{1 \text{ min}} = 2.59 \times 10^5 \text{ s}$

(c) $\dfrac{\$1.55}{\text{gal}} \times \dfrac{1 \text{ gal}}{4 \text{ qt}} \times \dfrac{1.057 \text{ qt}}{1 \text{ L}} = \dfrac{\$0.410}{\text{L}}$

(d) $\dfrac{5.0 \text{ pm}}{\text{ms}} \times \dfrac{1 \times 10^{-12} \text{ m}}{1 \text{ pm}} \times \dfrac{1 \text{ ms}}{1 \times 10^{-3} \text{ s}} = 5.0 \times 10^{-9}$ m/s

(e) $\dfrac{75.00 \text{ mi}}{\text{hr}} \times \dfrac{1.609 \text{ km}}{1 \text{ mi}} \times \dfrac{1000 \text{ m}}{1 \text{ km}} \times \dfrac{1 \text{ hr}}{60 \text{ min}} \times \dfrac{1 \text{ min}}{60 \text{ s}} = 33.52$ m/s

(f) $55.35 \text{ ft}^3 \times \dfrac{(12)^3 \text{ in}^3}{1 \text{ ft}^3} \times \dfrac{(2.54)^3 \text{ cm}^3}{1 \text{ in}^3} = 1.567 \times 10^6 \text{ cm}^3$

1.45   (a) $31 \text{ gal} \times \dfrac{4 \text{ qt}}{1 \text{ gal}} \times \dfrac{1 \text{ L}}{1.057 \text{ qt}} = 1.2 \times 10^2$ L

(b) $\dfrac{6 \text{ mg}}{\text{kg (body)}} \times \dfrac{1 \text{ kg}}{2.205 \text{ lb}} \times 150 \text{ lb} = 4 \times 10^2$ mg

(c) $\dfrac{254 \text{ mi}}{11.2 \text{ gal}} \times \dfrac{1.609 \text{ km}}{1 \text{ mi}} \times \dfrac{1 \text{ gal}}{4 \text{ qt}} \times \dfrac{1.057 \text{ qt}}{1 \text{ L}} = \dfrac{9.64 \text{ km}}{\text{L}}$

(d) $\dfrac{50 \text{ cups}}{1 \text{ lb}} \times \dfrac{1 \text{ qt}}{4 \text{ cups}} \times \dfrac{1 \text{ L}}{1.057 \text{ qt}} \times \dfrac{1000 \text{ mL}}{1 \text{ L}} \times \dfrac{1 \text{ lb}}{453.6 \text{ g}} = \dfrac{26 \text{ mL}}{\text{g}}$

1.47 $12.5 \text{ ft} \times 15.5 \text{ ft} \times 8.0 \text{ ft} = 1.6 \times 10^3 \text{ ft}^3$ (2 sig figs)

$1550 \text{ ft}^3 \times \dfrac{(1 \text{ yd})^3}{(3 \text{ ft})^3} \times \dfrac{(1 \text{ m})^3}{(1.0936)^3 \text{ yd}^3} \times \dfrac{10^3 \text{ dm}^3}{1 \text{ m}^3} \times \dfrac{1 \text{ L}}{1 \text{ dm}^3} \times \dfrac{1.19 \text{ g}}{\text{L}} \times \dfrac{1 \text{ kg}}{1000 \text{ g}} = 52$ kg air

1.49   A wire is a very long, thin cylinder of volume, $V = \pi r^2 h$, where h is the length of the wire and πr^2 is the cross-sectional area of the wire.

        Strategy: 1) Calculate total volume of copper in cm^3 from mass and density

                  2) h (length in cm) = $\dfrac{V}{\pi r^2}$

                  3) Change cm → ft

$150 \text{ lb Cu} \times \dfrac{453.6 \text{ g}}{1 \text{ lb Cu}} \times \dfrac{1 \text{ cm}^3}{8.94 \text{ g}} = 7.61 \times 10^3 \text{ cm}^3$

$r = d/2 = 8.25 \text{ mm} \times \dfrac{1 \text{ cm}}{10 \text{ mm}} \times \dfrac{1}{2} = 0.413$ cm

$h = \dfrac{V}{\pi r^2} = \dfrac{7610.7 \text{ cm}^3}{\pi (0.4125)^2 \text{ cm}^2} = 1.42 \times 10^4$ cm

$1.4237 \times 10^4 \text{ cm} \times \dfrac{1 \text{ in}}{2.54 \text{ cm}} \times \dfrac{1 \text{ ft}}{12 \text{ in}} = 467$ ft

Additional Exercises

1.51   Composition is the contents of a substance, the kinds of elements that are present and their relative amounts. Structure is the arrangement of these contents.

1.54   Any sample of vitamin C has the same relative amount of carbon and oxygen; the ratio of oxygen to carbon in the isolated sample is the same as the ratio in synthesized vitamin C.

$$\frac{2.00\,g\,O}{1.50\,g\,C} = \frac{x\,g\,O}{6.35\,g\,C}; \quad x = \frac{(2.00\,g\,O)(6.35\,g\,C)}{1.50\,g\,C} = 8.47\,g\,O$$

This illustrates the *law of constant composition*.

1.56   (a) $\dfrac{m}{s^2}$ (b) $\dfrac{kg \cdot m}{s^2}$ (c) $\dfrac{kg \cdot m}{s^2} \times m = \dfrac{kg \cdot m^2}{s^2}$

      (d) $\dfrac{kg \cdot m}{s^2} \times \dfrac{1}{m^2} = \dfrac{kg}{m \cdot s^2}$ (e) $\dfrac{kg \cdot m^2}{s^2} \times \dfrac{1}{s} = \dfrac{kg \cdot m^2}{s^3}$

1.59 $K = {}^\circ C + 273.15; \quad K = -268.9{}^\circ C + 273.15 = 4.3\,K$

      ${}^\circ F = 9/5\,({}^\circ C) + 32; \quad {}^\circ F = 9/5\,(-268.9) + 32 = -452.0{}^\circ F$

      (We consider 32 to be exact, so the result has 4 significant figures, as does the data.)

1.62   (a) $25.83 \times 10^9\,lb \times \dfrac{453.6\,g}{1\,lb} = 1.172 \times 10^{13}\,g\,NaOH$

      (b) $1.17165 \times 10^{13}\,g \times \dfrac{1\,cm^3}{2.130\,g} \times \dfrac{1\,m^3}{(100)^3\,cm^3} \times \dfrac{1\,km^3}{(1000)^3\,m^3} = 5.501 \times 10^{-3}\,km^3$

1.65   There are 209.1 degrees between the freezing and boiling points on the Celsius (C) scale and 100 degrees on the glycol (G) scale. Also, $-11.5{}^\circ C = 0{}^\circ G$. By analogy with ${}^\circ F$ and ${}^\circ C$,

      ${}^\circ G = \dfrac{100}{209.1}\,({}^\circ C + 11.5)$ or ${}^\circ C = \dfrac{209.1}{100}\,({}^\circ G) - 11.5$

      These equations correctly relate the freezing point and boiling point of ethylene glycol on the two scales.

      f.p. of H_2O: ${}^\circ G = \dfrac{100}{209.1}\,(0{}^\circ C + 11.5) = 5.50{}^\circ G$

      b.p. of H_2O: ${}^\circ G = \dfrac{100}{209.1}\,(100{}^\circ C + 11.5) = 53.3{}^\circ G$

1.68   (a) $2.4 \times 10^5\,mi \times \dfrac{1.609\,km}{1\,mi} \times \dfrac{1000\,m}{1\,km} = 3.9 \times 10^8\,m$

      (b) $2.4 \times 10^5\,mi \times \dfrac{1.609\,km}{1\,mi} \times \dfrac{1\,hr}{2.4 \times 10^3\,km} \times \dfrac{60\,min}{1\,hr} \times \dfrac{60\,s}{1\,min} = 5.8 \times 10^5\,s$

1.70   (a) $\dfrac{\$2000}{acre \cdot ft} \times \dfrac{1\,acre}{4840\,yd^2} \times \dfrac{3\,ft}{1\,yd} \times \dfrac{(1.094\,yd)^3}{(1\,m)^3} \times \dfrac{(1\,m)^3}{(10\,dm)^3} \times \dfrac{(1\,dm)^3}{1\,L} =$

         $\$1.6 \times 10^{-3}/L$ or $0.16\cent\,/\,L$

(b) $\dfrac{\$2000}{acre \cdot ft} \times \dfrac{1\ acre \cdot ft}{2\ households \cdot year} \times \dfrac{1\ year}{365\ days} \times 1\ household = \dfrac{\$2.74}{day}$

1.72 $9.64\ g\ ethanol \times \dfrac{1\ cm^3}{0.789\ g\ ethanol} = 12.2\ cm^3$, volume of cylinder

$V = \pi r^2 h; \ \ r = (V/\pi h)^{1/2} = \left[\dfrac{12.218\ cm^3}{\pi \times 15.0\ cm}\right]^{1/2} = 0.509\ cm$

$d = 2r = 1.02\ cm$

1.76   The densities are:

carbon tetrachloride (methane, tetrachloro) - $1.5940\ g/cm^3$

hexane - $0.6603\ g/cm^3$

benzene - $0.87654\ g/cm^3$

methylene iodide (methane, diiodo) - $3.3254\ g/cm^3$

The volume of a sphere is $4/3\ \pi\ r^3$.

$V = 4/3\ \pi\ (0.56)^3\ cm^3 = 0.74\ cm^3; \ \ density = \dfrac{2.00\ g}{0.7356\ cm^3} = 2.7\ g/cm^3$

The marble will float on the methylene iodide only.

2 Atoms, Molecules, and Ions

Atomic Theory and Atomic Structure

2.1   Postulate 4 of the atomic theory is the *law of constant composition*. It states that the relative number and kinds of atoms in a compound are constant, regardless of the source. Therefore, 1.0 g of pure water should always contain the same relative amounts of hydrogen and oxygen, no matter where or how the sample is obtained.

2.3   (a) $\dfrac{17.37 \text{ g oxygen}}{15.20 \text{ g nitrogen}} = \dfrac{1.143 \text{ g O}}{1 \text{ g N}}$; 1.143/1.143 = 1.0; 1.0 × 2 = 2

                  $\dfrac{34.74 \text{ g oxygen}}{15.20 \text{ g nitrogen}} = \dfrac{2.286 \text{ g O}}{1 \text{ g N}}$; 2.286/1.143 = 2.0; 2.0 × 2 = 4

                  $\dfrac{43.43 \text{ g oxygen}}{15.20 \text{ g nitrogen}} = \dfrac{2.857 \text{ g O}}{1 \text{ g N}}$; 2.857/1.143 = 2.5; 2.5 × 2 = 5

     (b)   These masses of oxygen per one gram nitrogen are in the ratio of 2:4:5 and thus obey the *law of multiple proportions*.

2.5   Evidence that cathode rays were negatively charged particles was (1) that electric and magnetic fields deflected the rays in the same way they would deflect negatively charged particles and (2) that a metal plate exposed to cathode rays acquired a negative charge.

2.7   The droplets contain different charges because there may be 1, 2, 3 or more excess electrons on the droplet. The electronic charge is likely to be the lowest common factor in all the observed charges. Assuming this is so, we calculate the apparent electronic charge from each drop as follows:

A: $1.60 \times 10^{-19} / 1 = 1.60 \times 10^{-19}$ C

B: $3.15 \times 10^{-19} / 2 = 1.58 \times 10^{-19}$ C

C: $4.81 \times 10^{-19} / 3 = 1.60 \times 10^{-19}$ C

D: $6.31 \times 10^{-19} / 4 = 1.58 \times 10^{-19}$ C

The reported value is the average of these four values. Since each calculated charge has three significant figures, the average will also have three significant figures.

$(1.60 \times 10^{-19}$ C $+ 1.58 \times 10^{-19}$ C $+ 1.60 \times 10^{-19}$ C $+ 1.58 \times 10^{-19}$ C$) / 4 = 1.59 \times 10^{-19}$ C

2.9   The Be nuclei have a much smaller volume and positive charge than the Au nuclei; the charge repulsion between the alpha particles and the Be nuclei will be less, and there will be fewer direct hits because the Be nuclei have an even smaller volume than the Au nuclei. Fewer alpha particles will be scattered in general and fewer will be strongly back scattered.

2.11   (a) $2.4 \text{ Å} \times \dfrac{1 \times 10^{-8} \text{ cm}}{1 \text{ Å}} \times \dfrac{1 \text{ m}}{100 \text{ cm}} \times \dfrac{1 \text{ nm}}{1 \times 10^{-9} \text{ m}} = 0.24 \text{ nm}$

$2.4 \text{ Å} \times \dfrac{1 \times 10^{-10} \text{ m}}{1 \text{ Å}} \times \dfrac{1 \text{ pm}}{1 \times 10^{-12} \text{ m}} = 2.4 \times 10^2 \text{ or } 240 \text{ pm} \quad (1 \text{ Å} = 100 \text{ pm})$

(b) $1.0 \text{ cm} \times \dfrac{1 \text{ Å}}{1 \times 10^{-8} \text{ cm}} \times \dfrac{1 \text{ Cr atom}}{2.4 \text{ Å}} = 4.2 \times 10^7 \text{ Cr atoms}$

2.13   p = protons, n = neutrons, e = electrons

(a) ^{40}Ar has 18 p, 22 n, 18 e   (b) ^{55}Mn has 25 p, 30 n, 25 e

(c) ^{65}Zn has 30 p, 35 n, 30 e   (d) ^{79}Se has 34 p, 45 n, 34 e

(e) ^{184}W has 74 p, 110 n, 74 e   (f) ^{235}U has 92 p, 143 n, 92 e

2.15

Symbol	^{39}K	^{55}Mn	^{112}Cd	^{137}Ba	^{207}Pb
Protons	19	25	48	56	82
Neutrons	20	30	64	81	125
Electrons	19	25	48	56	82
Mass no.	39	55	112	137	207

2.17   (a) $^{23}_{11}$Na   (b) $^{51}_{23}$V   (c) $^{4}_{2}$He   (d) $^{37}_{17}$Cl   (e) $^{24}_{12}$Mg

The Periodic Table; Molecules and Ions

2.19   (a) Ag (metal)   (b) He (nonmetal)   (c) P (nonmetal)   (d) Cd (metal)
(e) Ca (metal)   (f) Br (nonmetal)   (g) As (metalloid)

2.21   (a) K, alkali metals (metal)   (b) I, halogens (nonmetal)   (c) Mg, alkaline earth metals (metal) (d) Ar, noble gases (nonmetal)   (e) S, chalcogens (nonmetal)

2.23   A structural formula contains the most information. It shows the total number and kinds of atoms (the molecular formula, from which the empirical formula can be deduced) and how these atoms are connected.

2.25   (a) 6   (b) 6   (c) 12

2.27 (a) C_2H_6O

$$H-\underset{\underset{H}{|}}{\overset{\overset{H}{|}}{C}}-O-\underset{\underset{H}{|}}{\overset{\overset{H}{|}}{C}}-H$$

(b) C_2H_6O

$$H-\underset{\underset{H}{|}}{\overset{\overset{H}{|}}{C}}-\underset{\underset{H}{|}}{\overset{\overset{H}{|}}{C}}-O-H$$

(c) CH_4O

$$H-\underset{\underset{H}{|}}{\overset{\overset{H}{|}}{C}}-O-H$$

(d) PCl_3

$$Cl-\underset{\underset{Cl}{|}}{P}-Cl$$

2.29 CH: C_2H_2, C_6H_6

CH$_2$: C_2H_4, C_3H_6, C_4H_8

NO$_2$: N_2O_4, NO_2

2.31

Symbol	$^{31}P^{3-}$	$^{40}Ca^{2+}$	$^{51}V^{2+}$	$^{79}Se^{2-}$	$^{59}Ni^{2+}$
Protons	15	20	23	34	28
Neutrons	16	20	28	45	31
Electrons	18	18	21	36	26
Net Charge	3-	2+	2+	2-	2+

2.33 (a) Al^{3+} (b) Ca^{2+} (c) S^{2-} (d) I^- (e) Cs^+

2.35 (a) GaF_3, gallium(III) fluoride   (b) LiH, lithium hydride
      (c) AlI_3, aluminum iodide   (d) K_2S, potassium sulfide

2.37 (a) $CaBr_2$ (b) NH_4Cl (c) $Al(C_2H_3O_2)_3$ (d) K_2SO_4 (e) $Mg_3(PO_4)_2$

2.39 Molecular (all elements are nonmetals): (a) B_2H_6 (b) CH_3OH (f) NOCl   (g) NF_3

      Ionic (formed by a cation and an anion, usually contains a metal cation): (c) $LiNO_3$,
      (d) Sc_2O_3, (e) CsBr, (h) Ag_2SO_4

Naming Iorganic Compounds

2.41 (a) ClO_2^- (b) Cl^- (c) ClO_3^- (d) ClO_4^- (e) ClO^-

2.43 (a) aluminum fluoride   (b) iron(II) hydroxide (ferrous hydroxide)
      (c) copper(II) nitrate (cupric nitrate)   (d) barium perchlorate   (e) lithium phosphate
      (f) mercury(I) sulfide (mercurous sulfide)   (g) calcium acetate   (h) chromium(III) carbonate
      (chromic carbonate)   (i) potassium chromate   (j) ammonium sulfate

2.45 (a) Cu_2O (b) K_2O_2 (c) $Al(OH)_3$ (d) $Zn(NO_3)_2$ (e) Hg_2Br_2 (f) $Fe_2(CO_3)_3$ (g) NaBrO

2.47 (a) bromic acid   (b) hydrobromic acid   (c) phosphoric acid   (d) HClO   (e) HIO_3
      (f) H_2SO_3

2.49   (a) sulfur hexafluoride   (b) iodine pentafluoride   (c) xenon trioxide   (d) N_2O_4 (e) HCN
       (f) P_4S_6

2.51   (a) $ZnCO_3$, ZnO, CO_2 (b) HF, SiO_2, SiF_4, H_2O (c) SO_2, H_2O, H_2SO_3
       (d) H_3P (or PH_3)   (e) $HClO_4$, Cd, $Cd(ClO_4)_2$ (f) VBr_3

Additional Exercises

2.54   (a)   Most of the volume of an atom is empty space in which electrons move. Most alpha
                particles passed through this space. The path of the massive alpha particle would
                not be significantly altered by interaction with a "puny" electron.

       (b)   Most of the mass of an atom is contained in a very small, dense area called the
                nucleus. The few alpha particles that hit the massive, positively charged gold nuclei
                were strongly repelled and essentially deflected back in the direction they came from.

2.57   (a)   5 significant figures. $^1H^+$ is a bare proton with mass 1.0073 amu. ^1H is a hydrogen
                atom, with 1 proton and 1 electron. The mass of the electron is 5.486×10^{-4} or
                0.0005486 amu. Thus the mass of the electron is significant in the fourth decimal
                place or fifth significant figure in the mass of ^1H.

       (b)   Mass of ^1H = 1.0073 amu   (proton)
                            0.0005486 amu   (electron)
                            1.0078 amu   (We have not rounded up to 1.0079 since
                                              49 < 50 in the final sum.)

                Mass % of electron = $\dfrac{\text{mass of e}^-}{\text{mass of } ^1H} \times 100 = \dfrac{5.486 \times 10^{-4} \text{ amu}}{1.0078 \text{ amu}} \times 100 = 0.05444\%$

2.60   (a) an alkali metal - K   (b) an alkaline earth metal - Ca   (c) a noble gas - Ar
       (d) a halogen - Br   (e) a metalloid - Ge   (f) a nonmetal in 1A - H
       (g) a metal that forms a 3+ ion - Al   (h) a nonmetal that forms a 2- ion - O
       (i) an element that resembles Al - Ga

2.62

Symbol	^{106}Pd	^{122}Sb^{3-}	^{184}W^{3+}	^{207}Pb^{2+}	^{232}Th^{4+}
Protons	46	51	74	82	90
Neutrons	60	71	110	125	142
Electrons	46	54	71	80	86
Net Charge	0	3-	2+	2+	4+

2.65   (a) perbromate   (b) selenite   (c) AsO_4^{3-} (d) $HTeO_4^-$

2.68   (a) potassium nitrate   (b) sodium carbonate   (c) calcium oxide
       (d) hydrochloric acid   (e) magnesium sulfate   (f) magnesium hydroxide

3 Stoichiometry: Calculations with Chemical Formulas and Equations

Balancing Chemical Equations

3.1 (a) In balancing chemical equations, the *law of conservation of mass*, that atoms are neither created nor destroyed during the course of a reaction, is observed. This means that the **number** and **kinds** of atoms on both sides of the chemical equation must be the same.

(b) gases - (g); liquids - (l); solids - (s); aqueous solutions - (aq)

(c) P_4 indicates that there are four phosphorus atoms bound together by chemical bonds into a single molecule; 4P denotes four separate phosphorus atoms.

3.3 (a) $\textbf{1 } NH_4NO_3(s) \rightarrow \textbf{1 } N_2O(g) + \textbf{2 } H_2O(l)$

(b) $\textbf{1 } La_2O_3(s) + \textbf{3 } H_2O(l) \rightarrow \textbf{2 } La(OH)_3(aq)$

(c) $\textbf{1 } Mg_3N_2(s) + \textbf{6 } H_2O(l) \rightarrow \textbf{3 } Mg(OH)_2(s) + \textbf{2 } NH_3(aq)$

(d) $\textbf{1 } NCl_3(aq) + \textbf{3 } H_2O(l) \rightarrow \textbf{1 } NH_3(aq) + \textbf{3 } HOCl(aq)$

(e) $\textbf{1 } Al(OH)_3(s) + \textbf{3 } HNO_3(aq) \rightarrow \textbf{1 } Al(NO_3)_3(aq) + \textbf{3 } H_2O(l)$

(f) $\textbf{2 } C_6H_6(l) + \textbf{15 } O_2(g) \rightarrow \textbf{12 } CO_2(g) + \textbf{6 } H_2O(l)$

(g) $\textbf{4 } CH_3NH_2(g) + \textbf{9 } O_2(g) \rightarrow \textbf{4 } CO_2(g) + \textbf{10 } H_2O(g) + \textbf{2 } N_2(g)$

3.5 (a) $\textbf{2}NH_3(g) + \textbf{2}Na(l) \rightarrow \textbf{2}NaNH_2(s) + H_2(g)$

(b) $Zn(s) + H_2SO_4(aq) \rightarrow ZnSO_4(aq) + H_2(g)$

(c) $\textbf{2}KNO_3(s) \overset{\Delta}{\rightarrow} \textbf{2}KNO_2(s) + O_2(g)$

(d) $PCl_3(l) + \textbf{3}H_2O(l) \rightarrow H_3PO_3(aq) + \textbf{3}HCl(aq)$

(e) $Cu(s) + \textbf{2}H_2SO_4(aq) \rightarrow CuSO_4(aq) + SO_2(g) + \textbf{2}H_2O(l)$

Patterns of Chemical Reactivity

3.7 (a) $C_6H_{12}(l) + \textbf{9}O_2(g) \rightarrow \textbf{6}CO_2(g) + \textbf{6}H_2O(l)$

(b) $C_2H_5OC_2H_5(l) + \textbf{6}O_2(g) \rightarrow \textbf{4}CO_2(g) + \textbf{5}H_2O(l)$

(c) $\textbf{2}Cs(s) + \textbf{2}H_2O(l) \rightarrow \textbf{2}CsOH(aq) + H_2(g)$

(d) $Mg(s) + Br_2(l) \rightarrow MgBr_2(s)$

3.9   (a) $C_5H_6O(l) + 6O_2(g) \rightarrow 5CO_2(g) + 3H_2O(l)$ combustion

       (b) $2H_2O_2(l) \rightarrow 2H_2O(l) + O_2(g)$ decomposition

       (c) $CaO(s) + H_2O(l) \rightarrow Ca(OH)_2(aq)$ combination

       (d) $6Li(s) + N_2(g) \rightarrow 2Li_3N(s)$ combination

3.11   (a) $2K(s) + 2NH_3(l) \rightarrow 2KNH_2(solv) + H_2(g)$

       (b) $4CH_3NO_2(g) + 7O_2(g) \rightarrow 4NO_2(g) + 4CO_2(g) + 6H_2O(l)$

       (c) $CH_4(g) + 4F_2(g) \rightarrow CF_4(g) + 4HF(g)$

Atomic and Molecular Weights

3.13   (a) $^{12}_{6}C$ (b)   An atomic mass unit is exactly 1/12 of the mass of one atom of $^{12}_{6}C$, or 1.66054×10^{-24} g.

3.15   The **average** mass of a boron atom is given by the sum of the mass of each isotope present in the sample times its **fractional** abundance.

      average atomic mass = 0.1978(10.013 amu) + 0.8022(11.009 amu)
      (atomic weight, AW) = 1.981 amu + 8.831 amu = 10.812 amu

3.17   FW in amu to 1 decimal place (see Sample Exercise 3.5)

       (a) P_2O_3: 2(31.0) + 3(16.0) = 110.0 amu

       (b) $BaSO_4$: 1(137.3) + 1(32.1) + 4(16.0) = 233.4 amu

       (c) $Mg(C_2H_3O_2)_2$: 1(24.3) + 4(12.0) + 6(1.0) + 4(16.0) = 142.3 amu

       (d) $(NH_4)_2Cr_2O_7$: 2(14.0) + 8(1.0) + 2(52.0) + 7(16.0) = 252.0 amu

       (e) HNO_3: 1(1.0) + 1(14.0) + 3(16.0) = 63.0 amu

       (f) Li_2CO_3: 2(6.9) + 1(12.0) + 3(16.0) = 73.8 amu

       (g) Si_2Cl_6: 2(28.1) + 6(35.5) = 269.2 amu

3.19   Calculate the formula weight, then the mass % of oxygen in the compound.

       (a) SO_3: FW = 1(32.1) + 3(16.0) = 80.1 amu

$$\% \, O = \frac{3(16.0) \text{ amu}}{80.1 \text{ amu}} \times 100 = 59.9\%$$

       (b) CH_3COOH: FW = 2(12.0) + 4(1.0) + 2(16.0) = 60.0 amu

$$\% \, O = \frac{2(16.0) \text{ amu}}{60.0 \text{ amu}} \times 100 = 53.3\%$$

       (c) $Ca(NO_3)_2$: FW = 1(40.1) + 2(14.0) + 6(16.0) = 164.1 amu

$$\% \, O = \frac{6(16.0) \text{ amu}}{164.1 \text{ amu}} \times 100 = 58.5\%$$

       (d) $(NH_4)_2SO_4$: FW = 132.1 amu (Exercise 3.18(b))

$$\% \, O = \frac{4(16.0) \text{ amu}}{132.1 \text{ amu}} \times 100 = 48.4\%$$

3.21   (a) CO_2: FW = 1(12.0) + 2(16.0) = 44.0 amu

%C = $\dfrac{12.0 \text{ amu}}{44.0 \text{ amu}}$ × 100 = 27.3%

(b) C_2H_6: FW = 2(12.0) + 6(1.0) = 30.0

%C = $\dfrac{2(12.0) \text{ amu}}{30.0 \text{ amu}}$ × 100 = 80.0%

(c) CH_3OH: FW = 1(12.0) + 4(1.0) + 1(16.0) = 32.0 amu

%C = $\dfrac{12.0 \text{ amu}}{32.0 \text{ amu}}$ × 100 = 37.5%

(d) $CS(NH_2)_2$: FW = 1(12.0) + 1(32.1) + 2(14.0) + 4(1.0) = 76.1 amu

%C = $\dfrac{12.0 \text{ amu}}{76.1 \text{ amu}}$ × 100 = 15.8%

3.23   (a)   Three peaks: ^1H - ^1H, ^1H - ^2H, ^2H - ^2H

(b) ^1H - ^1H = 2(1.00783) = 2.01566 amu

^1H - ^2H = 1.00783 + 2.01411 = 3.02194 amu

^2H - ^2H = 2(2.01411) = 4.02822 amu

[The mass ratios are 1 : 1.49923 : 1.99846 or 1 : 1.5 : 2]

(c) ^1H - ^1H is largest, because there is the greatest chance that two atoms of the more abundant isotope will combine.

^2H - ^2H is the smallest, because there is the least chance that two atoms of the less abundant isotope will combine.

The Mole

3.25   (a)   A *mole* is the amount of matter that contains as many objects as the number of atoms in exactly 12 g of ^{12}C.

(b)   6.022 × 10^{23}. This is the number of objects in a mole of anything.

(c)   The formula weight of a substance in amu has the same numerical value as the molar mass expressed in grams.

3.27   (a) $\dfrac{12 \text{ H atoms}}{6 \text{ C atoms}} = \dfrac{2 \text{ H}}{1 \text{ C}}$ × 4.0 × 10^{22}C atoms = 8.0 × 10^{22} H atoms

(b) $\dfrac{1 \ C_6H_{12}O_6 \text{ molecule}}{6 \text{ C atoms}}$ × 4.0 × 10^{22} C atoms = 6.67 × 10^{21}

= 6.7 × 10^{21} $C_6H_{12}O_6$ molecules

(c)   6.67 × 10^{21} $C_6H_{12}O_6$ molecules × = 1.1 × 10^{-2} mol $C_6H_{12}O_6$

(d)   1 mole of $C_6H_{12}O_6$ weighs 180.0 g (Sample Exercise 3.8)

$$1.1 \times 10^{-2} \text{ mol } C_6H_{12}O_6 \times \frac{180.0 \text{ g } C_6H_{12}O_6}{1 \text{ mol}} = 2.0 \text{ g } C_6H_{12}O_6$$

3.29   (a)   molar mass: $63.55 + 2(14.01) + 6(16.00) = 187.57$ g

        (b) $0.120 \text{ mol } Cu(NO_3)_2 \times \frac{187.57 \text{ g}}{1 \text{ mol}} = 22.5 \text{ g } Cu(NO_3)_2$

        (c) $3.15 \text{ g } Cu(NO_3)_2 \times \frac{1 \text{ mol}}{187.57 \text{ g}} = 1.68 \times 10^{-2} \text{ mol}$

        (d) $1.25 \text{ mg} \times \frac{1 \times 10^{-3} \text{g}}{1 \text{ mg}} \times \frac{1 \text{ mol}}{187.57 \text{ g}} \times \frac{6.022 \times 10^{23} \text{molecules}}{\text{mol}} \times \frac{2 \text{ N atoms}}{1 \text{ molecule}}$

$$= 8.03 \times 10^{18} \text{ N atoms}$$

3.31   (a) $0.00150 \text{ mol } SO_2 \times \frac{64.06 \text{ g } SO_2}{1 \text{ mol } SO_2} = 0.0961 \text{ g } SO_2$

        (b) $2.98 \times 10^{21} \text{ Ar atom} \times \frac{1 \text{ mol}}{6.022 \times 10^{23} \text{ atoms}} \times \frac{39.95 \text{ g Ar}}{\text{mol Ar}} = 0.198 \text{ g Ar}$

        (c) $1.05 \times 10^{20} \text{ molecules} \times \frac{1 \text{ mol}}{6.022 \times 10^{23} \text{ molecules}} \times \frac{194.2 \text{ g } C_8H_{10}N_4O_2}{1 \text{ mol caffeine}}$

$$= 3.39 \times 10^{-2} \text{ g } C_8H_{10}N_4O_2$$

3.33   (a) $0.0666 \text{ mol } C_3H_8 \times \frac{6.022 \times 10^{23} \text{ molecules}}{1 \text{ mol}} = 4.01 \times 10^{22} \text{ } C_3H_8 \text{ molecules}$

        (b) $50.0 \text{ mg } C_8H_9O_2N \times \frac{1 \times 10^{-3} \text{ g}}{1 \text{ mg}} \times \frac{1 \text{ mol } C_8H_9O_2N}{151.2 \text{ g } C_8H_9O_2N} \times \frac{6.022 \times 10^{23} \text{ molecules}}{1 \text{ mol}}$

$$= 1.99 \times 10^{20} \text{ } C_8H_9O_2N \text{ molecules}$$

        (c) $10.5 \text{ g } C_{12}H_{22}O_{11} \times \frac{1 \text{ mol } C_{12}H_{22}O_{11}}{342.3 \text{ g } C_{12}H_{22}O_{11}} \times \frac{6.022 \times 10^{23} \text{ molecules}}{1 \text{ mol}}$

$$= 1.85 \times 10^{22} \text{ } C_{12}H_{22}O_{11} \text{ molecules}$$

3.35   Strategy: volume H_2O → mass H_2O → moles H_2O → molecules H_2O

$$350 \text{ ft} \times 99 \text{ ft} \times 6.0 \text{ ft} \times \frac{12^3 \text{ in}^3}{1 \text{ ft}^3} \times \frac{2.54^3 \text{ cm}^3}{\text{in}^3} \times \frac{1 \text{ mL}}{1 \text{ cm}^3} = 5.887 \times 10^9 = 5.9 \times 10^9 \text{ mL}$$

$$5.887 \times 10^9 \text{ mL} \times \frac{1.0 \text{ g}}{\text{mL}} \times \frac{1 \text{ mol } H_2O}{18.02 \text{ g } H_2O} \times 6.022 \times 10^{23} = 2.0 \times 10^{32} \text{ } H_2O \text{ molecules}$$

3.37 $\dfrac{2.05 \times 10^{-6} \text{ g C}_2\text{H}_3\text{Cl}}{1 \text{ L}} \times \dfrac{1 \text{ mol C}_2\text{H}_3\text{Cl}}{62.50 \text{ g C}_2\text{H}_3\text{Cl}} = 3.280 \times 10^{-8} = 3.28 \times 10^{-8} \text{ mol C}_2\text{H}_3\text{Cl/L}$

$\dfrac{3.280 \times 10^{-8} \text{ mol C}_2\text{H}_3\text{Cl}}{1 \text{ L}} \times \dfrac{6.022 \times 10^{23} \text{ molecules}}{1 \text{ mol}} = 1.97 \times 10^{16} \text{ molecules/L}$

Empirical Formulas

3.39   An *empirical formula* gives the relative number and kind of each atom in a compound, but a *molecular formula* gives the actual number of each kind of atom, and thus the molecular weight.

3.41   (a)   Find the **simplest ratio of moles** by dividing by the smallest number of moles present.

             0.0130 mol C / 0.0065 = 2
             0.039   mol H / 0.0065   = 6
             0.0065 mol O / 0.0065 = 1

             The empirical formula is C_2H_6O.

      (b)   Calculate the moles of each element present, then the simplest ratio of moles.

             $11.66 \text{ g Fe} \times \dfrac{1 \text{ mol Fe}}{55.85 \text{ g Fe}} = 0.2088 \text{ mol Fe};\ 0.2088 / 0.2088 = 1$

             $5.01 \text{ g O} \times \dfrac{1 \text{ mol O}}{16.00 \text{ g O}} = 0.3131 \text{ mol O};\ 0.3131 / 0.2088 \approx 1.5$

             Multiplying by two, the integer ratio is 2 Fe : 3 O; the empirical formula is Fe_2O_3.

      (c)   Assume 100 g sample, calculate moles of each element, find the simplest ratio of moles.

             $40.0 \text{ g C} \times \dfrac{1 \text{ mol C}}{12.01 \text{ g C}} = 3.33 \text{ mol C};\ 3.33 / 3.33 = 1$

             $6.7 \text{ g H} \times \dfrac{1 \text{ mol H}}{1.008 \text{ mol H}} = 6.65 \text{ mol H};\ 6.65 / 3.33 \approx 2$

             $53.3 \text{ g O} \times \dfrac{1 \text{ mol O}}{16.00 \text{ mol O}} = 3.33 \text{ mol O};\ 3.33 / 3.33 = 1$

             The empirical formula is CH_2O.

3.43   The procedure in all these cases is to assume 100 g of sample, calculate the number of moles of each element present in that 100 g, then obtain the ratio of moles as smallest whole numbers.

(a) $32.79 \text{ g Na} \times \dfrac{1 \text{ mol Na}}{22.99 \text{ g Na}} = 1.426 \text{ mol Na}; \quad 1.426 / 0.4826 \approx 3$

         $13.02 \text{ g Al} \times \dfrac{1 \text{ mol Al}}{26.98 \text{ g Al}} = 0.4826 \text{ mol Al}; \quad 0.4826 / 0.4826 = 1$

         $54.19 \text{ g F} \times \dfrac{1 \text{ mol F}}{19.00 \text{ g F}} = 2.852 \text{ mol F}; \quad 2.852 / 0.4826 \approx 6$

         The empirical formula is Na_3AlF_6.

(b) $62.1 \text{ g C} \times \dfrac{1 \text{ mol C}}{12.01 \text{ g C}} = 5.17 \text{ mol C}; \quad 5.17 / 0.864 \approx 6$

         $5.21 \text{ g H} \times \dfrac{1 \text{ mol H}}{1.008 \text{ g H}} = 5.17 \text{ mol O}; \quad 5.17 / 0.864 \approx 6$

         $12.1 \text{ g N} \times \dfrac{1 \text{ mol N}}{14.01 \text{ g N}} = 0.864 \text{ mol N}; \quad 0.864 / 0.864 = 1$

         $20.7 \text{ g O} \times \dfrac{12 \text{ mol O}}{16.00 \text{ g O}} = 1.29 \text{ mol O}; \quad 1.29 \times 0.864 \approx 1.5$

         Multiplying by two, the formula is $C_{12}H_{12}N_2O_3$.

3.45   (a)   Calculate the empirical formula weight (FW) and its ratio with the molar mass.

         $FW = 12.01 + 1.01 = 13.02; \quad \dfrac{MM}{FW} = \dfrac{78}{13} = 6$

         The subscripts of the empirical formula are multiplied by 6. The molecular formula is C_6H_6.

     (b) $FW = 14.01 + 2(16.00) = 46.01; \quad \dfrac{MM}{FW} = \dfrac{92.02}{46.01} = 2$

         The molecular formula is N_2O_4.

3.47   Assume 100 g in the following problems.

     (a) $38.7 \text{ g C} \times \dfrac{1 \text{ mol C}}{12.01 \text{ g C}} = 3.22 \text{ mol C}; \quad 3.22 / 3.22 = 1$

         $9.7 \text{ g H} \times \dfrac{1 \text{ mol C}}{1.008 \text{ g H}} = 9.62 \text{ mol H}; \quad 9.62 / 3.22 \approx 3$

         $51.6 \text{ g O} \times \dfrac{1 \text{ mol O}}{16.00 \text{ g O}} = 3.23 \text{ mol O}; \quad 3.23 / 3.23 \approx 1$

         Thus, CH_3O, formula weight = 31. If the molar mass is 62.1 amu, a factor of 2 gives the molecular formula $C_2H_6O_2$.

     (b) $49.5 \text{ g C} \times \dfrac{1 \text{ mol C}}{12.01 \text{ g C}} = 4.12 \text{ mol C}; \quad 4.12 / 1.03 \approx 4$

         $5.15 \text{ g H} \times \dfrac{1 \text{ mol H}}{1.008 \text{ g H}} = 5.11 \text{ mol H}; \quad 5.11 / 1.03 \approx 5$

$$28.9 \text{ g N} \times \frac{1 \text{ mol N}}{14.01 \text{ g N}} = 2.06 \text{ mol N}; \quad 2.06 / 1.03 \approx 2$$

$$16.5 \text{ g O} \times \frac{1 \text{ mol O}}{16.00 \text{ g O}} = 1.03 \text{ mol O}; \quad 1.03 / 1.03 = 1$$

Thus, $C_4H_5N_2O$, FW = 97. If the molar mass is about 195, a factor of 2 gives the molecular formula $C_8H_{10}N_4O_2$.

3.49   We can calculate the mass of O by subtraction, but first the masses of C and H in the sample must be found.

$$6.32 \times 10^{-3} \text{ g CO}_2 \times \frac{12.01 \text{ g C}}{44.01 \text{ g CO}_2} = 1.725 \times 10^{-3} \text{ g C} = 1.73 \text{ mg C}$$

$$2.58 \times 10^{-3} \text{ g H}_2O \times \frac{2.016 \text{ g H}}{18.02 \text{ g H}_2O} = 2.886 \times 10^{-4} \text{ g H} = 0.289 \text{ mg H}$$

mass of O = 2.78 mg sample - (1.725 mg C + 0.289 mg H) = 0.77 mg O

$$1.725 \times 10^{-3} \text{ g C} \times \frac{1 \text{ mol C}}{12.01 \text{ g C}} = 1.44 \times 10^{-4} \text{ mol C}; \quad 1.44 \times 10^{-4} / 4.81 \times 10^{-5} \approx 3$$

$$2.886 \times 10^{-4} \text{ g C} \times \frac{1 \text{ mol H}}{1.008 \text{ g H}} = 2.86 \times 10^{-4} \text{ mol H}; \quad 2.86 \times 10^{-4} / 4.81 \times 10^{-5} \approx 6$$

$$7.7 \times 10^{-4} \text{ g O} \times \frac{1 \text{ mol O}}{16.00 \text{ g H}} = 4.81 \times 10^{-5} \text{ mol O}; \quad 4.81 \times 10^{-5} / 4.81 \times 10^{-5} = 1$$

The empirical formula is C_3H_6O.

3.51   The reaction involved is $Na_2CO_3 \cdot xH_2O(s) \rightarrow Na_2CO_3(s) + xH_2O(g)$.

Calculate the mass of H_2O lost and then the mole ratio of Na_2CO_3 and H_2O.

g H_2O lost = 2.558 g sample - 0.948 g Na_2CO_3 = 1.610 g H_2O

$$0.948 \text{ g Na}_2CO_3 \times \frac{1 \text{ mol Na}_2CO_3}{106.0 \text{ g Na}_2CO_3} = 0.00894 \text{ mol Na}_2CO_3$$

$$1.610 \text{ g H}_2O \times \frac{1 \text{ mol H}_2O}{18.02 \text{ g H}_2O} = 0.08935 \text{ mol H}_2O$$

$$\frac{\text{mol H}_2O}{\text{mol Na}_2CO_3} = \frac{0.08935}{0.00894} = 9.99; \quad x = 10.$$

The formula is $Na_2CO_3 \cdot \underline{\mathbf{10}} \text{ H}_2O$.

Calculations Based on Chemical Equations

3.53   The mole ratios implicit in the coefficients of a balanced chemical equation are essential for solving stoichiometry problems. If the equation is not balanced, the mole ratios will be incorrect and lead to erroneous calculated amounts of reactants and/or products.

3.55   **Apply the mole ratio** of moles CO_2/moles C_2H_5OH to calculate moles CO_2 produced. This is the heart of every stoichiometry problem.

$$C_2H_5OH(l) + 3O_2(g) \rightarrow 2CO_2(g) + 3H_2O(l)$$

(a) $3.00 \text{ mol } C_2H_5OH \times \dfrac{2 \text{ mol } CO_2}{1 \text{ mol } C_2H_5OH} = 6.00 \text{ mol } CO_2$

(b)   g $C_2H_5OH \longrightarrow$ mol $C_2H_5OH \xrightarrow[\text{ratio}]{\text{mole}}$ mol $CO_2 \longrightarrow$ g CO_2

$3.00 \text{ g } C_2H_5OH \times \dfrac{1 \text{ mol } C_2H_5OH}{46.07 \text{ } C_2H_5OH} \times \dfrac{2 \text{ mol } CO_2}{1 \text{ mol } C_2H_5OH} \times \dfrac{44.01 \text{ g } CO_2}{1 \text{ mol } CO_2} = 5.73 \text{ g } CO_2$

3.57   (a) $CaH_2(s) + 2H_2O(l) \rightarrow Ca(OH)_2(aq) + 2H_2(g)$

(b) $10.0 \text{ g } H_2 \times \dfrac{1 \text{ mol } H_2}{2.016 \text{ g } H_2} \times \dfrac{1 \text{ mol } CaH_2}{2 \text{ mol } H_2} \times \dfrac{42.10 \text{ g } CaH_2}{1 \text{ mol } CaH_2} = 104 \text{ g } CaH_2$

3.59 $C_6H_{12}O_6 \rightarrow 2C_2H_5OH(aq) + 2CO_2(g)$

(a) $0.330 \text{ mol } C_6H_{12}O_6 \times \dfrac{2 \text{ mol } CO_2}{1 \text{ mol } C_6H_{12}O_6} = 0.660 \text{ mol } CO_2$

(b) $2.00 \text{ mol } C_2H_5OH \times \dfrac{1 \text{ mol } C_6H_{12}O_6}{2 \text{ mol } C_2H_5OH} \times \dfrac{180.2 \text{ g } C_6H_{12}O_6}{1 \text{ mol } C_6H_{12}O_6} = 180 \text{ g } C_6H_{12}O_6$

(c) $2.00 \text{ g } C_2H_5OH \times \dfrac{1 \text{ mol } C_2H_5OH}{46.07 \text{ g } C_2H_5OH} \times \dfrac{2 \text{ mol } CO_2}{2 \text{ mol } C_2H_5OH} \times \dfrac{44.01 \text{ g } CO_2}{1 \text{ mol } CO_2} = 1.91 \text{ g } CO_2$

3.61   (a) $2NaN_3(s) \rightarrow 2Na(s) + 3N_2(g)$

(b) $5.00 \text{ g } N_2 \times \dfrac{1 \text{ mol } N_2}{28.01 \text{ g } N_2} \times \dfrac{2 \text{ mol } NaN_3}{3 \text{ mol } N_2} \times \dfrac{65.01 \text{ g } NaN_3}{1 \text{ mol } NaN_3} = 7.74 \text{ g } NaN_3$

(c)   First determine how many g N_2 are in 10.0 ft^3, using the density of N_2.

$\dfrac{1.25 \text{ g}}{1 \text{ L}} \times \dfrac{1 \text{ L}}{1000 \text{ cm}^3} \times \dfrac{(2.54)^3 \text{ cm}^3}{1 \text{ in}^3} \times \dfrac{(12)^3 \text{ in}^3}{1 \text{ ft}^3} \times 10.0 \text{ ft}^3 = 354.0 = 354 \text{ g } N_2$

$354.0 \text{ g } N_2 \times \dfrac{1 \text{ mol } N_2}{28.01 \text{ g } N_2} \times \dfrac{2 \text{ mol } NaN_3}{3 \text{ mol } N_2} \times \dfrac{65.01 \text{ g } NaN_3}{1 \text{ mol } NaN_3} = 548 \text{ g } NaN_3$

Limiting Reactants; Theoretical Yields

3.63   (a)   The *limiting reactant* determines the maximum number of product moles resulting from a chemical reaction; any other reactant is an *excess reactant*.

         (b)   The limiting reactant regulates the amount of products because it is completely used up during the reaction; no more product can be made when one of the reactants is unavailable.

3.65   (a)   Each bicycle needs 2 wheels, 1 frame and 1 set of handlebars. A total of 5350 wheels corresponds to 2675 pairs of wheels. This is fewer than the number of frames but more than the number of handlebars. The 2655 handlebars determine that 2655 bicycles can be produced.

         (b)   3023 frames - 2655 bicycles = 368 frames left over

$$(2675 \text{ pairs of wheels} - 2655 \text{ bicycles}) \times \frac{2 \text{ wheels}}{\text{pair}} = 40 \text{ wheels left over}$$

         (c)   The handlebars are the "limiting reactant" in that they determine the number of bicycles that can be produced.

3.67 $SiO_2(s) + 3C(s) \rightarrow SiC(s) + 2CO(g)$

         (a)   Follow the approach in Sample Exercise 3.17.

$$3.00 \text{ g } SiO_2 \times \frac{1 \text{ mol } SiO_2}{60.09 \text{ g } SiO_2} \times \frac{1 \text{ mol SiC}}{1 \text{ mol } SiO_2} \times \frac{40.10 \text{ g SiC}}{1 \text{ mol SiC}} = 2.00 \text{ g SiC}$$

$$4.50 \text{ g C} \times \frac{1 \text{ mol C}}{12.01 \text{ g C}} \times \frac{1 \text{ mol SiC}}{3 \text{ mol C}} \times \frac{40.10 \text{ g SiC}}{1 \text{ mol SiC}} = 5.01 \text{ g SiC}$$

               The lesser amount, 2.00 g SiC, can be produced.

         (b) SiO_2 is the limiting reactant (it leads to the smaller amount of product) and C is the excess reactant.

               (c) $3.00 \text{ g } SiO_2 \times \dfrac{1 \text{ mol } SiO_2}{60.09 \text{ g } SiO_2} \times \dfrac{3 \text{ mol C}}{1 \text{ mol } SiO_2} \times \dfrac{12.01 \text{ g C}}{1 \text{ mol C}} = 1.80 \text{ g C}$

               4.50 g C initially present - 1.80 g C consumed = 2.70 g C remain

3.69 $H_2S(g) + 2NaOH(aq) \rightarrow Na_2S(s) + 2H_2O(l)$

$$1.50 \text{ g } H_2S \times \frac{1 \text{ mol } H_2S}{34.09 \text{ g } H_2S} \times \frac{1 \text{ mol } Na_2S}{1 \text{ mol } H_2S} \times \frac{78.05 \text{ g } Na_2S}{1 \text{ mol } Na_2S} = 3.43 \text{ g } Na_2S$$

$$1.65 \text{ g NaOH} \times \frac{1 \text{ mol NaOH}}{40.00 \text{ g NaOH}} \times \frac{1 \text{ mol } Na_2S}{2 \text{ mol NaOH}} \times \frac{78.05 \text{ g } Na_2S}{1 \text{ mol } Na_2S} = 1.61 \text{ g } Na_2S$$

         The lesser amount, 1.61 g Na_2S is formed.

3.71   Strategy: Write balanced equation; determine limiting reactant; calculate amounts of excess reactant remaining and products, based on limiting reactant.

$$2AgNO_3(aq) + Na_2CO_3(aq) \rightarrow Ag_2CO_3(s) + 2NaNO_3(aq)$$

Follow the method in Sample Exercise 3.16.

$$5.00 \text{ g AgNO}_3 \times \frac{1 \text{ mol AgNO}_3}{169.9 \text{ g AgNO}_3} \times \frac{1 \text{ mol Na}_2CO_3}{2 \text{ mol AgNO}_3} \times \frac{106.0 \text{ g Na}_2CO_3}{1 \text{ mol Na}_2CO_3} = 1.56 \text{ g Na}_2CO_3$$

5.00 g $AgNO_3$ will react completely with 1.56 g Na_2CO_3.

$AgNO_3$ is the limiting reactant and Na_2CO_3 is present in excess.

mass Na_2CO_3 remaining = 5.00 g initial - 1.56 g reacted = 3.44 g

$$5.00 \text{ g AgNO}_3 \frac{1 \text{ mol AgNO}_3}{169.9 \text{ g AgNO}_3} \times \frac{1 \text{ mol Ag}_2CO_3}{2 \text{ mol AgNO}_3} \times \frac{275.8 \text{ g Ag}_2CO_3}{1 \text{ mol Ag}_2NO_3} = 4.06 \text{ g Ag}_2CO_3$$

$$5.00 \text{ g AgNO}_3 \times \frac{1 \text{ mol AgNO}_3}{169.9 \text{ g AgNO}_3} \times \frac{2 \text{ mol NaNO}_3}{2 \text{ mol AgNO}_3} \times \frac{85.00 \text{ g NaNO}_3}{1 \text{ mol NaNO}_3} = 2.50 \text{ g NaNO}_3$$

After evaporation there will be no $AgNO_3$ remaining (limiting reactant), 3.44 g Na_2CO_3 (excess reactant), 4.06 g Ag_2CO_3 and 2.50 g $NaNO_3$. The sum of the amounts is 10.00 g; mass is conserved.

3.73   Strategy: Calculate theoretical (stoichiometric) yield and then the % yield.

$$2.00 \text{ g KClO}_3 \times \frac{1 \text{ mol KClO}_3}{122.6 \text{ g KClO}_3} \times \frac{3 \text{ mol O}_2}{2 \text{ mol KClO}_3} \times \frac{32.00 \text{ g O}_2}{1 \text{ mol O}_2} = 0.7830 = 0.783 \text{ g O}_2$$

$$\% \text{ yield} = \frac{\text{actual yield}}{\text{theoretical yield}} \times 100 = \frac{0.720 \text{ g O}_2}{0.783 \text{ g O}_2} \times 100 = 92.0\%$$

Two reasons might be uneven heating of the $KClO_3$ and thus incomplete reaction, or loss of O_2 before weighing.

Additional Exercises

3.75   (a) $C_4H_8O_2(l) + 5O_2(g) \rightarrow 4CO_2(g) + 4H_2O(l)$

(b) $Cu(OH)_2(s) \rightarrow CuO(s) + H_2O(g)$

(c) $Zn(s) + 2HCl(aq) \rightarrow ZnCl_2(aq) + H_2(g)$

3.78   (a)   4 peaks; there are 4 possible combinations of isotopes.

(b) $^1H^{35}Cl$ 1.0078 amu + 34.969 amu = 35.977 amu

$^2H^{35}Cl$ 2.0140 amu + 34.969 amu = 36.983 amu

$^1H^{37}Cl$ 1.0078 amu + 36.966 amu = 37.974 amu

$^2H^{37}Cl$ 2.0140 amu + 36.966 amu = 38.980 amu

(c)   The intensities will be the products of the natural abundances of the two isotopes.

$^1H^{35}Cl$ (99.985)(0.7553) = 75.52

$^2H^{35}Cl$ (0.015)(0.7553) = 0.011

$^1H^{37}Cl$ (99.985)(0.2447) = 24.47

$^2H^{37}Cl$ (0.015)(0.2447) = 0.0037

Dividing through by 0.0037, the relative intensities in order of increasing mass are 20,000 : 3.0 : 6600 : 1. The two ^2HCl peaks are very small relative to the other peaks.

(d)   Assuming the very small peaks due to ^2HX can be observed, element **X** has 4 naturally occurring isotopes.

3.81   (a) $\dfrac{5.342 \times 10^{-21}\,g}{1 \text{ molecule penicillin G}} \times \dfrac{6.0221 \times 10^{23} \text{ molecules}}{1 \text{ mol}} = 3217$ g/mol penicillin G

       (b)   1.00 g hemoglobin (hem) contains 3.40×10^{-3} g Fe.

$\dfrac{1.00\,g \text{ hem}}{3.40 \times 10^{-3}\,g \text{ Fe}} \times \dfrac{55.85\,g \text{ Fe}}{1 \text{ mol Fe}} \times \dfrac{4 \text{ mol Fe}}{1 \text{ mol hem}} = 6.57 \times 10^4$ g/mol hemoglobin

3.84   (a)   Let AW = the atomic weight of X.

According to the chemical reaction, moles XI_3 reacted = moles XCl_3 produced

0.5000 g XI_3 × 1 mol XI_3 / (AW + 389.70) g XI_3

$= 0.2360$ g $XCl_3 \times \dfrac{1 \text{ mol } XCl_3}{(AW + 106.35)\,g\ XCl_3}$

0.5000 (AW + 106.35) = 0.2360 (AW + 380.70)

0.5000 AW + 53.175 = 0.2360 AW + 89.845

0.2640 AW = 36.67;   AW = 138.9 g

       (b)   X is lanthanum, La, atomic number 57.

3.86   Since all the C in the vanillin must be present in the CO_2 produced, get g C from g CO_2.

2.43 g $CO_2 \times \dfrac{1 \text{ mol } CO_2}{44.01\,g\ CO_2} \times \dfrac{12.01\,g\ C}{1 \text{ mol C}} = 0.6631 = 0.663$ g C

Since all the H in vanillin must be present in the H_2O produced, get g H from g H_2O.

0.50 g $H_2O \times \dfrac{1 \text{ mol } H_2O}{18.02\,g\ H_2O} \times \dfrac{2 \text{ mol H}}{1 \text{ mol } H_2O} \times \dfrac{1.008\,g\ H}{1 \text{ mol H}} = 0.0559 = 0.056$ g H

Get g O by subtraction. (Since the analysis was performed by combustion, an unspecified amount of O_2 was a reactant, and thus not all the O in the CO_2 and H_2O produced came from vanillin.)

1.05 g vanillin - 0.663 g C - 0.056 g H = 0.331 g O

0.6631 g C × $\dfrac{1 \text{ mol C}}{12.01\,g\ C}$ = 0.0552 mol C; 0.0552 / 0.0207 = 2.67

$$0.0559 \text{ g H} \times \frac{1 \text{ mol H}}{1.008 \text{ g H}} = 0.0555 \text{ mol C}; \quad 0.0556 / 0.0207 = 2.68$$

$$0.331 \text{ g O} \times \frac{1 \text{ mol O}}{16.00 \text{ g O}} = 0.0207 \text{ mol O}; \quad 0.0207 / 0.0207 = 1.00$$

Multiplying the numbers above by **3** to obtain an integer ratio of moles, the empirical formula of vanillin is $C_8H_8O_3$.

3.89   Strategy: Because different sample sizes were used to analyze the different elements, calculate mass % of each element in the sample.

      i.   Calculate mass % C from g CO_2.

      ii.   Calculate mass % Cl from AgCl.

      iii.   Get mass % H by subtraction.

      iv.   Calculate mole ratios and the empirical formulas.

      i.

$$3.52 \text{ g CO}_2 \times \frac{1 \text{ mol CO}_2}{44.01 \text{ g CO}_2} \times \frac{1 \text{ mol C}}{1 \text{ mol CO}_2} \times \frac{12.01 \text{ g C}}{1 \text{ mol C}} = 0.9606 = 0.961 \text{ g C}$$

$$\frac{0.9606 \text{ g C}}{1.50 \text{ g sample}} \times 100 = 64.04 = 64.0\% \text{ C}$$

      ii.

$$1.27 \text{ g AgCl} \times \frac{1 \text{ mol AgCl}}{143.3 \text{ g AgCl}} \times \frac{1 \text{ mol Cl}}{1 \text{ mol AgCl}} \times \frac{35.45 \text{ g Cl}}{1 \text{ mol Cl}} = 0.3142 = 0.314 \text{ g Cl}$$

$$\frac{0.3142 \text{ g Cl}}{1.00 \text{ g sample}} \times 100 = 31.42 = 31.4\% \text{ Cl}$$

      iii.   % H = 100.0 - (64.04% C + 31.42% Cl) = 4.54 = 4.5% H

      iv.   Assume 100 g sample.

$$64.04 \text{ g C} \times \frac{1 \text{ mol C}}{12.01 \text{ g C}} = 5.33 \text{ mol C}; \quad 5.33 / 0.886 = 6.02$$

$$31.42 \text{ g Cl} \times \frac{1 \text{ mol Cl}}{35.45 \text{ g Cl}} = 0.886 \text{ mol Cl}; \quad 0.886 / 0.886 = 1.00$$

$$4.54 \text{ g H} \times \frac{1 \text{ mol H}}{1.008 \text{ g H}} = 4.50 \text{ mol H}; \quad 4.50 / 0.886 = 5.08$$

The empirical formula is probably C_6H_5Cl.

The subscript for H, 5.08, is relatively far from 5.00, but C_6H_5Cl makes chemical sense. More significant figures in the mass data are required for a more accurate mole ratio.

3.92 $2C_{57}H_{110}O_6$ + **163**O_2 → **114**CO_2 + **110**H_2O

molar mass of fat = 57(12.01) + 110(1.008) + 6(16.00) = 891.5

$$1.0 \text{ kg fat} \times \frac{1000 \text{ g}}{1 \text{ kg}} \times \frac{1 \text{ mol fat}}{891.5 \text{ g fat}} \times \frac{110 \text{ mol H}_2\text{O}}{2 \text{ mol fat}} \times \frac{18.02 \text{ g H}_2\text{O}}{1 \text{ mol H}_2\text{O}} \times \frac{1 \text{ kg}}{1000 \text{ g}} = 1.1 \text{ kg H}_2\text{O}$$

3.95   All of the O_2 is produced from $KClO_3$; get g $KClO_3$ from g O_2. All of the H_2O is produced from $KHCO_3$; get g $KHCO_3$ from g H_2O. The g H_2O produced also reveals the g CO_2 from the decomposition of $NaHCO_3$. The remaining CO_2 (13.2 g CO_2 - g CO_2 from $NaHCO_3$) is due to K_2CO_3 and g K_2CO_3 can be derived from it.

$$4.00 \text{ g } O_2 \times \frac{1 \text{ mol } O_2}{32.00 \text{ g } O_2} \times \frac{2 \text{ mol } KClO_3}{3 \text{ mol } O_2} \times \frac{122.6 \text{ g } KClO_3}{1 \text{ mol } KClO_3} = 10.22 = 10.2 \text{ g } KClO_3$$

$$1.80 \; H_2O \times \frac{1 \text{ mol } H_2O}{18.02 \text{ g } H_2O} \times \frac{2 \text{ mol } KHCO_3}{1 \text{ mol } H_2O} \times \frac{100.1 \text{ g } KHCO_3}{1 \text{ mol } KHCO_3} = 20.00 = 20.0 \text{ g } KHCO_3$$

$$1.80 \text{ g } H_2O \times \frac{1 \text{ mol } H_2O}{18.02 \text{ g } H_2O} \times \frac{2 \text{ mol } CO_2}{1 \text{ mol } H_2O} \times \frac{44.01 \text{ g } CO_2}{1 \text{ mol } CO_2} = 8.792 = 8.79 \text{ g } CO_2 \text{ from } KHCO_3$$

13.20 g CO_2 total - 8.792 CO_2 from $KHCO_3$ = 4.408 = 4.41 g CO_2 from K_2CO_3

$$4.408 \text{ g } CO_2 \times \frac{1 \text{ mol } CO_2}{44.01 \text{ g } CO_2} \times \frac{1 \text{ mol } K_2CO_3}{1 \text{ mol } CO_2} \times \frac{138.2 \text{ g } K_2CO_3}{1 \text{ mol } K_2CO_3} = 13.84 = 13.8 \text{ g } K_2CO_3$$

100.0 g mixture - 10.22 g $KClO_3$ - 20.00 g $KHCO_3$ - 13.84 g K_2CO_3 = 56.0 g KCl

Integrative Exercises

3.98   Strategy: volume of alloy sphere $\xrightarrow{\text{density}}$ mass of alloy $\xrightarrow{\text{mass \% Cu}}$ mass of Cu

$$V = 4/3 \; \pi \; r^3 = 4/3 \; \pi \; (2.0)^3 \text{ in}^3 \times \frac{(2.54)^3 \text{ cm}^3}{1 \text{ in}^3} = 549 = 5.5 \times 10^2 \text{ cm}^3$$

$$549 \text{ cm}^3 \times \frac{10.3 \text{ g sterling}}{1 \text{ cm}^3} \times \frac{7.5 \text{ g Cu}}{100 \text{ g sterling}} \times \frac{1 \text{ mol Cu}}{63.55 \text{ g Cu}} \times \frac{6.022 \times 10^{23} \text{ atoms}}{\text{mol}}$$

$$= 4.0 \times 10^{24} \text{ g Cu atoms}$$

3.100   Strategy: $m^3 \longrightarrow L \xrightarrow{\text{density}} g \; C_8H_{18} \; g \xrightarrow[\text{ratio}]{\text{mole}} g \; CO_2 \longrightarrow kg \; CO_2$

$$0.15 \text{ m}^3 \; C_8H_{18} \times \frac{10^3 \text{ dm}^3}{1 \text{ m}^3} \times \frac{1 \text{ L}}{1 \text{ dm}^3} \times \frac{1000 \text{ mL}}{1 \text{ L}} \times \frac{0.69 \text{ g } C_8H_{18}}{1 \text{ mL } C_8H_{18}} = 1.035 \times 10^5$$

$$= 1.0 \times 10^5 \text{ g } C_8H_{18}$$

$$2C_8H_{18}(l) + 25O_2(g) \rightarrow 16CO_2(g) + 18H_2O(l)$$

$$1.035 \times 10^5 \text{ g } C_8H_{18} \times \frac{1 \text{ mol } C_8H_{18}}{114.2 \text{ g } C_8H_{18}} \times \frac{16 \text{ mol } CO_2}{2 \text{ mol } C_8H_{18}} \times \frac{44.01 \text{ g } CO_2}{1 \text{ mol } CO_2} \times \frac{1 \text{ kg}}{1000 \text{ g}}$$

$$= 3.2 \times 10^2 \text{ kg } CO_2$$

3.102   (a) $S(s) + O_2(g) \rightarrow SO_2(g)$; $SO_2(g) + CaO(s) \rightarrow CaSO_3(s)$

(b) $\dfrac{2000 \text{ tons coal}}{\text{day}} \times \dfrac{2000 \text{ lb}}{1 \text{ ton}} \times \dfrac{1 \text{ kg}}{2.20 \text{ lb}} \times \dfrac{1000 \text{ g}}{1 \text{ kg}} \times \dfrac{0.028 \text{ g S}}{1 \text{ g coal}} \times \dfrac{1 \text{ mol S}}{32.1 \text{ g S}}$

$\times \dfrac{1 \text{ mol } SO_2}{1 \text{ mol S}} \times \dfrac{1 \text{ mol } CaSO_3}{1 \text{ mol } SO_2} \times \dfrac{120 \text{ g } CaSO_3}{1 \text{ mol } CaSO_3} \times \dfrac{1 \text{ kg } CaSO_3}{1000 \text{ g } CaSO_3}$

$$= 1.9 \times 10^5 \text{ kg } CaSO_3/\text{day}$$

This corresponds to over 200 tons of $CaSO_3$ per day as a waste product.

4 Aqueous Reactions and Solution Stoichiometry

Solution Composition: Molarity

4.1 Concentration is the **ratio** of the amount of solute present in a certain quantity of solvent or solution. This ratio remains constant regardless of how much solution is present. Thus, concentration is an intensive property. The absolute concentration does depend on the amount of solute present, but once this ratio is established, it doesn't vary with the volume of solution present.

4.3 The second solution is 5 times as concentrated as the first. An equal volume of the more concentrated solution will contain 5 times as much solute (5 times the number of moles and also 5 times the mass) as the 0.50 M solution. Thus, the mass of solute in the 2.50 M solution is 5 × 4.5 g = 22.5 g.

Mathematically:

$$\frac{\dfrac{2.50 \text{ mol solute}}{1 \text{ L solution}}}{\dfrac{0.50 \text{ mol solute}}{1 \text{ L solution}}} = \frac{x \text{ grams solute}}{4.5 \text{ g solute}}$$

$$\frac{2.50 \text{ mol solute}}{0.50 \text{ mol solute}} = \frac{x \text{g solute}}{4.5 \text{ g solute}}; \quad 5.0(4.5 \text{ g solute}) = 23 \text{ g solute}$$

The result has 2 sig figs; 22.5 rounds to 23 g solute

4.5 (a) $M = \dfrac{\text{mol solute}}{\text{L solution}}; \dfrac{0.0345 \text{ mol NH}_4\text{Cl}}{400 \text{ mL}} \times \dfrac{1000 \text{ mL}}{1 \text{ L}} = 0.0863 \ M \text{ NH}_4\text{Cl}$

(b) $\text{mol} = M \times \text{L}; \dfrac{2.20 \text{ mol HNO}_3}{1 \text{ L}} \times 0.0350 \text{ L} = 0.0770 \text{ mol HNO}_3$

(c) $L = \dfrac{\text{mol}}{M}; \dfrac{0.125 \text{ mol KOH}}{1.50 \text{ mol KOH/L}} = 0.0833 \text{ L or } 83.3 \text{ mL of } 1.50 \ M \text{ KOH}$

4.7 $M = \dfrac{\text{mol}}{\text{L}}; \text{mol} = \dfrac{\text{g}}{\text{MM}}$ (MM is the symbol for molar mass in this manual.)

(a) $\dfrac{0.150 \ M \text{ KBr}}{1 \text{ L}} \times 0.250 \text{ L} \times \dfrac{119.0 \text{ g KBr}}{1 \text{ mol KBr}} = 4.46 \text{ g KBr}$

(b) $4.75 \text{ g Ca(NO}_3)_2 \times \dfrac{1 \text{ mol Ca(NO}_3)_2}{164.1 \text{ g Ca(NO}_3)_2} \times \dfrac{1}{0.200 \text{ L}} = 0.145 \text{ } M \text{ Ca(NO}_3)_2$

(c) $5.00 \text{ g Na}_3\text{PO}_4 \times \dfrac{1 \text{ mol Na}_3\text{PO}_4}{163.9 \text{ g Na}_3\text{PO}_4} \times \dfrac{1 \text{ L}}{1.50 \text{ mol Na}_3\text{PO}_4} \times \dfrac{1000 \text{ mL}}{1 \text{ L}}$

$$= 20.3 \text{ mL solution}$$

4.9   The number of moles of sucrose needed is

$\dfrac{0.150 \text{ mol}}{1 \text{ L}} \times 0.125 \text{ L} = 0.01875 = 0.0188 \text{ mol}$

Weigh out $0.01875 \text{ mol C}_{12}\text{H}_{22}\text{O}_{11} \times \dfrac{342.3 \text{ g C}_{12}\text{H}_{22}\text{O}_{11}}{1 \text{ mol C}_{12}\text{H}_{22}\text{O}_{11}} = 6.42 \text{ g C}_{12}\text{H}_{22}\text{O}_{11}$

Add this amount of solid to a 125 mL volumetric flask, dissolve in a small volume of water, and add water to the mark on the neck of the flask. Agitate thoroughly to ensure total mixing.

4.11   Calculate the moles of solute present in the final 400 mL of 0.100 M $C_{12}H_{22}O_{11}$ solution:

moles $C_{12}H_{22}O_{11} = M \times L = \dfrac{0.100 \text{ mol C}_{12}\text{H}_{22}\text{O}_{11}}{1 \text{ L}} \times 0.400 \text{ L} = 0.0400 \text{ mol C}_{12}\text{H}_{22}\text{O}_{11}$

Calculate the volume of 1.50 M glucose solution that would contain 0.0400 mol $C_{12}H_{22}O_{11}$:

$L = \text{moles}/M$; $0.0400 \text{ mol C}_{12}\text{H}_{22}\text{O}_{11} \times \dfrac{1 \text{ L}}{1.50 \text{ mol C}_{12}\text{H}_{22}\text{O}_{11}} = 0.02667 = 0.0267 \text{ L}$

$0.02667 \text{ L} \times \dfrac{1000 \text{ mL}}{1 \text{ L}} = 26.7 \text{ mL}$

Thoroughly rinse, clean and fill a 50 mL buret with the 1.50 M $C_{12}H_{22}O_{11}$. Dispense 26.7 mL of this solution into a 400 mL volumetric container, add water to the mark and mix thoroughly. (26.7 mL is a difficult volume to measure with a pipette.)

4.13   Calculate the mass of acetic acid, $HC_2H_3O_2$, present in 10.0 mL of the pure liquid.

$10.00 \text{ mL acetic acid} \times \dfrac{1.049 \text{ g acetic acid}}{1 \text{ mL acetic acid}} = 10.49 \text{ g acetic acid}$

$10.49 \text{ g HC}_2\text{H}_3\text{O}_2 \times \dfrac{1 \text{ mol HC}_2\text{H}_3\text{O}_2}{60.05 \text{ g HC}_2\text{H}_3\text{O}_2} = 0.17469 = 0.1747 \text{ mol HC}_2\text{H}_3\text{O}_2$

$M = \text{mol/L} = \dfrac{0.17469 \text{ mol HC}_2\text{H}_3\text{O}_2}{0.1000 \text{ L solution}} = 1.747 \text{ } M \text{ HC}_2\text{H}_3\text{O}_2$

Electrolytes

4.15   Tap water contains enough dissolved electrolytes to conduct a significant amount of electricity. Thus, water can complete a circuit between an electrical appliance and our body, producing a shock.

4.17   (a)   HF -- weak   (b) C_2H_5OH -- non   (c) NH_3 -- weak

          (d) $KClO_3$ -- strong   (e) $Cu(NO_3)_2$ -- strong

4.19   (a) $0.14\ M\ Na^+$, $0.14\ M\ OH^-$

          (b) $0.25\ M\ Ca^{2+}$, $0.50\ M\ Br^-$

          (c) $0.25\ M$ (CH_3OH is a molecular solute)

          (d) $M_2 = M_1 V_1 / V_2$, where V_2 is the total solution volume.

$$K^+:\ \frac{0.20\ M \times 0.050\ L}{0.075\ L} = 0.133 = 0.13\ M$$

ClO_3^-: concentration ClO_3^- = concentration Na^+ = 0.13 M

$$SO_4^{2-}:\ \frac{0.20\ M \times 0.0250\ L}{0.075\ L} = 0.0667 = 0.067\ M\ SO_4^{2-}$$

Na^+: concentration Na^+ = 2 × concentration SO_4^{2-} = 0.13 M

4.21 $KCl \rightarrow K^+ + Cl^-$; 0.20 M KCl = 0.20 M K^+

$K_2Cr_2O_7 \rightarrow 2\ K^+ + Cr_2O_7^{2-}$; 0.15 M $K_2Cr_2O_7$ = 0.30 M K^+

$K_3PO_4 \rightarrow 3\ K^+ + PO_4^{3-}$; 0.080 M K_3PO_4 = 0.24 M K^+

0.15 M $K_2Cr_2O_7$ has the highest K^+ concentration.

Acids, Bases and Salts

4.23   Since the solution does conduct some electricity, but less than an equimolar NaCl solution (a strong electrolyte) the unknown solute must be a weak electrolyte. The weak electrolytes in the list of choices are NH_3 and H_3PO_3; since the solution is acidic, the unknown must be **H_3PO_3.**

4.25   (a)   A *monoprotic acid* has one ionizable (acidic) H and a *diprotic acid* has two.

          (b)   A *strong acid* is completely ionized in aqueous solution whereas only a fraction of *weak acid* molecules are ionized.

      (c)   An *acid* is an H^+ donor, a substance that increases the concentration of H^+ in aqueous solution. A *base* is an H^+ acceptor and thus increases the concentration of OH^- in aqueous solution.

4.27   In aqueous solution, HNO_3 exists entirely as H^+ and NO_3^- ions; the single arrow denotes complete ionization. HCN is only dissociated to a small extent; the double arrow denotes a mixture of H^+ ions, CN^- ions and neutral undissociated HCN molecules in solution.

4.29   (a) $2HBr(aq) + Ca(OH)_2(aq) \rightarrow CaBr_2(aq) + 2H_2O(l)$

          (b) $Cu(OH)_2(s) + 2HClO_4(aq) \rightarrow Cu(ClO_4)_2(aq) + 2H_2O(l)$

          (c) $2Fe(OH)_3(s) + 3H_2SO_4(aq) \rightarrow Fe_2(SO_4)_3(aq) + 6H_2O(l)$

Ionic Equations; Metathesis Reactions

4.31   (a) $Pb^{2+}(aq) + SO_4^{2-}(aq) \rightarrow PbSO_4(s)$; spectators: Na^+, NO_3^-

          (b) $2Al(s) + 6H^+(aq) \rightarrow 2Al^{3+}(aq) + 3H_2(g)$; spectator: Cl^-

          (c) $FeO(s) + 2H^+(aq) \rightarrow H_2O(l) + Fe^{2+}(aq)$; spectator: ClO_4^-

4.33   The driving force in a metathesis reaction is the formation of a product that removes ions from solution. This means that at least some reactant species present as ions exist in a different form in the products, as indicated by the net ionic equation.

        Driving forces in Exercise 4.32:

        (a) $H_2O(l)$ -- nonelectrolyte

        (b) $H_2O(l)$, $CO_2(g)$ -- nonelectrolyte and gas, respectively

        (c) $Cu(OH)_2(s)$ -- precipitate

4.35   Follow the guidelines in Table 4.3.

        (a) $NiCl_2$ - soluble   (b) Ag_2S - insoluble

        (c) Cs_3PO_4 - soluble (Cs^+ is an alkali metal cation)

        (d) $SrCO_3$ - insoluble   (e) $(NH_4)_2SO_4$ - soluble

4.37 Br^- and NO_3^- can be ruled out because the Ba^{2+} salts are soluble. (Actually all NO_3^- salts are soluble.) CO_3^{2-} forms insoluble salts with the three cations given; it must be the anion in question.

4.39   (a) $Ba^{2+}(aq) + HSO_4^-(aq) \rightarrow BaSO_4(s) + H^+(aq)$

          (b)   No reaction. Both components are soluble electrolytes, as are possible metathesis reaction products.

          (c) $2Ag^+(aq) + CO_3^{2-}(aq) \rightarrow Ag_2CO_3(s)$

          (d) $H^+(aq) + OH^-(aq) \rightarrow H_2O(l)$

          (e) $HC_2H_3O_2(aq) + OH^-(aq) \rightarrow H_2O(l) + C_2H_3O_2^-(aq)$

          (f) $Pb^{2+}(aq) + SO_4^{2-}(aq) \rightarrow PbSO_4(s)$

Oxidation - Reduction Reactions

4.41   Corrosion was one of the first oxidation-reduction processes to be studied in detail. During corrosion, a metal reacts with some environmental agent, usually oxygen, to form a metal compound. Thus, reaction of a metal with oxygen causes the metal to lose electrons; reaction with oxygen is logically called *oxidation*. Eventually the term oxidation was generalized to mean any process where a substance loses electrons, whether or not the substance is a metal or oxygen is a reactant.

4.43   The most easily oxidized metals are near the bottom of groups on the left side of the chart, especially groups 1A and 2A. The least easily oxidized metals are on the lower right of the transition metals, particularly those near the bottom of groups 8B and 1B.

4.45   (a) $2HCl(aq) + Ni(s) \rightarrow NiCl_2(aq) + H_2(g)$; $Ni(s) + 2H^+(aq) \rightarrow Ni^{2+}(aq) + H_2(g)$

        (b) $H_2SO_4(aq) + Fe(s) \rightarrow FeSO_4(aq) + H_2(g)$; $Fe(s) + 2H^+(aq) \rightarrow Fe^{2+}(aq) + H_2(g)$

        (c) $2HBr(aq) + Zn(s) \rightarrow ZnBr_2(aq) + H_2(g)$; $Zn(s) + 2H^+(aq) \rightarrow Zn^{2+}(aq) + H_2(g)$

        (d) $2HC_2H_3O_2(aq) + Mg(s) \rightarrow Mg(C_2H_3O_2)_2(aq) + H_2(g)$;

               $Mg(s) + 2HC_2H_3O_2(aq) \rightarrow Mg^{2+}(aq) + 2C_2H_3O_2^-(aq) + H_2(g)$

4.47   (a) $2Al(s) + 3NiCl_2(aq) \rightarrow 2AlCl_3(aq) + 3Ni(s)$

        (b) $Ag(s) + Pb(NO_3)_2(aq) \rightarrow$ no reaction

        (c) $2Cr(s) + 3NiSO_4(aq) \rightarrow Cr_2(SO_4)_3(aq) + 3Ni(s)$

        (d) $Mn(s) + 2HBr(aq) \rightarrow MnBr_2(aq) + H_2(g)$

        (e) $H_2(g) + CuCl_2(aq) \rightarrow Cu(s) + 2HCl(aq)$

        (f) $Ba(s) + 2H_2O(l) \rightarrow Ba(OH)_2(aq) + H_2(g)$

               (The most active metals can displace H^+ from H_2O as well as from acids.)

4.49   (a)   i. $Zn(s) + Cd^{2+}(aq) \rightarrow Cd(s) + Zn^{2+}(aq)$

               ii. $Cd(s) + Ni^{2+}(aq) \rightarrow Ni(s) + Cd^{2+}(aq)$

        (b)   According to Table 4.5, the most active metals are most easily oxidized, and Zn is more active than Ni. Observation (i) indicates that Cd is less active than Zn; (ii) indicates that Cd is more active than Ni. Cd is between Zn and Ni on the activity series.

        (c)   Place an iron strip in $CdCl_2(aq)$. If Cd(s) is deposited, Cd is less active than Fe; if there is no reaction, Cd is more active than Fe. Do the same test with Co if Cd is less active than Fe or with Cr if Cd is more active than Fe.

3

Solution Stoichiometry; Titrations

4.51   (a)   Write the balanced equation for the reaction in question:

$HClO_4(aq) + NaOH(aq) \rightarrow NaClO_4(aq) + H_2O(l)$

Calculate the moles of the known substance, in this case NaOH.

moles NaOH = $M \times L = \dfrac{0.0875 \text{ mol NaOH}}{1 \text{ L}} \times 0.0500 \text{ L} = 0.004375$

$= 0.00438 \text{ mol NaOH}$

Apply the mole ratio (mol unknown/mol known) from the chemical equation.

$0.004375 \text{ mol NaOH} \times \dfrac{1 \text{ mol } HClO_4}{1 \text{ mol NaOH}} = 0.004375 \text{ mol } HClO_4$

Calculate the desired quantity of unknown, in this case the volume of 0.115 M $HClO_4$ solution.

L = mol/M;   L = $0.004375 \text{ mol } HClO_4 \times \dfrac{1 \text{ L}}{0.115 \text{ mol HCl}} = 0.0380 \text{ L} = 38.0 \text{ mL}$

  (b)   Following the procedure outlined in part (a):

$2HCl(aq) + Mg(OH)_2(s) \rightarrow MgCl_2(aq) + 2H_2O(l)$

$2.87 \text{ g } Mg(OH)_2 \times \dfrac{1 \text{ mol } Mg(OH)_2}{58.33 \text{ g } Mg(OH)_2} = 0.04920 = 0.04920 \text{ mol } Mg(OH)_2$

$0.0492 \text{ mol } Mg(OH)_2 \times \dfrac{2 \text{ mol HCl}}{1 \text{ mol } Mg(OH)_2} = 0.0984 \text{ mol HCL}$

L = mol/M = $0.09840 \text{ mol HCl} \times \dfrac{1 \text{ L HCl}}{0.128 \text{ mol HCl}} = 0.769 \text{ L} = 769 \text{ mL}$

  (c) $AgNO_3(aq) + KCl(aq) \rightarrow AgCl(s) + KNO_3(aq)$

$785 \text{ mg KCl} \times \dfrac{1 \times 10^{-3} \text{ g}}{1 \text{ mg}} \times \dfrac{1 \text{ mol KCl}}{74.55 \text{ g KCl}} \times \dfrac{1 \text{ mol } AgNO_3}{1 \text{ mol KCl}} = 0.01053$

$= 0.0105 \text{ mol } AgNO_3$

$M = \text{mol/L} = \dfrac{0.01053 \text{ mol } AgNO_3}{0.0258 \text{ L}} = 0.408 \ M \ AgNO_3$

  (d) $HCl(aq) + KOH(aq) \rightarrow KCl(aq) + H_2O(l)$

$\dfrac{0.108 \text{ mol HCl}}{1 \text{ L}} \times 0.0453 \text{ L} \times \dfrac{\text{mol KOH}}{\text{mol HCl}} \times \dfrac{56.11 \text{ g KOH}}{1 \text{ mol KOH}} = 0.275 \text{ g KOH}$

4.53   See Exercise 4.51(a) for a more detailed approach.

$\dfrac{6.0 \text{ mol } H_2SO_4}{1 \text{ L}} \times 0.035 \text{ L} \times \dfrac{2 \text{ mol } NaHCO_3}{1 \text{ mol } H_2SO_4} \times \dfrac{84.01 \text{ g } NaHCO_3}{1 \text{ mol } NaHCO_3} = 35 \text{ g } NaHCO_3$

4.55   The neutralization reaction here is:

$2HBr(aq) + Ca(OH)_2(aq) \rightarrow CaBr_2(aq) + 2H_2O(l)$

$0.0488 \text{ L HBr soln} \times \dfrac{5.00 \times 10^{-2} \text{ mol HBr}}{1 \text{ L soln}} \times \dfrac{1 \text{ mol Ca(OH)}_2}{2 \text{ mol HBr}} \times \dfrac{1}{0.100 \text{ L of Ca(OH)}_2}$

$= 1.220 \times 10^{-2} = 1.22 \times 10^{-2} \ M \text{ Ca(OH)}_2$

From the molarity of the saturated solution, we can calculate the gram solubility of $Ca(OH)_2$ in 100 mL of H_2O.

$0.100 \text{ L soln} \times \dfrac{1.220 \times 10^{-2} \text{ mol Ca(OH)}_2}{1 \text{ L soln}} \times \dfrac{74.10 \text{ g Ca(OH)}_2}{1 \text{ mol Ca(OH)}_2}$

$= 0.0904 \text{ g Ca(OH)}_2 \text{ in 100 mL soln}$

Additional Exercises

4.57   (a) $0.0500 \text{ L soln} \times \dfrac{0.200 \text{ mol NaCl}}{1 \text{ L soln}} = 1.00 \times 10^{-2} \text{ mol NaCl}$

$0.1000 \text{ L soln} \times \dfrac{0.100 \text{ mol NaCl}}{1 \text{ L soln}} = 1.00 \times 10^{-2} \text{ mol NaCl}$

Total moles NaCl $= 2.00 \times 10^{-2}$, total volume = 0.0500 L + 0.1000 L + 0.1500 L

Molarity $= \dfrac{2.00 \times 10^{-2} \text{ mol}}{0.150 \text{ L}} = 0.133 \ M$

(b) $0.0245 \text{ L soln} \times \dfrac{1.50 \text{ mol NaOH}}{1 \text{ L soln}} = 0.03675 = 0.0368 \text{ mol NaOH}$

$0.0250 \text{ L soln} \times \dfrac{0.850 \text{ mol NaOH}}{1 \text{ L soln}} = 0.017425 = 0.0174 \text{ mol NaOH}$

Total moles NaOH = 0.054175 = 0.0542, total volume = 0.0450 L

Molarity $= \dfrac{0.054175 \text{ mol NaOH}}{0.0450 \text{ L}} = 1.20 \ M$

4.60 Na^+ must replace the total + charge due to Ca^{2+} and Mg^{2+}. Think of this as moles of charge rather than moles of particles.

$\dfrac{0.010 \text{ mol Ca}^{2+}}{1 \text{ L water}} \times 1.0 \times 10^3 \text{ L} \times \dfrac{2 \text{ mol + charge}}{1 \text{ mol Ca}^{2+}} = 20 \text{ mol of + charge}$

$\dfrac{0.0050 \text{ mol Mg}^{2+}}{1 \text{ L water}} \times 1.0 \times 10^3 \text{ L} \times \dfrac{2 \text{ mol + charge}}{1 \text{ mol Mg}^{2+}} = 10 \text{ mol of + charge}$

30 moles of + charge must be replaced; 30 mol Na^+ are needed.

4.63   The two precipitates formed are due to AgCl(s) and $SrSO_4$(s). Since no precipitate forms on addition of hydroxide ion to the remaining solution, the other two possibilities, Ni^{2+} and Mn^{2+}, are absent.

4.66   A metal on Table 4.4 is able to displace the metal cations below it from their compounds. That is, zinc will reduce the cations below it to their metals.

     (a) $Zn(s) + Na^+(aq) \rightarrow$ no reaction

     (b) $Zn(s) + Pb^{2+}(aq) \rightarrow Zn^{2+}(aq) + Pb(s)$

     (c) $Zn(s) + Mg^{2+}(aq) \rightarrow$ no reaction

     (d) $Zn(s) + Fe^{2+}(aq) \rightarrow Zn^{2+}(aq) + Fe(s)$

     (e) $Zn(s) + Cu^{2+}(aq) \rightarrow Zn^{2+}(aq) + Cu(s)$

     (f) $Zn(s) + Al^{3+}(aq) \rightarrow$ no reaction

4.68 $H_2C_4H_4O_6 + 2OH^-(aq) \rightarrow C_4H_4O_6^{2-}(aq) + 2H_2O(l)$

$$0.02262 \text{ L NaOH soln} \times \frac{0.2000 \text{ mol NaOH}}{1 \text{ L}} \times \frac{1 \text{ mol } H_2C_4H_4O_6}{2 \text{ mol NaOH}} \times \frac{1}{0.04000 \text{ L } H_2C_4H_4O_6}$$

$$= 0.05655 \ M \ H_2C_4H_4O_6 \text{ soln}$$

Integrative Exercises

4.70   Strategy: $M \times L = \text{mol } Na_3PO_4 \rightarrow \text{mol } Na^+ \rightarrow Na^+$ ions

$$\frac{0.0100 \text{ mol } Na_3PO_4}{1 \text{ L solution}} \times 1.00 \text{ mL} \times \frac{1 \text{ L}}{1000 \text{ mL}} \times \frac{3 \text{ mol } Na^+}{1 \text{ mol } Na_3PO_4} \times \frac{6.022 \times 10^{23} \ Na^+ \text{ ions}}{1 \text{ mol } Na^+}$$

$$= 1.81 \times 10^{19} \ Na^+ \text{ ions}$$

4.72 $Ba^{2+}(aq) + SO_4^{2-}(aq) \rightarrow BaSO_4(s)$

$$0.4123 \text{ g } BaSO_4 \times \frac{137.3 \text{ g Ba}}{233.4 \text{ g } BaSO_4} = 0.2425 \text{ g Ba}$$

$$\text{mass \%} = \frac{\text{g Ba}}{\text{g sample}} \times 100 = \frac{0.24254 \text{ g Ba}}{6.977 \text{ g sample}} \times 100 = 3.476\% \text{ Ba}$$

4.75 $0.0250 \text{ L soln} \times \dfrac{0.102 \text{ mol } Ag^+}{1 \text{ L soln}} \times \dfrac{1 \text{ mol } Ag_3AsO_4}{3 \text{ mol } Ag^+} \times \dfrac{1 \text{ mol As}}{1 \text{ mol } Ag_3AsO_4} \times \dfrac{74.92 \text{ g As}}{1 \text{ mol As}}$

$$= 0.06368 = 0.0637 \text{ g As}$$

$$\text{mass percent} = \frac{0.06368 \text{ g As}}{1.22 \text{ g sample}} \times 100 = 5.22\% \text{ As}$$

5 Thermochemistry

Nature of Energy

5.1   (a)   *Heat* is the transfer of energy due to a difference in temperature.

       (b)   Heat is transferred from one system to another until the two systems are at the same temperature.

5.3   (a)   Gravity; work is done because the force of gravity is opposed and the pencil is lifted.

       (b)   Mechanical force; work is done because the force of the coiled spring is opposed as the spring is compressed over a distance.

5.5   (a)   Since $1 \text{ J} = 1 \text{ kg} \cdot \text{m}^2/\text{s}^2$, convert g $\rightarrow$ kg to obtain E_k in joules.

$$E_k = 1/2 \ mv^2 = 1/2 \ \times 45 \text{ g} \ \times \frac{1 \text{ kg}}{1000 \text{ g}} \ \times \left(\frac{61 \text{ m}}{1 \text{ s}} \right)^2 = \frac{84 \text{ kg} \cdot \text{m}^2}{1 \text{ s}^2} = 84 \text{ J}$$

       (b) $83.72 \text{ J} \ \times \ \dfrac{1 \text{ cal}}{4.184 \text{ J}} = 20 \text{ cal}$

       (c)   As the ball hits the tree, its speed (and hence its kinetic energy) drops to zero. Most of the kinetic energy is transferred to the potential energy of a slightly deformed golf ball, some is absorbed by the tree and some is released as heat. As the ball bounces off the tree, its potential energy is reconverted to kinetic energy.

5.7   Find: J/Btu

      Given: heat capacity of water = 1 Btu/lb $\cdot$ °F

      Know: heat capacity of water = 4.184 J/g $\cdot$ °C

      Strategy: $\dfrac{\text{J}}{\text{g} \cdot \text{°C}} \rightarrow \dfrac{\text{J}}{\text{lb} \cdot \text{°F}} \rightarrow \dfrac{\text{J}}{\text{Btu}}$

      This strategy requires changing °F to °C. Since this involves the magnitude of a degree on each scale, rather than a specific temperature, the 32 in the temperature relationship is not needed.

      100 °C = 180 °F;   5 °C = 9 °F

$$\frac{4.184 \text{ J}}{\text{g} \cdot \text{°C}} \ \times \ \frac{453.6 \text{ g}}{\text{lb}} \ \times \ \frac{5 \text{ °C}}{9 \text{ °F}} \ \times \ \frac{1 \text{ lb} \cdot \text{°F}}{1 \text{ Btu}} = 1054 \text{ J/Btu}$$

First Law of Thermodynamics

5.9   (a)   In thermodynamics, the *system* is the well-defined part of the universe whose energy changes are being studied.

       (b)   A closed system can exchange heat but not mass with its surroundings.

5.11   (a)   In any chemical or physical change, energy can be neither created nor destroyed, but it can be changed in form.

       (b)   The total *internal energy* (E) of a system is the sum of all the kinetic and potential energies of the system components.

       (c)   The internal energy of a system increases when work is done on the system by the surroundings and/or when heat is transferred to the system from the surroundings (the system is heated).

5.13   In each case, evaluate q and w in the expression $\Delta E = q + w$. For an exothermic process, q is negative; for an endothermic process, q is positive.

       (a)   q is positive because the system gains heat and w is negative because the system does work. $\Delta E = +327$ kJ - 430 kJ = -103 kJ. The process is endothermic.

       (b) $\Delta E = -1.15$ kJ - 934 J = -1.15 kJ - 0.934 kJ = -2.08 kJ The process is exothermic.

       (c)   q is negative because the system loses heat and w is positive because the work is done on the system. $\Delta E = -245$ J + 97 J = -148 J. The process is exothermic.

5.15   (a)   A *state function* is a property of a system that depends only on the physical state (pressure, temperature, etc.) of the system, not on the route used by the system to get to the current state.

       (b)   Internal energy and enthalpy <u>are</u> state functions; work <u>is not</u> a state function.

       (c)   Temperature is a state function; regardless of how hot or cold the sample has been, the temperature depends only on its present condition.

Enthalpy

5.17   (a)   When a process occurs under constant external pressure, the enthalpy change (ΔH) equals the amount of heat transferred. $\Delta H = q_p$.

       (b)   No. Enthalpy is a state function, so it is totally defined by the current conditions (state) of the system, not the history of how the system arrived at its current state.

       (c) $\Delta H = q_p$. If the system absorbs heat, q and ΔH are positive and the enthalpy of the system increases.

5.19   (a) $CH_3OH(l) + 3/2\ O_2(g) \rightarrow 2\ H_2O(l) + CO_2(g)$

               $\Delta H = -726.7$ kJ

        (b) $\underline{CH_3OH(l) + 3/2\ O_2(g)}$

                 $\Delta H = -726.7$ kJ

                 $\downarrow$

                 $\overline{2\ H_2O(l) + CO_2(g)}$

5.21   (a)   Exothermic (ΔH is negative).

        (b) 5.6 g Na $\times \dfrac{1\ \text{mol}}{22.99\ \text{g}} \times \dfrac{-821.8\ \text{kJ}}{2\ \text{mol Na}} = 1.0 \times 10^2$ kJ

        (c) 16.5 kJ $\times \dfrac{2\ \text{mol NaCl}}{821.8\ \text{kJ}} \times \dfrac{58.44\ \text{g NaCl}}{1\ \text{mol NaCl}} = 2.35$ g NaCl produced

        (d) $2NaCl(s) \rightarrow 2Na(s) + Cl_2(g)$ $\Delta H = +821.8$ kJ

           This is the reverse of the reaction given above, so the sign of ΔH is reversed.

           44.1g NaCl $\times \dfrac{1\ \text{mol NaCl}}{58.44\ \text{g NaCl}} \times \dfrac{821.8\ \text{kJ}}{2\ \text{mol NaCl}} = 310$ kJ absorbed

5.23   (a) 0.715 mol $O_2 \times \dfrac{-89.4\ \text{kJ}}{3\ \text{mol }O_2} = -21.3$ kJ

        (b) 6.14 g KCl $\times \dfrac{1\ \text{mol KCl}}{74.55\ \text{g KCl}} \times \dfrac{-89.4\ \text{kJ}}{2\ \text{mol KCl}} = -3.68$ kJ

        (c) 12.3 g $KClO_3 \times \dfrac{1\ \text{mol }KClO_3}{122.6\text{g }KClO_3} \times \dfrac{+89.4\ \text{kJ}}{2\ \text{mol }KClO_3} = +4.48$ kJ (sign of ΔH reversed)

5.25   Enthalpy of $H_2O(s) < H_2O(l) < H_2O(g)$. Heat must be added to convert s $\rightarrow$ l $\rightarrow$ g.

5.27   At constant pressure, $\Delta E = \Delta H - P\Delta V$. In order to calculate ΔE, more information about the conditions of the reaction must be known. For an ideal gas at constant pressure and temperature, $P\Delta V = RT\Delta n$. The values of either P and ΔV or T and Δn must be known to calculate ΔE from ΔH.

5.29   q = -135 kJ (heat is given off by the system), w = -63 kJ (work is done by the system).

      ΔE = q + w = -135 kJ - 63 kJ = -198 kJ. ΔH = q = -135 kJ (at constant pressure).

5.31   (a) $2CO_2(g) + 3H_2O(g) \rightarrow C_2H_6(g) + 7/2\ O_2(g)$ $\Delta H = +1430$ kJ

        (b) $2C_2H_6(g) + 7O_2(g) \rightarrow 4CO_2(g) + 6H_2O(g)$ $\Delta H = 2(-1430)$ kJ = -2860 kJ

        (c)   The exothermic forward reaction is more likely to be thermodynamically favored.

(d)   Vaporization (liquid → gas) is endothermic so the reverse process, condensation (gas → liquid) is exothermic. If the product was $H_2O(l)$, the reaction would be more exothermic and ΔH would have a larger negative value.

Calorimetry

The specific heat of water to four significant figures, **4.184 J/g • K**, will be used in many of the following exercises; temperature units of K and °C will be used interchangeably.

5.33   (a) $\dfrac{4.184\ J}{1\ g \cdot K}$ or $\dfrac{4.184\ J}{1\ g \cdot °C}$ (b) $\dfrac{265\ g\ H_2O \times 4.184\ J}{1\ g \cdot °C} = \dfrac{1.11 \times 10^3\ J}{°C}$ = 1.11 kJ/°C

       (c)   1.00 kg $H_2O \times \dfrac{1000\ g}{1\ kg} \times \dfrac{4.184\ J}{1\ g \cdot °C} \times \dfrac{1\ kJ}{1000\ J} \times 25.0\ °C$ = 105 kJ

5.35   156 g Si $\times \dfrac{0.702\ J}{g \cdot K} \times (37.5\ °C - 25.0\ °C) = 1.37 \times 10^3\ J$

5.37   Since the temperature of the water increases, the dissolving process is exothermic and the sign of ΔH is negative. The heat lost by the NaOH(s) dissolving equals the heat gained by the solution.

Calculate the heat gained by the solution. The temperature change is 37.8 - 21.6 = 16.2°C. The total mass of solution is (100.0 g H_2O + 6.50 g NaOH) = 106.5 g.

106.5 g solution $\times \dfrac{4.184\ J}{1\ g \cdot °C} \times 16.2°C \times \dfrac{1\ kJ}{1000\ J}$ = 7.219 = 7.22 kJ

This is the amount of heat lost when 6.50 g of NaOH dissolves.

The heat loss per mole NaOH is

$\dfrac{-7.219\ kJ}{6.50\ g\ NaOH} \times \dfrac{40.00\ g\ NaOH}{1\ mol\ NaOH}$ = -44.4 kJ/mol $\Delta H = q_p$ = -44.4 kJ/mol NaOH

5.39 $q_{bomb} = -q_{rxn}$; ΔT = 30.57°C - 23.44°C = 7.13°C

$q_{bomb} = \dfrac{7.854\ kJ}{1°C} \times 7.13°C$ = 56.00 = 56.0 kJ

At constant volume, $q_V = \Delta E$. ΔE and ΔH are very similar.

$\Delta E_{rxn} = q_{rxn} = -q_{bomb} = \dfrac{-56.0\ kJ}{2.20\ g\ C_6H_4O_2}$ = -25.454 = -25.5 kJ/g $C_6H_4O_2$

$\Delta E_{rxn} = \dfrac{-25.454\ kJ}{1\ g\ C_6H_4O_2} \times \dfrac{108.1\ g\ C_6H_4O_2}{1\ mol\ C_6H_4O_2}$ = -2.75 × 10³ kJ/mol $C_6H_4O_2$

5.41 (a) $C_{total} = 2.500$ g glucose $\times \dfrac{15.57 \text{ kJ}}{1 \text{ g glucose}} \times \dfrac{1}{2.70°C} = 14.42 = 14.4$ kJ/°C

(b) $C_{H_2O} = 2.700$ kg $H_2O \times \dfrac{4.184 \text{ kJ}}{1 \text{ kg} \cdot °C} = 11.30$ kJ/°C

$C_{empty\ calorimeter} = \dfrac{14.42 \text{ kJ}}{1°C} - \dfrac{11.30 \text{ kJ}}{1°C} = 3.12 = 3.1$ kJ/°C

(c) $q = 2.500$ g glucose $\times \dfrac{15.57 \text{ kJ}}{1 \text{ g glucose}} = 38.93$ kJ produced

$C_{H_2O} = 2.000$ kg $H_2O \times \dfrac{4.184 \text{ kJ}}{1 \text{ kg} \cdot °C} = 8.368$ kJ/°C

$C_{total} = \dfrac{8.368 \text{ kJ}}{1°C} + \dfrac{3.12 \text{ kJ}}{1°C} = 11.49 = 11.5$ kJ/°C

38.98 kJ $= \dfrac{11.49 \text{ kJ}}{°C} \times \Delta T; \quad \Delta T = 3.39°C$

Hess's Law

5.43   If a reaction can be described as a series of steps, ΔH for the reaction is the sum of the enthalpy changes for each step. As long as we can describe a route where ΔH for each step is known, ΔH for any process can be calculated.

5.45 (a)
$$\begin{array}{lll} A \rightarrow B & \Delta H = +30 \text{ kJ} \\ \underline{B \rightarrow C} & \underline{\Delta H = +60 \text{ kJ}} \\ A \rightarrow C & \Delta H = +90 \text{ kJ} \end{array}$$

(b)

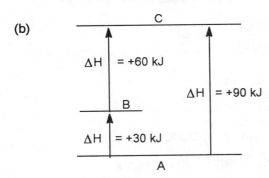

The process of A forming C can be described as A forming B and B forming C.

5.47
$$\begin{array}{ll} N_2(g) + 2O_2(g) \rightarrow 2NO_2(g) & \Delta H = +\ 67.6 \text{ kJ} \\ 2NO_2(g) \rightarrow O_2(g) + 2NO(g) & \Delta H = +113.2 \text{ kJ} \\ \hline \\ N_2(g) + O_2(g) \rightarrow 2NO(g) & \Delta H = +180.8 \text{ kJ} \end{array}$$

5.49

$$C_2H_4(g) \rightarrow 2\,H_2(g) + 2C(s) \qquad\qquad \Delta H = -52.3 \text{ kJ}$$
$$2C(s) + 4F_2(g) \rightarrow 2CF_4(g) \qquad\qquad \Delta H = 2(-680 \text{ kJ})$$
$$2H_2(g) + 2F_2(g) \rightarrow 4HF(g) \qquad\qquad \Delta H = 2(-537 \text{ kJ})$$

$$C_2H_4(g) + 6F_2(g) \rightarrow 2CF_4(g) + 4HF(g) \qquad \Delta H = -2.49 \times 10^3 \text{ kJ}$$

Enthalpies of Formation

5.51   (a)   *Standard conditions* for enthalpy changes are usually P = 1 atm and T = 298 K. For the purpose of comparison, standard enthalpy changes, $\Delta H°$, are tabulated for reactions at these conditions.

      (b)   *Enthalpy of formation*, ΔH_f, is the enthalpy change that occurs when a compound is formed from its component elements.

      (c)   *Standard enthalpy of formation*, $\Delta H_f^°$, is the enthalpy change that accompanies formation of one mole of a substance from elements in their standard states.

5.53   (a) $H_2(g) + O_2(g) \rightarrow H_2O_2(g)$ $\Delta H_f^° = -136.10 \text{ kJ}$

      (b) $N_2(g) + 1/2\,O_2(g) \rightarrow N_2O(g)$ $\Delta H_f^° = +81.6 \text{ kJ}$

      (c) $Pb(s) + C(s, gr) + 3/2\,O_2(g) \rightarrow PbCO_3(s)$ $\Delta H_f^° = -699.1 \text{ kJ}$

      (d) $Na(s) + 1/2\,H_2(g) + C(s, gr) + 3/2\,O_2(g) \rightarrow NaHCO_3(s)$ $\Delta H_f^° = -947.7 \text{ kJ}$

5.55   Use heats of formation to calculate $\Delta H°$ for the combustion of butane.

$$C_4H_{10}(l) + 13/2\,O_2(g) \rightarrow 4CO_2(g) + 5H_2O(l)$$

$$\Delta H_{rxn}^° = 4\Delta H_f^°\ CO_2(g) + 5\Delta H_f^°\ H_2O(l) - \Delta H_f^°\ C_4H_{10}(l)$$

$$\Delta H_{rxn}^° = 4(-393.5 \text{ kJ}) + 5(-285.83 \text{ kJ}) - (-147.6 \text{ kJ}) = -2855.6 = -2856 \text{ kJ/mol } C_4H_{10}$$

$$1.0 \text{ g } C_4H_{10} \times \frac{1 \text{ mol } C_4H_{10}}{58.123 \text{ g } C_4H_{10}} \times \frac{-2855.6 \text{ kJ}}{1 \text{ mol } C_4H_{10}} = 49 \text{ kJ}$$

5.57   (a) $\Delta H_{rxn}^° = \Delta H_f^°\ N_2O_4(g) - 2\Delta H_f^°\ NO_2(g)$

              $= 9.66 \text{ kJ} - 2(33.84 \text{ kJ}) = -58.02 \text{ kJ}$

      (b) $\Delta H_{rxn}^° = \Delta H_f^°\ CaO(s) + \Delta H_f^°\ CO_2(g) - \Delta H_f^°\ CaCO_3(s)$

              $= -635.5 \text{ kJ} + (-393.5 \text{ kJ}) - (-1207.1 \text{ kJ}) = +178.1 \text{ kJ}$

      (c) $\Delta H_{rxn}^° = 2\Delta H_f^°\ NH_3(g) - \Delta H_f^°\ N_2(g) - 3\Delta H_f^°\ H_2(g)$

              $= 2(-46.19 \text{ kJ}) - 0 - 3(0) = -92.38 \text{ kJ}$

      (d) $\Delta H_{rxn}^° = 2\Delta H_f^°\ FeCl_3(s) + 3\Delta H_f^°\ H_2O(g) - \Delta H_f^°\ Fe_2O_3(s) - 6\Delta H_f^°\ HCl(g)$

              $= 2(-400 \text{ kJ}) + 3(-241.82 \text{ kJ}) - (-822.16 \text{ kJ}) - 6(-92.30 \text{ kJ}) = -150 \text{ kJ}$

5.59 $\Delta H^{\circ}_{rxn} = 3\Delta H^{\circ}_{f} \; CO_2(g) + 3\Delta H^{\circ}_{f} \; H_2O(l) - \Delta H^{\circ}_{f} \; C_3H_6O(l)$

$-1790 \text{ kJ} = 3(-393.5 \text{ kJ}) + 3(-285.83 \text{ kJ}) - \Delta H^{\circ}_{f} \; C_3H_6O(l)$

$\Delta H^{\circ}_{f} \; C_3H_6O(l) = -248 \text{ kJ}$

5.61   (a) $C_8H_{18}(l) + 25/2 \; O_2(g) \rightarrow 8CO_2(g) + 9H_2O(g) \qquad \Delta H^{\circ} = -5069 \text{ kJ}$

   (b) $8C(s, gr) + 9H_2(g) \rightarrow C_8H_{18}(l) \quad \Delta H^{\circ}_{f} = ?$

   (c) $\Delta H^{\circ}_{rxn} = 8\Delta H^{\circ}_{f} \; CO_2(g) + 9\Delta H^{\circ}_{f} \; H_2O(g) - \Delta H^{\circ}_{f} \; C_8H_{18}(l) - 25/2 \; \Delta H^{\circ}_{f} \; O_2(g)$

   $-5069 \text{ kJ} = 8(-393.5 \text{ kJ}) + 9(-241.82 \text{ kJ}) - \Delta H^{\circ}_{f} \; C_8H_{18}(l) - 0$

   $\Delta H^{\circ}_{f} \; C_8H_{18}(l) = 8(-393.5 \text{ kJ}) + 9(-241.82 \text{ kJ}) + 5069 \text{ kJ} = -255 \text{ kJ}$

5.63 $2B(s) + 3/2 \; O_2(g) \rightarrow B_2O_3(s) \qquad\qquad \Delta H^{\circ} = 1/2(-2509.1 \text{ kJ})$

   $3H_2(g) + 3/2 \; O_2(g) \rightarrow 3H_2O(l) \qquad\quad \Delta H^{\circ} = 3/2(-571.7 \text{ kJ})$

   $B_2O_3(s) + 3H_2O(l) \rightarrow B_2H_6(g) + 3O_2(g) \quad \Delta H^{\circ} = -(-2147.5 \text{ kJ})$

   $\qquad 2B(s) + 3H_2(g) \rightarrow B_2H_6(g) \qquad\qquad \Delta H^{\circ}_{f} = +35.4 \text{ kJ}$

Foods and Fuels

5.65   Calculate the Cal (kcal) due to each nutritional component of the Campbell's® soup, then sum.

   $9 \text{ g carbohydrates} \times \dfrac{17 \text{ kJ}}{1 \text{ g carbohydrate}} = 153 \text{ or } 2 \times 10^2 \text{ kJ}$

   $1 \text{ g protein} \times \dfrac{17 \text{ kJ}}{1 \text{ g protein}} = 17 \text{ or } 0.2 \times 10^2 \text{ kJ}$

   $7 \text{ g fat} \times \dfrac{38 \text{ kJ}}{1 \text{ g fat}} = 266 \text{ or } 3 \times 10^2 \text{ kJ}$

   total energy = 153 kJ + 17 kJ + 266 kJ = 436 or 4×10^2 kJ

   $436 \text{ kJ} \times \dfrac{1 \text{ kcal}}{4.184 \text{ kJ}} \times \dfrac{1 \text{ Cal}}{1 \text{ kcal}} = 104 \text{ or } 1 \times 10^2 \text{ Cal/serving}$

5.67 $16.0 \text{ g } C_6H_{12}O_6 \times \dfrac{1 \text{ mol } C_6H_{15}O_6}{180.2 \text{ g } C_6H_{12}O_6} \times \dfrac{2812 \text{ kJ}}{\text{mol } C_6H_{12}O_6} \times \dfrac{1 \text{ Cal}}{4.184 \text{ kJ}} = 59.7 \text{ Cal}$

5.69   Propyne: $C_3H_4(g) + 4O_2(g) \rightarrow 3CO_2(g) + 2H_2O(g)$

   (a) $\Delta H = 3(-393.5 \text{ kJ}) + 2(-241.82 \text{ kJ}) - (185.4 \text{ kJ}) = -1849.5 = -1850 \text{ kJ/mol } C_3H_4$

   (b) $\dfrac{-1849.5 \text{ kJ}}{1 \text{ mol } C_3H_4} \times \dfrac{1 \text{ mol } C_3H_4}{40.065 \text{ g } C_3H_4} \times \dfrac{1000 \text{ g } C_3H_4}{1 \text{ kg } C_3H_4} = -4.616 \times 10^4 \text{ kJ/kg } C_3H_4$

Propylene: $C_3H_6(g) + 9/2\ O_2(g) \rightarrow 3CO_2(g) + 3H_2O(g)$

(a) $\Delta H = 3(-393.5\ kJ) + 3(-241.82\ kJ) - (20.4\ kJ) = -1926.4 = -1926\ kJ/mol\ C_3H_6$

(b) $\dfrac{-1926.4\ kJ}{1\ mol\ C_3H_6} \times \dfrac{1\ mol\ C_3H_6}{42.080\ g\ C_3H_6} \times \dfrac{1000\ g\ C_3H_6}{1\ kg\ C_3H_6} = -4.578 \times 10^4\ kJ/kg\ C_3H_6$

Propane: $C_3H_8(g) + 5O_2(g) \rightarrow 3CO_2(g) + 4H_2O(g)$

(a) $\Delta H = 3(-393.5\ kJ) + 4(-241.82\ kJ) - (-103.8\ kJ) = -2044.0 = -2044\ kJ/mol\ C_3H_8$

(b) $\dfrac{-2044.0\ kJ}{1\ mol\ C_3H_8} \times \dfrac{1\ mol\ C_3H_8}{44.096\ g\ C_3H_8} \times \dfrac{1000\ g\ C_3H_8}{1\ kg\ C_3H_8} = -4.635 \times 10^4\ kJ/kg\ C_3H_8$

(c)   These three substances yield nearly identical quantities of heat per unit mass, but propane is marginally higher than the other two.

Additional Exercises

5.72 $w = \Delta E - q;$ $\Delta E = +1.32\ kJ,$ $q = +1.55\ kJ;$ $w = 1.32\ kJ - 1.55\ kJ = -0.23\ kJ$

The negative sign for w indicates that work is done by the system on the surroundings.

5.76   Find the heat capacity of 1000 gal H_2O.

$$C_{H_2O} = 1000\ gal\ H_2O \times \dfrac{4\ qt}{1\ gal} \times \dfrac{1\ L}{1.06\ qt} \times \dfrac{1 \times 10^3\ cm^3}{1\ L} \times \dfrac{1\ g}{1\ cm^3} \times \dfrac{4.184\ J}{1\ g \cdot {}^\circ C}$$

$$= 1.58 \times 10^7\ J/{}^\circ C = 1.58 \times 10^4\ kJ/{}^\circ C;\ \text{then,}$$

$$\dfrac{1.58 \times 10^7\ J}{1\ ^\circ C} \times \dfrac{1\ ^\circ C \cdot g}{0.85\ J} \times \dfrac{1\ kg}{1 \times 10^3\ g} \times \dfrac{1\ brick}{1.8\ kg} = 1.0 \times 10^4\ \text{or 10,000 bricks}$$

5.79   (a)   From the mass of benzoic acid that produces a certain temperature change, we can calculate the heat capacity of the calorimeter.

$$\dfrac{0.235\ g\ benzoic\ acid}{1.642\ ^\circ C\ change\ observed} \times \dfrac{26.38\ kJ}{1\ g\ benzoic\ acid} = \dfrac{3.78\ kJ}{^\circ C}$$

Now we can use this experimentally determined heat capacity with the data for caffeine.

$$\dfrac{1.525\ ^\circ C\ rise}{0.265\ g\ caffeine} \times \dfrac{3.78\ kJ}{1\ ^\circ C} \times \dfrac{194.2\ g\ caffeine}{1\ mol\ caffeine} = 4.22 \times 10^3\ kJ/mol\ caffeine$$

(b)   The overall uncertainty is approximately equal to the sum of the uncertainties due to each effect. The uncertainty in the mass measurement is 0.235/0.001 or 0.265/0.001, about 1 part in 235 or 1 part in 265. The uncertainty in the temperature measurements is 1.642/0.002 or 1.525/0.002, about 1 part in 820 or 1 part in 760. Thus the uncertainty in heat of combustion from each measurement is

$$\dfrac{4220}{235} = 18\ kJ;\quad \dfrac{4220}{265} = 16\ kJ;\quad \dfrac{4220}{820} = 5\ kJ;\quad \dfrac{4220}{760} = 6\ kJ$$

The sum of these uncertainties is 45 kJ. In fact, the overall uncertainty is less than this because independent errors in measurement do tend to partially cancel.

5.82   (a) $C_6H_{12}O_6(s) + 6O_2(g) \rightarrow 6CO_2(g) + 6H_2O(l)$

$\Delta H^{\circ}_{rxn} = 6\Delta H^{\circ}_f\ CO_2(g) + 6\Delta H^{\circ}_f\ H_2O(l) - \Delta H^{\circ}_f\ C_6H_{12}O_6(s)\ - 6\Delta H^{\circ}_f\ O_2(g)$

$\quad\quad = 6(-393.5\ kJ) + 6(-285.83\ kJ) - (-1260\ kJ) - 6(0)$

$\quad\quad = -2816\ kJ/mol\ C_6H_{12}O_6$

$C_{12}H_{22}O_{11}(s) + 12O_2(g) \rightarrow 12CO_2(g) + 11H_2O(l)$

$\Delta H^{\circ}_{rxn} = 12\Delta H^{\circ}_f\ CO_2(g) + 11\Delta H^{\circ}_f\ H_2O(l) - \Delta H^{\circ}_f\ C_{12}H_{22}O_{11}(s) - 12\Delta H^{\circ}_f\ O_2(g)$

$\quad\quad = 12(-393.5\ kJ) + 11(-285.83\ kJ) - (-2221\ kJ) - 12(0)$

$\quad\quad = -5645\ kJ/mol\ C_{12}H_{22}O_{11}$

(b) $\dfrac{-2816\ kJ}{1\ mol\ C_6H_{12}O_6} \times \dfrac{1\ mol\ C_6H_{12}O_6}{180.2\ g\ C_6H_{12}O_6} = \dfrac{15.63\ kJ}{1\ g\ C_6H_{12}O_6} \rightarrow 16\ kJ/g\ C_6H_{12}O_6$

$\dfrac{-5645\ kJ}{1\ mol\ C_{12}H_{22}O_{11}} \times \dfrac{1\ mol\ C_{12}H_{22}O_{11}}{342.3\ g\ C_{12}H_{22}O_{11}} = \dfrac{16.49\ kJ}{1\ g\ C_{12}H_{22}O_{11}} \rightarrow 16\ kJ/g\ C_{12}H_{22}O_{11}$

(c)   The average fuel value of carbohydrates (Section 5.8) is 17 kJ/g. These two carbohydrates have fuel values (16 kJ/g) slightly lower but in line with this average. (More complex carbohydrates supply more energy and raise the average value.)

5.86   **1,3-butadiene**, C_4H_6, MM = 54.092 g/mol

(a) $C_4H_6(g) + 11/2\ O_2(g) \rightarrow 4CO_2(g) + 3H_2O(l)$

$\Delta H^{\circ}_{rxn} = 4\Delta H^{\circ}_f\ CO_2(g) + 3\Delta H^{\circ}_f\ H_2O(l) - \Delta H^{\circ}_f\ C_4H_6(g) + 11/2\ \Delta H^{\circ}_f\ O_2(g)$

$\quad\quad = 4(-393.5\ kJ) + 3(-285.83\ kJ) - 111.9\ kJ + 11/2\ (0) = -2543.4\ kJ/mol\ C_4H_6$

(b) $\dfrac{-2543.4\ kJ}{1\ mol\ C_4H_6} \times \dfrac{1\ mol\ C_4H_6}{54.092\ g} = 47.020 \rightarrow 47\ kJ/g$

(c)   % H $= \dfrac{6(1.008)}{54.092} \times 100 = 11.18\%$ H

1-butene, C_4H_8, MM = 56.108 g/mol

(a) $C_4H_8(g) + 6O_2(g) \rightarrow 4CO_2(g) + 4H_2O(l)$

$\Delta H^{\circ}_{rxn} = 4\Delta H^{\circ}_f\ CO_2(g) + 4\Delta H^{\circ}_f\ H_2O(l) - \Delta H^{\circ}_f\ C_4H_8(g) - 6\Delta H^{\circ}_f\ O_2(g)$

$\quad\quad = 4(-393.5\ kJ) + 4(-285.83\ kJ) - 1.2\ kJ - 6(0) = -2718.5\ kJ/mol\ C_4H_8$

(b) $\dfrac{-2718.5\ kJ}{1\ mol\ C_4H_8} \times \dfrac{1\ mol\ C_4H_8}{56.108\ g\ C_4H_8} = 48.451 \rightarrow 48\ kJ/g$

(c)   % H $= \dfrac{8(1.008)}{56.108} \times 100 = 14.37\%$ H

n-butane, $C_4H_{10}(g)$, MM = 58.124 g/mol

(a) $C_4H_{10}(g) + 13/2 \, O_2(g) \rightarrow 4CO_2(g) + 5H_2O(l)$

$\Delta H^{\circ}_{rxn} = 4\Delta H^{\circ}_f \, CO_2(g) + 5\Delta H^{\circ}_f \, H_2O(l) - \Delta H^{\circ}_f \, C_4H_{10}(g) - 13/2 \, \Delta H^{\circ}_f \, O_2(g)$

     $= 4(-393.5 \text{ kJ}) + 5(-285.83 \text{ kJ}) - (-124.7 \text{ kJ}) - 3/2 \, (0) = -2878.5 \text{ kJ/mol } C_4H_{10}$

(b) $\dfrac{-2878.5 \text{ kJ}}{1 \text{ mol } C_4H_{10}} \times \dfrac{1 \text{ mol } C_4H_{10}}{58.124 \text{ g } C_4H_{10}} = 49.523 \rightarrow 50 \text{ kJ/g}$

(c)   % H = $\dfrac{10(1.008)}{58.124} \times 100 = 17.34\%$ H

(d)   It is certainly true that as the mass % H increases, the fuel value (kJ/g) of the hydrocarbon increases, given the same number of C atoms. A graph of the data in parts (b) and (c) (see below) suggests that mass % H and fuel value are directly proportional when the number of C atoms is constant.

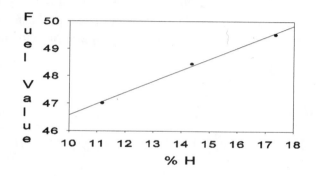

Integrative Exercises

5.89   (a)   At constant pressure, $\Delta E = \Delta H - P\Delta V$. The difference between ΔE and ΔH is the expansion work done on or by the system. If all reactants and products are liquids or solids, very little volume change is possible. $P\Delta V$ is essentially zero and $\Delta E \approx \Delta H$.

         (b)   The reaction that produces the insoluble gas will have a greater difference between ΔE and ΔH, because the gas causes a significant $+\Delta V$, while the aqueous weak electrolyte does not.

5.92   (a) $AgNO_3(aq) + NaCl(aq) \rightarrow NaNO_3(aq) + AgCl(s)$
                net ionic equation: $Ag^+(aq) + Cl^-(aq) \rightarrow AgCl(s)$
                $\Delta H^{\circ} = \Delta H^{\circ}_f \, AgCl(s) - \Delta H^{\circ}_f \, Ag^+(aq) - \Delta H^{\circ}_f \, Cl^-(aq)$
                $\Delta H^{\circ} = -127.0 \text{ kJ} - (105.90 \text{ kJ}) - (-167.2 \text{ kJ}) = -65.7 \text{ kJ}$

(b) $\Delta H°$ for the complete molecular equation will be the same as $\Delta H°$ for the net ionic equation. $Na^+(aq)$ and $NO_3^-(aq)$ are spectator ions; they appear on both sides of the chemical equation. Since the overall enthalpy change is the enthalpy of the products minus the enthalpy of the reactants, the contributions of the spectator ions cancel.

(c) $\Delta H° = \Delta H_f°$ $NaNO_3(aq) + \Delta H_f°$ $AgCl(s) - \Delta H_f°$ $AgNO_3(aq) - \Delta H_f°$ $NaCl(aq)$

$\Delta H_f°$ $AgNO_3(aq) = \Delta H_f°$ $NaNO_3(aq) + \Delta H_f°$ $AgCl(s) - \Delta H_f°$ $NaCl(aq) - \Delta H°$

$\Delta H_f°$ $AgNO_3(aq) = -446.2$ kJ $+ (-127.0$ kJ$) - (-407.1$ kJ$) - (-65.7$ kJ$)$

$\Delta H_f°$ $AgNO_3(aq) = -100.4$ kJ/mol

6 Electronic Structure of Atoms

Radiant Energy

6.1   (a)   meters (m)   (b)   1/seconds (s^{-1})   (c)   meters/second ($m \cdot s^{-1}$ or m/s)

6.3   (a)   False. Only a small fraction of electromagnetic radiation (visible light, 400-700 nm) is visible.
         (b)   True
         (c)   False. Ultraviolet light has shorter wavelengths than does visible light.

6.5   Wavelength of (a) gamma rays < (d) yellow (visible) light < (e) red (visible) light < (b) 93.1 MHz FM (radio) waves < (c) 680 kHz or 0.680 MHz AM (radio) waves

6.7   (a) $\nu = c/\lambda$; $\dfrac{2.998 \times 10^8 \text{ m}}{\text{s}} \times \dfrac{1}{0.589 \text{ pm}} \times \dfrac{1 \text{ pm}}{1 \times 10^{-12} \text{ m}} = 5.09 \times 10^{20} \text{ s}^{-1}$

         (b) $\lambda = c/\nu$; $\dfrac{2.998 \times 10^8 \text{ m}}{\text{s}} \times \dfrac{1 \text{ s}}{5.11 \times 10^{11}} = 5.87 \times 10^{-4} \text{ m}$ (587 µm)

         (c)   No. The radiation in (a) is gamma rays and in (b) is infrared. Neither is visible to humans.

         (d) $6.54 \text{ s} \times \dfrac{2.998 \times 10^8 \text{ m}}{\text{s}} = 1.96 \times 10^9 \text{ m}$

6.9 $\nu = c/\lambda$; $\dfrac{2.998 \times 10^8 \text{ m}}{1 \text{ s}} \times \dfrac{1}{436 \text{ nm}} \times \dfrac{1 \text{ nm}}{1 \times 10^{-9} \text{ m}} = 6.88 \times 10^{14} \text{ s}^{-1}$
The color is blue. (See Figure 24.23.)

Quantized Energy and Photons

6.11   (a)   *Quantization* means that energy can only be absorbed or emitted in specific amounts or multiples of these amounts. This minimum amount of energy is called a quantum and is equal to a constant times the frequency of the radiation absorbed or emitted. **E = hν**.

         (b)   In everyday activities, we deal with macroscopic objects such as our bodies or our cars, which gain and lose total amounts of energy much larger than a single quantum, hν. The gain or loss of the relatively minuscule quantum of energy is unnoticed.

6.13   (a) $E = h\nu = hc/\lambda = 6.626 \times 10^{-34}\ \text{J}\cdot\text{s} \times \dfrac{2.998 \times 10^8\ \text{m}}{1\ \text{s}} \times \dfrac{1}{645\ \text{nm}} \times \dfrac{1\ \text{nm}}{1 \times 10^{-9}\ \text{m}}$

$$= 3.08 \times 10^{-19}\ \text{J}$$

(b) $E = h\nu = 6.626 \times 10^{-34}\ \text{J}\cdot\text{s} \times \dfrac{2.85 \times 10^{12}}{1\ \text{s}} = 1.89 \times 10^{-21}\ \text{J}$

(c) $\lambda = hc/E = 6.626 \times 10^{-34}\ \text{J}\cdot\text{s} \times \dfrac{2.998 \times 10^8\ \text{m}}{1\ \text{s}} \times \dfrac{1}{8.23 \times 10^{-19}\ \text{J}} = 2.41 \times 10^{-7}\ \text{m}$

$$= 241\ \text{nm}$$

This radiation is in the ultraviolet region.

6.15   (a) $E = hc/\lambda = 6.626 \times 10^{-34}\ \text{J}\cdot\text{s} \times \dfrac{2.998 \times 10^8\ \text{m}}{1\ \text{s}} \times \dfrac{1}{3.3\ \mu\text{m}} \times \dfrac{1\ \mu\text{m}}{1 \times 10^{-6}\ \text{m}}$

$$= 6.0 \times 10^{-20}\ \text{J}$$

$E = hc/\lambda = 6.626 \times 10^{-34}\ \text{J}\cdot\text{s} \times \dfrac{2.998 \times 10^8\ \text{m}}{1\ \text{s}} \times \dfrac{1}{0.154\ \text{nm}} \times \dfrac{1\ \text{nm}}{1 \times 10^{-9}\ \text{m}}$

$$= 1.29 \times 10^{-15}\ \text{J}$$

(b)   The 3.3 μm photon is in the infrared and the 0.154 nm (1.54×10^{-10} m) photon is in the X-ray region; the X-ray photon has the greater energy.

6.17 $E_{photon} = hc/\lambda = \dfrac{6.626 \times 10^{-34}\ \text{J}\cdot\text{s}}{785\ \text{nm}} \times \dfrac{2.998 \times 10^8\ \text{m}}{1\ \text{s}} \times \dfrac{1\ \text{nm}}{1 \times 10^{-9}\ \text{m}} = 2.531 \times 10^{-19}$

$$= 2.53 \times 10^{-19}\ \text{J/photon}$$

$\dfrac{E_{total}}{E_{photon}} = \#\ \text{of photons};\ \ 31\ \text{J} \times \dfrac{1\ \text{photon}}{2.531 \times 10^{-19}\ \text{J}} = 1.2 \times 10^{20}\ \text{photons}$

6.19 $\dfrac{495 \times 10^3\ \text{J}}{\text{mol O}_2} \times \dfrac{1\ \text{mol}}{6.022 \times 10^{23}\ \text{photons}} = 8.220 \times 10^{-19} = 8.22 \times 10^{-19}\ \text{J/photon}$

$\lambda = hc/E = \dfrac{6.626 \times 10^{-34}\ \text{J}\cdot\text{s}}{8.220 \times 10^{-19}\ \text{J}} \times \dfrac{2.998 \times 10^8\ \text{m}}{1\ \text{s}} = 2.42 \times 10^{-7}\ \text{m} = 242\ \text{nm}$

According to Figure 6.4, this is ultraviolet radiation.

6.21   (a) $E = h\nu = 6.626 \times 10^{-34}\ \text{J}\cdot\text{s} \times 1.09 \times 10^{15}\ \text{s}^{-1} = 7.22 \times 10^{-19}\ \text{J}$

(b) $\lambda = c/\nu = \dfrac{2.998 \times 10^8\ \text{m}}{1\ \text{s}} \times \dfrac{1\ \text{s}}{1.09 \times 10^{15}} = 2.75 \times 10^{-7}\ \text{m} = 275\ \text{nm}$

(c) $E_{120} = hc/\lambda = 6.626 \times 10^{-34}\ \text{J}\cdot\text{s} \times \dfrac{2.998 \times 10^8\ \text{m}}{1\ \text{s}} \times \dfrac{1}{120\ \text{nm}} \times \dfrac{1\ \text{nm}}{1 \times 10^{-9}\ \text{m}}$

$$= 1.655 \times 10^{-18} = 1.66 \times 10^{-18}\ \text{J}$$

The excess energy of the 120 nm photon is converted into the kinetic energy of the emitted electron.

$E_k = E_{120} - E_{min} = 16.55 \times 10^{-19}$ J - 7.22 $\times 10^{-19}$ J = 9.3 $\times 10^{-19}$ J/electron

Bohr's Model; Matter Waves

6.23   When applied to atoms, the notion of quantized energies means that only certain energies can be gained or lost, only certain values of ΔE are allowed. The allowed values of ΔE are represented by the lines in the emission spectra of excited atoms.

6.25   An isolated electron is assigned an energy of zero; the closer the electron comes to the nucleus, the more negative its energy. Thus, as an electron moves closer to the nucleus, the energy of the electron decreases and the excess energy is emitted. Conversely, as an electron moves further from the nucleus, the energy of the electron increases and energy must be absorbed.

(a)   As the principle quantum number increases, the electron moves away from the nucleus and energy is **absorbed**.

(b)   A decrease in the radius of the orbit means the electron moves closer to the nucleus; energy is **emitted**.

(c)   Totally removing an electron from the atom requires that energy is **absorbed**.

6.27   (a) $\Delta E = R_H \left[\dfrac{1}{n_i^2} - \dfrac{1}{n_f^2} \right] = 2.18 \times 10^{-18}$ J (1/25 - 1/1) = -2.093 $\times 10^{-18}$ = -2.09 $\times 10^{-18}$ J

$\nu = E/h = \dfrac{2.093 \times 10^{-18} \text{ J}}{6.626 \times 10^{-34} \text{ J} \cdot \text{s}} = 3.158 \times 10^{15} = 3.16 \times 10^{15}$ s^{-1}

$\lambda = c/\nu = \dfrac{2.998 \times 10^8 \text{ m}}{1 \text{ s}} \times \dfrac{1 \text{ s}}{3.158 \times 10^{15}} = 9.49 \times 10^{-8}$ m

Since the sign of ΔE is negative, radiation is emitted.

(b) $\Delta E = 2.18 \times 10^{-18}$ J(1/36 - 1/4) = -4.844 $\times 10^{-19}$ = -4.84 $\times 10^{19}$ J

$\nu = \dfrac{4.844 \times 10^{-19} \text{ J}}{6.626 \times 10^{-34} \text{ J} \cdot \text{s}} = 7.311 \times 10^{14} = 7.31 \times 10^{14}$ s^{-1}; $\lambda = \dfrac{2.998 \times 10^8 \text{ m/s}}{7.311 \times 10^{14}/\text{s}}$

$= 4.10 \times 10^{-7}$ m

Visible radiation is emitted.

(c) $\Delta E = 2.18 \times 10^{-18}$ J$(1/16 - 1/25) = 4.095 \times 10^{-20} = 4.91 \times 10^{-20}$ J

$$\nu = \frac{4.095 \times 10^{-20} \text{ J}}{6.626 \times 10^{-34} \text{ J}\cdot\text{s}} = 7.403 \times 10^{13} = 7.40 \times 10^{13} \text{ s}^{-1}; \; \lambda = \frac{2.998 \times 10^8 \text{ m/s}}{7.403 \times 10^{13} /\text{s}}$$

$$= 4.05 \times 10^{-6} \text{ m}$$

Radiation is absorbed.

6.29   (a)   Only lines with $n_f = 2$ represent ΔE values and wavelengths that lie in the visible portion of the spectrum. Lines with $n_f = 1$ have larger ΔE values and shorter wavelengths that lie in the ultraviolet. Lines with $n_f > 2$ have smaller ΔE values and lie in the lower energy longer wavelength regions of the electromagnetic spectrum.

     (b) $n_i = 3, n_f = 2; \; \Delta E = R_H \left[\dfrac{1}{n_i^2} - \dfrac{1}{n_f^2} \right] = 2.18 \times 10^{-18}$ J $(1/9 - 1/4)$

$$\lambda = hc/E = \frac{6.626 \times 10^{-34} \text{ J}\cdot\text{s} \times 2.998 \times 10^8 \text{ m/s}}{2.18 \times 10^{-18} \text{ J } (1/9 - 1/4)} = 6.56 \times 10^{-7} \text{ m}$$

This is the yellow line at 656 nm.

$n_i = 4, n_f = 2; \; \lambda = hc/E = \dfrac{6.626 \times 10^{-34} \text{ J}\cdot\text{s} \times 2.998 \times 10^8 \text{ m/s}}{2.18 \times 10^{-18} \text{ J } (1/16 - 1/4)} = 4.86 \times 10^{-7} \text{ m}$

This is the blue-green (cyan) line at 486 nm.

$n_i = 5, n_f = 2; \; \lambda = hc/E = \dfrac{6.626 \times 10^{-34} \text{ J}\cdot\text{s} \times 2.998 \times 10^8 \text{ m/s}}{2.18 \times 10^{-18} \text{ J } (1/25 - 1/4)} = 4.34 \times 10^{-7} \text{ m}$

This is the blue line at 434 nm.

6.31   (a)   93.8 nm $\times \dfrac{1 \times 10^{-9} \text{ m}}{1 \text{ nm}} = 9.38 \times 10^{-8}$ m; this line is in the ultraviolet region.

     (b)   Only lines with $n_f = 1$ have a large enough ΔE to lie in the ultraviolet region. Solve Equation 6.6 for n_i, recalling that ΔE is negative for emission.

$$- hc/\lambda = R_H \left[\frac{1}{n_i^2} - \frac{1}{n_f^2} \right]; \; -\frac{hc}{\lambda \times R_H} = \left[\frac{1}{n_i^2} - 1 \right]; \; \frac{1}{n_i^2} = 1 - \frac{hc}{\lambda \times R_H}$$

$$n_i^2 = \left(1 - \frac{hc}{\lambda \times R_H} \right)^{-1}; \; n_i = \left(1 - \frac{hc}{\lambda \times R_H} \right)^{-1/2}$$

$$n_i = \left(1 - \frac{6.626 \times 10^{-34} \text{ J}\cdot\text{s} \times 2.998 \times 10^8 \text{ m/s}}{9.38 \times 10^{-8} \text{ m} \times 2.18 \times 10^{-18} \text{ J}} \right)^{-1/2} = 6 \; (n \text{ values must be integers})$$

$n_i = 6, n_f = 1$

6.33 $\lambda = \dfrac{h}{mv}$; $1\ J = \dfrac{1\ kg \cdot m^2}{s^2}$; Change mass to kg and velocity to m/s in each case.

(a) $\dfrac{60\ km}{1\ hr} \times \dfrac{1000\ m}{1\ km} \times \dfrac{1\ hr}{60\ min} \times \dfrac{1\ min}{60\ s} = 16.67 = 17$ m/s

$\lambda = \dfrac{6.626 \times 10^{-34}\ kg \cdot m^2 \cdot s}{1\ s^2} \times \dfrac{1}{85\ kg} \times \dfrac{1\ s}{16.67\ m} = 4.7 \times 10^{-37}$ m

(b) $50\ g \times \dfrac{1\ kg}{1000\ g} = 0.050$ kg

$\lambda = \dfrac{6.626 \times 10^{-34}\ kg \cdot m^2 \cdot s}{1\ s^2} \times \dfrac{1}{0.050\ kg} \times \dfrac{1\ s}{400\ m} = 3.3 \times 10^{-35}$ m

(c) We need to calculate the mass of a single Li atom in kg.

$\dfrac{6.94\ g\ Li}{1\ mol\ Li} \times \dfrac{1\ kg}{1000\ g} \times \dfrac{1\ mol}{6.022 \times 10^{23}\ Li\ atoms} = 1.152 \times 10^{-26} = 1.15 \times 10^{-26}$ kg

$\lambda = \dfrac{6.626 \times 10^{-34}\ kg \cdot m^2 \cdot s}{1\ s^2} \times \dfrac{1}{1.152 \times 10^{-26}\ kg} \times \dfrac{1\ s}{6.5 \times 10^5\ m} = 8.8 \times 10^{-14}$ m

6.35 $v = h/m\lambda$; $\lambda = 0.88\ \text{Å} \times \dfrac{1 \times 10^{-10}\ m}{1\ \text{Å}} = 8.8 \times 10^{-11}$ m; $m = 1.67 \times 10^{-27}$ kg

$v = \dfrac{6.626 \times 10^{-34}\ kg \cdot m^2 \cdot s}{1\ s^2} \times \dfrac{1}{1.67 \times 10^{-27}\ kg} \times \dfrac{1\ s}{8.8 \times 10^{-11}\ m} = 4.5 \times 10^3$ m/s

6.37 The uncertainty principle states that there is a limit to how precisely we can know the simultaneous position and momentum of an electron. The more precisely we know the position the greater the uncertainty in the momentum, and vice versa. In the Bohr model, electrons move in exact spherical paths that have known energy, implying that the position and momentum of an electron can be known exactly and simultaneously. This violates the uncertainty principle and is the reason that the Bohr model is an inadequate description of the electronic structure of atoms.

Quantum Mechanics and Atomic Orbitals

6.39 In the Bohr model, the electron is treated as a small object that moves about the nucleus in circular orbits. A Bohr orbit specifies the exact path and energy of the electron. In the quantum-mechanical model, the wave properties of the electron are considered; any attempt to describe the exact path of an electron is inconsistent with the Heisenberg uncertainty principle. The quantum mechanical model is a statistical model that tells us the probability of finding an electron in certain regions around the nucleus. Thus, quantum mechanics would give the probability of finding the electron at 0.53 Å and this probability would always be less than 100%.

6.41   (a) $n = 4, l = 3, 2, 1, 0$ (b) $l = 2, m_l = -2, -1, 0, 1, 2$

6.43   (a)   2, 1, 1;   2, 1, 0;   2, 1 -1

       (b)   5, 2, 2;   5, 2, 1;   5, 2, 0;   5, 2, -1;   5, 2, -2

6.45   (a)   permissible, 2p   (b)   forbidden, for $l = 0$, m_l can only equal 0
       (c)   permissible, 4d   (d)   forbidden, for n = 3, the largest l value is 2

6.47   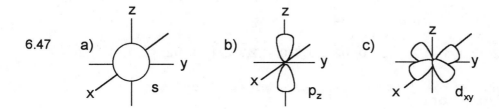

6.49   (a)   The 2s and 3s orbitals of a hydrogen atom have the same overall spherical shape, but the 3s orbital has a larger radial extension and one more node than the 2s orbital. Since the 3s orbital is "larger" than the 2s, there is a greater probability of finding an electron further from the nucleus in the 3s orbital.

       (b)   The shapes of the hydrogen 2s and 2p orbitals are quite different (spherical vs dumbbell), while the average distance from the nucleus of an electron occupying either orbital is similar.

       (c)   In the hydrogen atom, orbitals with the same n value are degenerate and energy increases with increasing n value. Thus, 2s and 2p have the same energy and 3s is at a higher energy.

Many-Electron Atoms; Electron Spin

6.51   (a)   In the hydrogen atom, orbitals with the same principle quantum number, n, are degenerate.

       (b)   In a many-electron atom, orbitals with the same principle and azimuthal quantum numbers, n and l, are degenerate.

6.53   The 2p electron in boron is shielded from the full charge of the nucleus by the 2s electrons. (Both the 2s and 2p electrons are shielded by the 1s electrons). Thus, the 2p electron experiences a smaller nuclear charge than the 2s electrons.

6.55   A 2p electron in Ne experiences a greater effective nuclear charge. The shielding experienced by a 2p electron in the two atoms is similar, so the electron in the atom with the larger Z ($Z_{Ne} = 10$, $Z_O = 8$) experiences the larger effective nuclear charge.

6.57 (a) +1/2, - 1/2

(b) Electrons with opposite spins are affected differently by a strong inhomogeneous magnetic field. An apparatus similar to that in Figure 6.25 can be used to distinguish electrons with opposite spins.

(c) The Pauli exclusion principle states that no two electrons can have the same four quantum numbers. Two electrons in a 1s orbital have the same, n, l and m_l values. They must have different m_s values.

6.59 (a) 10 (b) 2 (c) 6 (d) 14

6.61 2, 1, 1, 1/2; 2, 1, 1, -1/2; 2, 1, 0, 1/2; 2, 1, 0, -1/2; 2, 1, -1, 1/2; 2, 1, -1, -1/2

Electron Configurations

6.63 (a) Each box represents an orbital.

(b) Electron spin is represented by the direction of the half-arrows.

(c) No. The electron configuration of Be is $1s^2 2s^2$. There are no electrons in subshells that have degenerate orbitals, so Hund's rule is not used.

6.65 (a) Rb - $[Kr]5s^1$ (b) Se - $[Ar]4s^2 3d^{10} 4p^4$ (c) Zn - $[Ar]4s^2 3d^{10}$

(d) V - $[Ar]4s^2 3d^3$ (e) Pb - $[Xe]6s^2 4f^{14} 5d^{10} 6p^2$ (f) Yb - $[Xe]6s^2 4f^{14}$

6.67 (a) As [↑↓] 4s [↑↓][↑↓][↑↓][↑↓][↑↓] 3d [↑][↑][↑] 4p 3 unpaired electrons

(b) Te [↑↓] 5s [↑↓][↑↓][↑↓][↑↓][↑↓] 4d [↑↓][↑][↑] 5p 2 unpaired electrons

(c) Sn [↑↓] 5s [↑↓][↑↓][↑↓][↑↓][↑↓] 4d [↑][↑][ ] 5p 2 unpaired electrons

(d) Ag [↑] 5s [↑↓][↑↓][↑↓][↑↓][↑↓] 4d 1 unpaired electron

(e) Nb [↑↓] $5s^2$ [↑][↑][↑][ ][ ] 4d 3 unpaired electrons

6.69 (a) Mg (b) Al (c) Cr (d) Te

Additional Exercises

6.73   The round trip is 4.8×10^5 mi. Light travels at 2.998×10^8 m/s.

$$4.8 \times 10^5 \text{ mi} \times \frac{1.6093 \text{ km}}{1 \text{ mi}} \times \frac{1000 \text{ m}}{1 \text{ km}} \times \frac{1 \text{ s}}{2.998 \times 10^8 \text{ m}} = 2.6 \text{ s}$$

6.75   Find the energy of one photon.

$$E = hc/\lambda = \frac{6.626 \times 10^{-34} \text{ J} \cdot \text{s} \times 2.998 \times 10^8 \text{ m/s}}{6300 \text{ Å}} \times \frac{1 \text{ Å}}{1 \times 10^{-10} \text{ m}} = 3.153 \times 10^{-19} \text{ J}$$

$$1 \text{ W} = 1 \text{ J/s}, \; 1 \text{ W} \cdot \text{s} = 1 \text{ J}$$

$$5.0 \text{ mW} \times \frac{1 \times 10^{-3} \text{ W}}{1 \text{ mW}} \times \frac{1 \text{ J}}{1 \text{ W} \cdot \text{s}} \times 2.0 \text{ min} \times \frac{60 \text{ s}}{1 \text{ min}} \times \frac{1 \text{ photon}}{3.153 \times 10^{-19} \text{ J}} = 1.9 \times 10^{18} \text{ photons}$$

6.78 $\dfrac{8.6 \times 10^{-13} \text{ C}}{1 \text{ s}} \times \dfrac{1 e^-}{1.602 \times 10^{-19} \text{ C}} \times \dfrac{1 \text{ photon}}{1 e^-} = 5.368 \times 10^6 = 5.4 \times 10^6 \text{ photons/s}$

$$\frac{E}{\text{photon}} = hc/\lambda = \frac{6.626 \times 10^{-34} \text{ J} \cdot \text{s}}{550 \text{ nm}} \times \frac{2.998 \times 10^8 \text{ m}}{1 \text{ s}} \times \frac{1 \text{ nm}}{1 \times 10^{-9} \text{ m}} \times \frac{5.368 \times 10^6 \text{ photon}}{\text{s}}$$

$$= 1.9 \times 10^{-12} \text{ J/s}$$

6.81   **(a)** $\Delta E = R_H \left[\dfrac{1}{n_i^2} - \dfrac{1}{n_f^2} \right] = 2.18 \times 10^{-18} \text{ J} \left(\dfrac{1}{1} - \dfrac{1}{\infty} \right) = 2.18 \times 10^{-18} \text{ J/atom}$

$$2.18 \times 10^{-18} \text{ J/atom} \times \frac{6.022 \times 10^{23} \text{ atoms}}{\text{mol}} \times \frac{1 \text{ kJ}}{1000 \text{ J}} = 1.31 \times 10^3 \text{ kJ/mol}$$

        (b) $\lambda = hc/E = \dfrac{6.626 \times 10^{-34} \text{ J} \cdot \text{s} \times 2.998 \times 10^8 \text{ m/s}}{2.18 \times 10^{-18} \text{ J}} = 9.11 \times 10^{-8} \text{ m} = 91.1 \text{ nm}$

        (c) ΔE is positive, so light is absorbed. This is reasonable, since energy must be supplied to overcome the electrostatic attraction of the electron for the nucleus.

        (d) $\Delta E = R_H \left[\dfrac{1}{2^2} - \dfrac{1}{\infty} \right] = 2.18 \times 10^{-18} \text{ J} \, (1/4) = 5.45 \times 10^{-19} \text{ J}$

6.83 $\lambda = h/mv$; $v = h/m\lambda$. $\lambda = 0.711 \text{ Å} \times \dfrac{1 \times 10^{-10} \text{ m}}{1 \text{ Å}} = 7.11 \times 10^{-11}$ m; $m_e = 9.1094 \times 10^{-31}$ kg

$v = \dfrac{6.626 \times 10^{-31} \text{ J} \cdot \text{s}}{9.1094 \times 10^{-31} \text{ kg} \times 7.11 \times 10^{-11} \text{ m}} \times \dfrac{1 \text{ kg} \cdot \text{m}^2/\text{s}^2}{1 \text{ J}} = 1.02 \times 10^7$ m/s

6.86   (a) l (b) n and l (c) m_s (d) m_l

6.90   (d) 3p < (a) 4s < (b) 3d = (c) 3d

6.94   (a) O - excited   (b) Br - ground   (c) P - excited   (d) In - ground

7 Periodic Properties of the Elements

Periodic Table; Electron Shells; Atomic Radii

7.1   Mendeleev insisted that elements with similar chemical and physical properties be placed within a family or column of the table. Since many elements were as yet undiscovered, Mendeleev left blanks. He predicted properties for the "blanks" based on properties of other elements in the family.

7.3   The quantum mechanical model describes electron structure in terms of the probability of finding electrons in some volume element of space. On a plot of radial electron density (the electron density on the surface of a sphere whose radius is a certain distance from the nucleus) there are certain radii with high electron densities. The number of these maxima corresponds to the number of "electron shells" for a particular atom as proposed by Lewis. This is also the number of principal quantum levels in the atom. In a multielectron atom, the total of the electron densities of all orbitals or subshells in a principal quantum level is roughly spherical, corresponding to the spherical electron shell.

7.5   Krypton has a larger nuclear charge (Z = 36) than argon (Z = 18). The shielding of the n = 3 shells by the 1s and 2s electrons in the two atoms is approximately equal, so the n = 3 electrons in Kr experience a greater effective nuclear charge and are thus situated closer to the nucleus.

7.7   Since the quantum mechanical description of the atom does not specify the exact location of electrons, there is no specific distance from the nucleus where the last electron can be found. Rather, the electron density decreases gradually as the distance from the nucleus increases. There is no quantum mechanical "edge" of an atom.

7.9   The distance between Cr atoms in Cr metal will be 2 times the atomic radius of Cr, 2 × 1.25 Å = 2.50 Å.

7.11   From atomic radii, As-I = 1.21 Å + 1.33 Å = 2.54 Å. This is very close to the experimental value of 2.55 Å.

7.13   (a)   Atomic radii **decrease** moving from left to right across a row and (b) **increase** from top to bottom within a group.

        (c)   F < S < P <As. The order is unambiguous according to the trends of increasing atomic radius moving down a column and to the left in a row of the table.

7.15   (a)   The electrons in a He atom experience a nuclear charge of 2 and are drawn closer to the nucleus than the electron in H, which feels a nuclear charge of only 1.

         (b)   Even though Z = 10 for Ne and Z = 2 for He, the valence electrons in Ne are in $n = 2$ and those of He are closer to the nucleus in $n = 1$. Also, the 1 s electrons in Ne shield the valence electrons from the full nuclear charge. This shielding, coupled with the larger n value of the valence electrons in the Ne, means that its atomic radius is larger than that of He.

Ionization Energies;   Electron Affinities

7.17 $Te(g) \rightarrow Te^+(g) + 1e^-$; $Te^+(g) \rightarrow Te^{2+}(g) + 1e^-$; $Te^{2+}(g) \rightarrow Te^{3+}(g) + 1e^-$

7.19   The electron configuration of Li^+ is $1s^2$ or [He] and that of Be^+ is $[He]2s^1$. Be^+ has one more valence electron to lose while Li^+ has the stable noble gas configuration of He. It requires much more energy to remove a 1s core electron close to the nucleus of Li^+ than a 2s valence electron further from the nucleus of Be^+.

7.21   Moving from He to Rn in group 8A, first ionization energies decrease and atomic size increases, because the n value and thus distance of the valence electrons from the nucleus increases going down a group.

7.23   (a) Ne   (b) Mg   (c) Cr   (d) Br   (e) Ge

7.25   (a)   Na - In an isoelectronic series, all electronic effects (shielding and repulsion) are the same, so the particle with the smallest Z will have the smallest effective nuclear charge.

         (b) Si^{3+} - Si has the largest Z and effective nuclear charge.

         (c)   The greater the effective nuclear charge experienced by a valence electron, the larger the ionization energy for that electron. According to Table 7.2, I_1 for Na is 496 kJ/mol. I_4 for Si is 3230 kJ/mol.

7.27   Ionization energy: $Se(g)$ $\rightarrow$ $Se^+(g) + 1e^-$
                                    $[Ar]4s^23e^{10}4p^4$ $[Ar]4s^23d^{10}4p^3$

           Electron affinity: $Se(g) + 1e^- \rightarrow$ $Se^-(g)$
                                    $[Ar]4s^23d^{10}4p^4$ $[Ar]4s^23d^{10}4p^5$

7.29   F $+ 1e^- \rightarrow$ F^- ;   Ne $+ 1e^- \rightarrow$ Ne^-
       $[He]2s^22p^5$ $[He]2s^22p^6$ $[He]2s^22p^6$ $[He]2s^22p^63s^1$

           Adding an electron to F completes the $n = 2$ shell; F^- has a stable noble gas electron configuration and ΔE for the process is negative. An extra electron in Ne would occupy the higher energy 3s orbital; adding an electron increases the energy of the system and ΔE for the process is positive.

Properties of Metals and Nonmetals

7.31 $O_2 < Br_2 < K < Mg$. O_2 and Br_2 are nonmetals; at standard conditions, O_2 is a gas and Br_2 is a liquid, so O_2 has the lower melting point. Nonmetallic character increases going up and to the right on the chart. The greater the nonmetallic character, the lower the boiling point. O_2 is above but to the left of Br_2. As with other trends, the vertical difference is larger and determines the relationship. K and Mg are solid metals. K has the greater metallic character (further down and to the left on the chart) and the lower melting point.

7.33   Metallic character increases moving down a family and to the left in a period.

   (a)   Li   (b) Na   (c) Sn   (d) Al

7.35   Ionic: MgO, Li_2O, Y_2O_3;   molecular: SO_2, P_2O_5, N_2O, XeO_3

   Ionic compounds are formed by combining a metal and a nonmetal; molecular compounds are formed by two or more nonmetals.

7.37   When dissolved in water, an "acidic oxide" produces an acidic (pH < 7) solution. Oxides of nonmetals are acidic. Example: $SO_3(g)$. A "basic oxide" dissolved in water produces a basic (pH > 7) solution. Oxides of metals are basic. Example: CaO (quick lime).

7.39   (a) $CaO(s) + H_2O(l) \rightarrow Ca(OH)_2(aq)$

   (b) $CuO(s) + 2HNO_3(aq) \rightarrow Cu(NO_3)_2(aq) + H_2O(l)$

   (c) $SO_3(g) + H_2O(l) \rightarrow H_2SO_4(aq)$

   (d) $CO_2(g) + 2NaOH(aq) \rightarrow Na_2CO_3(aq) + H_2O(l)$

Group Trends in Metals and Nonmetals

7.41

	Na	**Mg**
(a)	[Ne] $3s^1$	[Ne] $3s^2$
(b)	+1	+2
(c)	+496 kJ/mol	+738 kJ/mol
(d)	very reactive	reacts with steam, but not $H_2O(l)$
(e)	1.86 Å	1.60 Å

   (b)   When forming ions, both adopt the stable configuration of Ne, but Na loses one electron and Mg two electrons to achieve this configuration.

   (c),(e)   The nuclear charge of Mg (Z = 12) is greater than that of Na, so it requires more energy to remove a valence electron with the same n value from Mg than Na. It also means that the 2s electrons of Mg are held closer to the nucleus, so the atomic radius (e) is smaller than that of Na.

   (d)   Mg is less reactive because it has a filled subshell and it has a higher ionization energy.

7.43   (a)   Ca and Mg are both metals; they tend to lose electrons and form cations when they react. Ca is more reactive because it has a lower ionization energy than Mg. The Ca valence electrons in the 4s orbital are less tightly held because they are farther from the nucleus and experience more shielding by core electrons than the 3s valence electrons of Mg.

(b)   K and Ca are both metals; they tend to lose electrons and form cations when they react. K is more reactive because it has a lower ionization energy. The 4s valence electron in K is less tightly held because it experiences a smaller nuclear charge ($Z = 19$ for K versus $Z = 20$ for Ca) with similar shielding effects than the 4s valence electrons of Ca.

7.45   (a) $2K(s) + 2H_2O(l) \rightarrow 2KOH(aq) + H_2(g)$

(b) $Ba(s) + 2H_2O(l) \rightarrow Ba(OH)_2(aq) + H_2(g)$

(c) $6Li(s) + N_2(g) \rightarrow 2Li_3N(s)$

(d) $2Mg(s) + O_2(g) \rightarrow 2MgO(s)$

7.47   H - $1s^1$; Li - [He] $2s^1$; F - [He] $2s^2 2p^6$. Like Li, H has only one valence electron, and its most common oxidation number is +1, which both H and Li adopt after losing the single valence electron. Like F, H needs only one electron to adopt the stable electron configuration of the nearest noble gas. Both H and F can exist in the -1 oxidation state, when they have gained an electron to complete their valence shells.

7.49

	F	**Cl**
(a)	[He] $2s^2 2p^5$	[Ne] $3s^2 3p^5$
(b)	-1	-1
(c)	1681 kJ/mol	1256 kJ/mol
(d)	reacts exothermically to form HF	reacts slowly to form HCl
(e)	-332 kJ/mol	-349 kJ/mol
(f)	0.72 Å	0.99 Å

(b)   F and Cl are in the same group, have the same valence electron configuration and common ionic charge.

(c),(f)   The $n = 2$ valence electrons in F are closer to the nucleus and more tightly held than the $n = 3$ valence electrons in Cl. Therefore, the ionization energy of F is greater, and the atomic radius is smaller.

(d)   In its reaction with H_2O, F is reduced; it gains an electron. Although the electron affinity, a gas phase single atom property, of F is less negative than that of Cl, the tendency of F to hold its own electrons (high ionization energy) coupled with a relatively large exothermic electron affinity makes it extremely susceptible to reduction and chemical bond formation. Cl is unreactive to water because it is less susceptible to reduction.

(e)   Although F has a larger Z_{eff} than Cl, its small atomic radius gives rise to large repulsions when an extra electron is added, so the overall electron affinity of F is smaller (less exothermic) than that of Cl.

(f)   The n = 2 valence electrons in F are closer to the nucleus so the atomic radius is smaller than that of Cl.

7.51   In 1962, N. Bartlett discovered that Xe, which has the lowest ionization energy of the nonradioactive Noble gases, would react with substances having a strong tendency to remove electrons, such as F_2. Thus, the term "inert" no longer described all the Group 8A elements. (Kr also reacts with F_2, but reactions of Ar, Ne and He are as yet unknown.)

7.53   (a) $2Li(s) + Cl_2(g) \rightarrow 2LiCl(s)$

(b) $S_8(s) + 8O_2(g) \rightarrow 8SO_2(g)$

(c) $2K I(l) \rightarrow 2K(l) + I_2(g)$

(d) $Cl_2(g) + H_2O(l) \rightarrow HCl(aq) + HOCl(aq)$

7.55   (a)   Valence electrons in F atoms experience a much greater attraction for the nucleus than those in I atoms, because they are much closer to the nucleus. Thus, F atoms have a much greater tendency to gain electrons and are more reactive toward relatively inert substances such as Xe.

(b) O_2 is the allotrope routinely found in the atmosphere (21% of air is O_2). O_3 is produced only under special conditions.

(c) F_2 is extremely reactive because it can remove electrons from almost any substance, so a special apparatus that will not react with F_2 and will not leak F_2 must be used to carry out reactions.

Additional Exercises

7.58   Up to Z = 83, there are four instances where atomic weights are reversed relative to atomic numbers: Ar and K; Co and Ni; Te and I; Ce and Pr.

7.60   According to Figure 7.5, the atomic radius of the most stable form of Fe at room temperature is 1.24 Å. This is derived from an Fe-Fe distance of 2.48 Å, which corresponds to α-iron.

7.62   (a)   Increasing atomic radius: F < O < C < Si ≈ Se
        F, O, C and Si can be ordered according to the trends for size and ionization energy. Se is difficult to place because it is both to the right and below Si; these two directions have conflicting trends. Also, it is in the fourth row and subject to the effects of the filling of the 3d subshell and accompanying increase in Z and Z_{eff}. Referring to the data in Figures 7.5 and 7.7, the atomic radii of Si and Se are approximately equal. The ionization energy of Se is greater than that of Si and less than that of C.

(b)   Increasing ionization energy: Si < Se < C < O < F

7.65   (a)   Y - $[Kr]5s^24d^1$.   Each succeeding ionization energy is larger than the previous one, due to increases in effective nuclear charge as electrons are removed (Exercise 7.64). I_1, I_2 and I_3 involve removing valence electrons and the size of the increases is similar. I_4 involves removing an inner-shell electron from the stable Kr core and requires significantly more energy than I_3.

        (b)   In Table 7.2, Al has the pattern of ionization energies most similar to yttrium. That is, I_1, I_2 and I_3 increase smoothly and there is a jump from I_3 to I_4. The actual values of I_2, I_3 and I_4 are greater for Al because the electrons being removed are closer to the nucleus and experience a greater effective nuclear charge.

7.68   (a)   The group 2B metals have complete (*n*-1)d subshells. An additional electron would occupy an *n*p subshell and be substantially shielded by both *n*s and (*n*-1) d electrons. Overall this is not a lower energy state than the neutral atom and a free electron.

        (b)   Valence electrons in Group 1B elements experience a relatively large effective nuclear charge due to the build-up in Z with the filling of the (*n*-1)d subshell. Thus, the electron affinities are large and negative. Group 1B elements are exceptions to the usual electron filling order and have the generic electron configuration $ns^1(n-1)d^{10}$. The additional electron would complete the *n*s subshell and experience repulsion with the other *n*s electrons. Going down the group, size of the *n*s subshell increases and repulsion effects decrease. That is, effective nuclear charge is greater going down the group because it is less diminished by repulsion, and electron affinities become more negative.

7.71   Since Xe reacts with F_2, and O_2 has approximately the same ionization energy as Xe, O_2 will probably react with F_2. Possible products would be O_2F_2, analogous to XeF_2, or OF_2.

$$O_2(g) + F_2(g) \rightarrow O_2F_2(g)$$

$$O_2(g) + 2F_2(g) \rightarrow 2OF_2(g)$$

Integrative Exercises

7.76   (a) $E = hc/\lambda$;   1 nm = 1×10^{-9} m;   58.4 nm = 58.4×10^{-9} m;

           1 eV•mol = 96.485 kJ/mol, 1 eV • mol = 96.485 kJ

$$E = \frac{6.626 \times 10^{-34} \text{ J•s} \times 2.998 \times 10^8 \text{ m/s}}{58.4 \times 10^{-9} \text{ m}} = 3.4015 \times 10^{-18} = 3.40 \times 10^{-18} \text{ J/photon}$$

$$\frac{3.4015 \times 10^{-18} \text{ J}}{\text{photon}} \times \frac{1 \text{ kJ}}{1000 \text{ J}} \times \frac{6.022 \times 10^{23} \text{ photons}}{\text{mol}} \times \frac{1 \text{ eV} \times \text{mol}}{96.485 \text{ kJ}} 124 = 21.230$$

$$= 21.2 \text{ eV}$$

        (b) $Hg(g) \rightarrow Hg^+(g) + 1e^-$

(c) $I_1 = E_{58.4} - E_K$ = 21.23 eV –10.75 eV = 10.48 = 10.5 eV

$$10.48 \text{ eV} \times \frac{96.485 \text{ kJ}}{1 \text{ eV} \cdot \text{mol}} = 1.01 \times 10^3 \text{ kJ/mol}$$

(d)   From Figures 7.6 and 7.7, iodine (I) appears to have the ionization energy closest to that of Hg, approximately 1000 kJ/mol. (Quantitative data is much easier to read on Figure 7.6.)

7.78   (a) Mg_3N_2

       (b) $Mg_3N_2(s) + 3H_2O(l) \rightarrow 3MgO(s) + 2NH_3(g)$
The driving force is the production of $NH_3(g)$.

       (c)   After the second heating, all the Mg is converted to MgO.
Calculate the initial mass Mg.

$$0.486 \text{ g MgO} \times \frac{24.305 \text{ g Mg}}{40.305 \text{ g MgO}} = 0.293 \text{ g Mg}$$

x = g Mg converted to MgO; y = g Mg converted to Mg_3N_2; x = 0.293 - y

$$\text{g MgO} = x \left(\frac{40.305 \text{ g MgO}}{24.305 \text{ g Mg}} \right); \text{ g Mg}_3\text{N}_2 = y \left(\frac{100.929 \text{ g Mg}_3\text{N}_2}{72.915 \text{ g Mg}} \right)$$

g MgO + g Mg_3N_2 = 0.470

$$(0.293 - y) \left(\frac{40.305}{24.305} \right) + y \left(\frac{100.929}{72.915} \right) = 0.470$$

$$(0.293 - y)(1.6583) + y(1.3842) = 0.470$$

$$-1.6583 \, y + 1.3842 \, y = 0.470 - 0.48588$$

$$-0.2741 \, y = -0.016$$

$$y = 0.05794 = 0.058 \text{ g Mg in Mg}_3\text{N}_2$$

$$\text{g Mg}_3\text{N}_2 = 0.05794 \text{ g Mg} \times \frac{100.929 \text{ g Mg}_3\text{N}_2}{72.915 \text{ g Mg}} = 0.0802 = 0.080 \text{ g Mg}_3\text{N}_2$$

$$\text{mass \% Mg}_3\text{N}_2 = \frac{0.0802 \text{ g Mg}_3\text{N}_2}{0.470 \text{ g (MgO + Mg}_3\text{N}_2)} \times 100 = 17\%$$

(The final mass % has 2 sig figs because the mass of Mg obtained from solving simultaneous equations has 2 sig figs.)

8 Basic Concepts of Chemical Bonding

Lewis Symbols and Ionic Bonding

8.1   (a)   Valence electrons are those that take part in chemical bonding, those in the outermost electron shell of the atom. This usually means the electrons beyond the core noble gas configuration of the atom, although it is sometimes only the outer shell electrons.

        (b)   C - [He] $2s^2 2p^2$ C has 4 valence electrons.
                    |_____|
            valence electrons

        (c) $1s^2 2s^2 2p^6 \; 3s^2 3p^1$ The atom (Al) has 3 valence electrons.
              |_____| |_____|
                [Ne]   valence electrons

8.3   (a) :Cl·   (b) Mg·   (c) :Br·   (d) :Ar: or A

8.5   Ca·   +   ·O: ⟶ Ca^{2+} + $\left[:\ddot{O}: \right]^{2-}$

8.7   (a) CaF_2 (b) Na_2S (c) Y_2O_3 (d) Li_3N

8.9   (a) Ba^{2+} : [Xe], noble gas configuration
        (b) Cl^- : $[Ne]3s^2 3p^6$ = [Ar], noble gas configuration
        (c) Te^{2-}: $[Kr]5s^2 4d^{10} 5p^6$ = [Xe], noble gas configuration
        (d) Cr^{2+} : $[Ar]3d^4$
        (e) Sc^{3+} : [Ar], noble gas configuration
        (f) Co^{3+} : $[Ar]3d^6$

8.11   (a)   *Lattice energy* is the energy required to totally separate one mole of solid ionic compound into its gaseous ions.

       (b)   The magnitude of the lattice energy depends on the magnitudes of the charges of the two ions, their radii and the arrangement of ions in the lattice. The main factor is the charges, because the radii of ions do not vary over a wide range.

8.13   Equation 8.4 predicts that as the oppositely charged ions approach each other, the energy of interaction will be large and negative. This more than compensates for the energy required to form Ca^{2+} and O^{2-} from the neutral atoms (see Figure 8.4 for the formation of NaCl).

8.15   NaF - 910 kJ/mol;   MgO - 3795 kJ/mol
         The two factors that affect lattice energies are charge and ionic radii. The Na-F and Mg-O separations are similar (Na^+ is larger than Mg^{2+}, but F^- is smaller than O^{2-}). The charges on Mg^{2+} and O^{2-} are twice those of Na^+ and F^-, so according to Equation 8.4, the lattice energy of MgO is approximately four times that of NaF.

8.17   (a) O^{2-} is smaller than S^{2-}; the closer approach of the oppositely charged ions in MgO leads to greater electrostatic attraction.

         (b)   The ions have 1+ and 1- charges in both compounds. However, the fact that the ionic radii are much smaller in LiF means that the ions can approach more closely, with a resultant increase in electrostatic attractive forces.

         (c)   The ions in CaO have 2+ and 2- charges as compared with 1+ and 1- charges in KF.

8.19 $RbCl(s) \rightarrow Rb^+(g) + Cl^-(g)$ ΔH (lattice energy) = ?

         By analogy to NaCl, Figure 8.4, the lattice energy is

$$\Delta H_{latt} = -\Delta H_f^\circ \, RbCl(s) + \Delta H_f^\circ \, Rb(g) + \Delta H_f^\circ \, Cl(g) + I_1 \, (Rb) + E \, (Cl)$$

$$= -(-430.5 \text{ kJ}) + 85.8 \text{ kJ} + 121.7 \text{ kJ} + 403 \text{ kJ} + (-349 \text{ kJ}) = +692 \text{ kJ}$$

         This value is smaller than that for NaCl (+788 kJ) because Rb^+ has a larger ionic radius than Na^+. This means that the value of d in the denominator of Equation 8.4 is larger for RbCl, and the potential energy of the electrostatic attraction is smaller.

Sizes of Ions

8.21   (a)   Electrostatic repulsions are reduced by removing an electron from a neutral atom, Z_{eff} increases, and the cation is smaller.

         (b)   The additional electrostatic repulsion produced by adding an electron to a neutral atom decreases the Z_{eff} of the valence electrons, causing them to be less tightly bound to the nucleus and the size of the anion to be larger.

         (c)   Going down a column, valence electrons are further from the nucleus and they experience greater shielding by core electrons. Inspite of an increase in Z, the size of particles with like charge increases.

8.23   (a)   In an isoelectronic series, the number of electrons in the particles is the same, but the values of Z are different.

     (b) Rb^+: **Kr**; Br^-: **Kr**; S^{2-}: **Ar**; Zn^{2+}: **Ni**

8.25 (a)   Since the number of electrons in an isoelectronic series is the same, repulsion and shielding effects do not vary for the different particles. As Z increases, Z_{eff} increases, the valence electrons are more strongly attracted to the nucleus and the size of the particle decreases.

     (b)   Because F^-, Ne and Na^+ have the same electron configuration, the 2p electron in the particle with the largest Z experiences the largest effective nuclear charge. A 2p electron in Na^+ experiences the greatest effective nuclear charge.

8.27 (a) $Li^+ < K^+ < Rb^+$ (b) $Mg^{2+} < Na^+ < Br^-$ (c) $K^+ < Ar < Cl^- < S^{2-}$ (d) $Ar < Cl < Cl^-$

Covalent Bonding, Electronegativity and Bond Polarity

8.29 (a)   A *covalent bond* is the bond formed when two atoms share one or more pairs of electrons.

     (b)   The ionic bonding in NaCl is due to strong electrostatic attraction between oppositely charged Na^+ and Cl^- ions. The covalent bonding in Cl_2 is due to sharing of a pair of electrons by two neutral chlorine atoms.

8.31 $H\cdot \; + \; H\cdot \; + \; H\cdot \; + \; \cdot\ddot{P}: \; \longrightarrow \; H{-}\underset{\underset{H}{|}}{\overset{\overset{H}{|}}{P}}:$

8.33 (a) $:\!\overset{\cdot\cdot}{O}\!\!=\!\!\underset{\cdot\cdot}{O}\!:$

     (b)   A double bond is required because there are not enough electrons to satisfy the octet rule with single bonds and unshared pairs.

     (c)   The greater the number of shared electron pairs between two atoms, the shorter the distance between the atoms. If O_2 has a double bond, the O-O distance will be shorter than the O-O single bond distance.

8.35 (a)   Electronegativity is the ability of an atom in a molecule (a bonded atom) to attract electrons to itself.

     (b)   The range of electronegativities on the Pauling scale is 0.7-4.0.

     (c)   Fluorine, F, is the most electronegative element.

8.37 Electronegativity increases going up and to the right of the periodic table.

     (a)   Cl   (b) B   (c) As   (d) Mg

8.39   The bonds in (a), (c) and (e) are polar because the atoms involved differ in electronegativity. The more electronegative element in each polar bond is: (a) Cl   (c) F   (e) O

Lewis Structure; Resonance Structures

8.41   Counting the **correct number of valence electrons** is the foundation of every Lewis structure.

(a)   Count valence electrons: $4 + (4 \times 1) = 8$ e$^-$, 4 e$^-$ pairs. Follow the procedure in Sample Exercise 8.7.

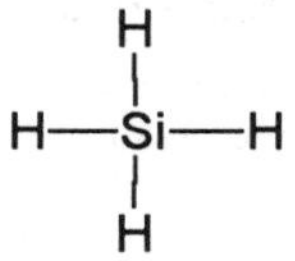

$$\begin{array}{c} H \\ | \\ H-Si-H \\ | \\ H \end{array}$$

(b)   Valence electrons: $[6 + (2 \times 7)] = 20$ e$^-$, 10 e$^-$ pairs.

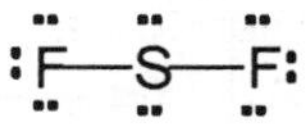

$$:\!\ddot{F}-\ddot{S}-\ddot{F}\!:$$

         i.   Place the S atom in the middle and connect each F atom with a single bond; this requires 2 e$^-$ pairs.

         ii.   Complete the octets of the F atoms with lone pairs of electrons; this requires an additional 6 e$^-$ pairs.

         iii.   The remaining 2 e$^-$ pairs complete the octet of the central S atom.

(c)   16 valence e$^-$, 8 e$^-$ pairs   (d)   32 valence e$^-$, 16 e$^-$ pairs

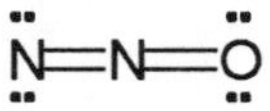

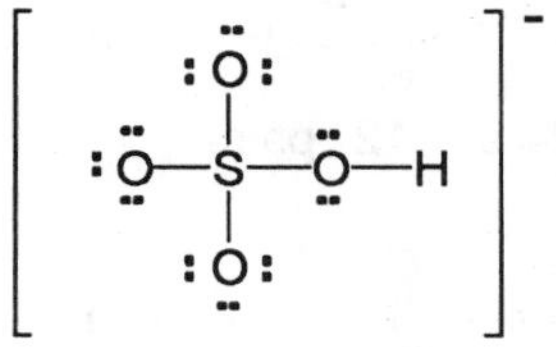

$$\ddot{N}=N=\ddot{O}$$

There are other
resonance structures.

$$\left[\begin{array}{c} :\ddot{O}: \\ | \\ :\ddot{O}-S-\ddot{O}-H \\ | \\ :\ddot{O}: \end{array}\right]^{-}$$

(e)   14 valence e$^-$, 6 e$^-$ pairs 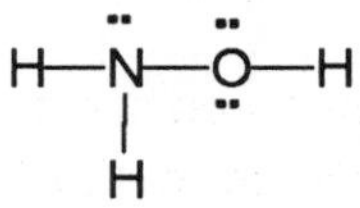   (Choose the Lewis structure that obeys the octet rule, Section 8.8.)

$$\begin{array}{c} H-\ddot{N}-\ddot{O}-H \\ | \\ H \end{array}$$

8.43   (a)   16 e$^-$, 8 e$^-$ pairs. The structure on the left minimizes formal charge.

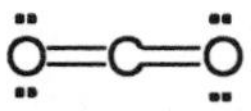

$$\ddot{O}=C=\ddot{O} \qquad\qquad :\ddot{O}-C\equiv O:$$
$$\;\;0\;\;\;\;0\;\;\;\;0 \qquad\qquad -1\;\;\;\;0\;\;+1$$

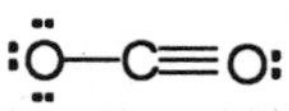

(b)   8 e⁻, 4 e⁻ pairs   (c)   32 e⁻, 16 e⁻ pairs 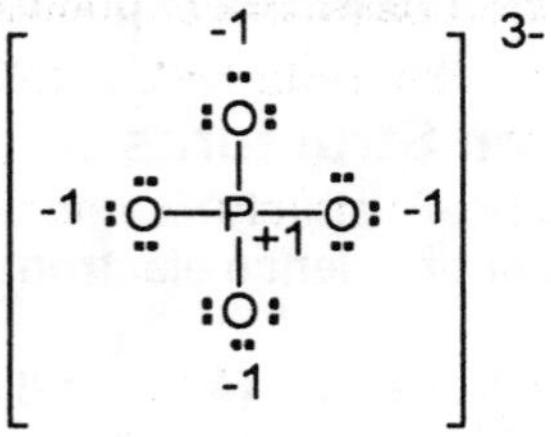   (d)   10 e⁻, 5 e⁻ pairs

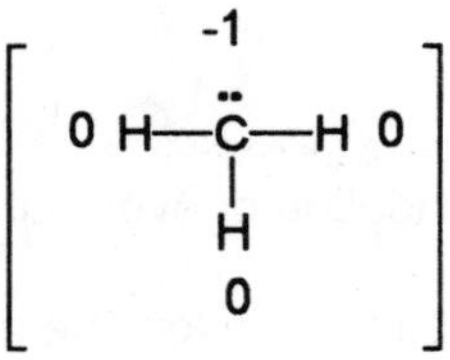

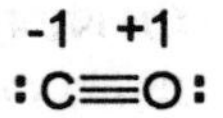

$$\left[\; 0\; \text{H}-\overset{-1}{\underset{\underset{\text{H}}{|}}{\text{C}}}-\text{H}\; 0\;\right]^{-}$$

$$\left[\;\underset{-1}{:\ddot{\text{O}}:}-\overset{-1}{\overset{:\ddot{\text{O}}:}{\underset{\underset{-1}{:\ddot{\text{O}}:}}{\overset{+1}{\text{P}}}}}-\overset{-1}{:\ddot{\text{O}}:}\;\right]^{3-}$$

$$\overset{-1\;\;+1}{:\text{C}\equiv\text{O}:}$$

8.45   (a)   18 e⁻, 9 e⁻ pairs

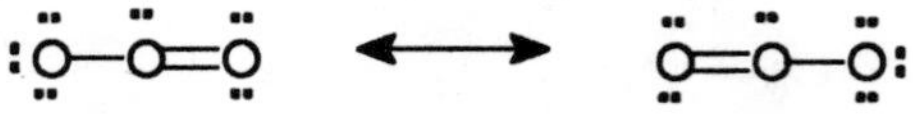

$$:\ddot{\text{O}}-\ddot{\text{O}}=\text{O} \longleftrightarrow \text{O}=\ddot{\text{O}}-\ddot{\text{O}}:$$

(b)   24 e⁻, 12 e⁻ pairs

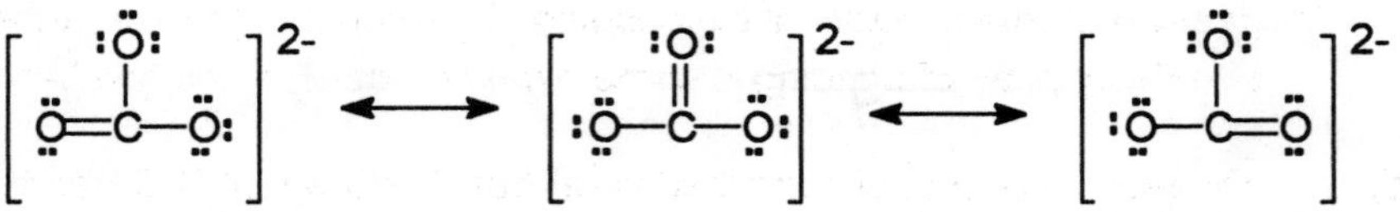

$$\left[\overset{:\ddot{\text{O}}:}{\ddot{\text{O}}=\text{C}-\ddot{\text{O}}:}\right]^{2-} \longleftrightarrow \left[\overset{:\ddot{\text{O}}:}{:\ddot{\text{O}}-\text{C}-\ddot{\text{O}}:}\right]^{2-} \longleftrightarrow \left[\overset{:\ddot{\text{O}}:}{:\ddot{\text{O}}-\text{C}=\ddot{\text{O}}}\right]^{2-}$$

(c)   18 e⁻, 9 e⁻ pairs

$$\left[\overset{:\ddot{\text{O}}:}{\text{H}-\text{C}-\ddot{\text{O}}:}\right]^{-} \longleftrightarrow \left[\overset{:\ddot{\text{O}}:}{\text{H}-\text{C}=\ddot{\text{O}}}\right]^{-}$$

(d)   24 e⁻, 12⁻ pairs

$$\overset{:\text{O}:}{:\ddot{\text{O}}-\text{S}-\ddot{\text{O}}:} \longleftrightarrow \overset{:\ddot{\text{O}}:}{\text{S}=} \longleftrightarrow \overset{:\ddot{\text{O}}:}{-\text{S}=\text{O}}$$

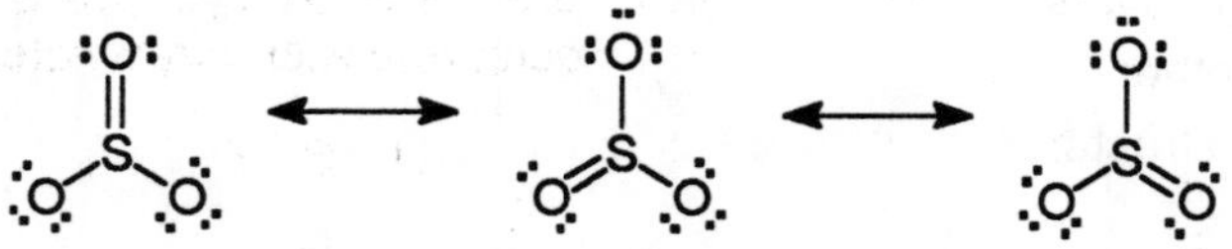

8.47   The Lewis structures are as follows:

5 e⁻ pairs   9 e⁻ pairs

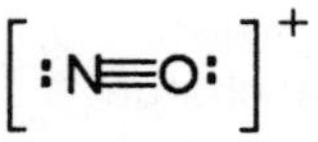

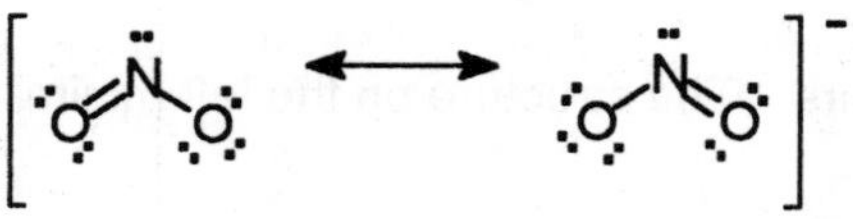

$$\left[:\text{N}\equiv\text{O}:\right]^{+} \qquad \left[\overset{\ddot{\text{N}}}{\ddot{\text{O}}\;\;\ddot{\text{O}}}\right]^{-} \longleftrightarrow \left[\overset{\ddot{\text{N}}}{\ddot{\text{O}}\;\;\ddot{\text{O}}}\right]^{-}$$

12 e⁻ pairs

$$\left[\overset{:\ddot{\text{O}}:}{:\ddot{\text{O}}-\text{N}-\ddot{\text{O}}:}\right]^{-} \quad \left[\overset{:\ddot{\text{O}}:}{:\ddot{\text{O}}-\text{N}-\ddot{\text{O}}:}\right]^{-} \quad \left[\overset{:\text{O}:}{:\ddot{\text{O}}-\text{N}-\ddot{\text{O}}:}\right]^{-}$$

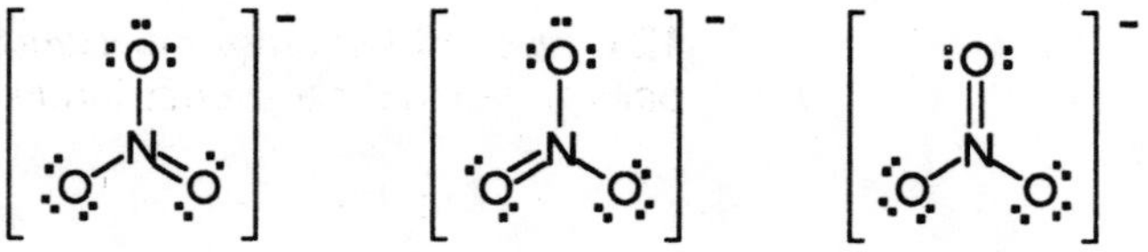

The average number of electron pairs in the N-O bond is 3.0 for NO^+, 1.5 for NO_2^- and 1.33 for NO_3^-. The more electron pairs shared between two atoms, the shorter the bond. Thus the N-O bond lengths vary in the order $NO^+ < NO_2^- < NO_3^-$.

8.49   (a)   Two equally valid Lewis structures can be drawn for benzene.

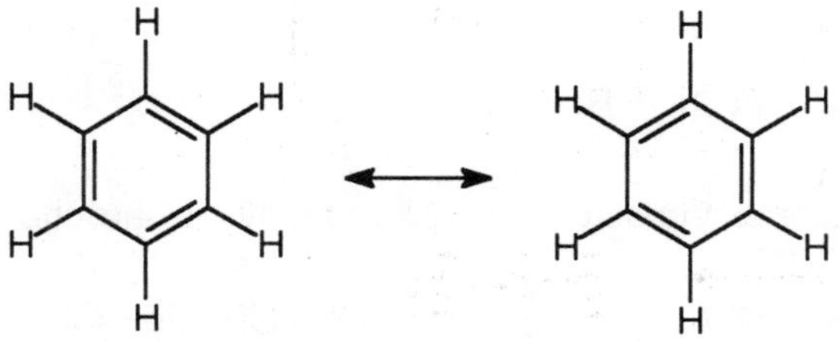

            Since the 6 C-C bond lengths are equal, neither of the individual Lewis structures gives an accurate picture of the bonding. The concept of resonance dictates that the true description of bonding is some hybrid or blend of the two Lewis structures.

      (b)   The resonance model described in (a) has 6 equivalent C-C bonds, each with some double bond character. That is, more than 1 pair but less than 2 pairs of electrons is involved in each C-C bond. We would expect the C-C bond length in benzene to be shorter than a single bond but longer than a double bond.

Exceptions to the Octet Rule

8.51   The most common exceptions to the octet rule are molecules with more than eight electrons around one or more atoms, usually the central atom.

8.53   (a)   26 e⁻, 13 e⁻ pairs   (b)   6 e⁻, 3 e⁻ pairs, impossible to satisfy octet rule with only 6 valence electrons

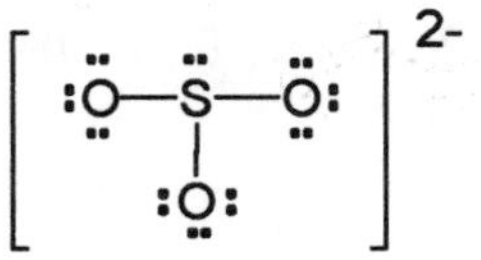

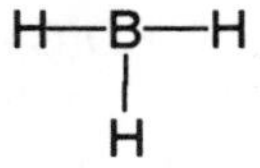

                                                     6 electrons around B

     c)   22 e⁻, 11 e⁻ pairs   (d)   48 e⁻ pairs, 24 e⁻ pairs

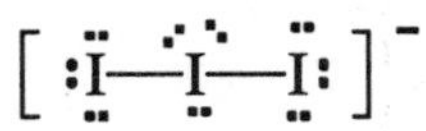

      10 e⁻ around central I

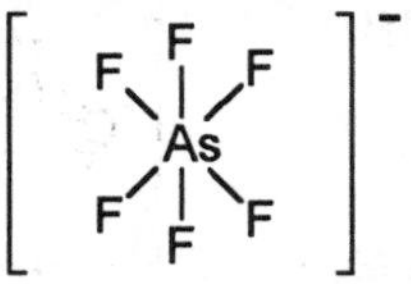

                                            12 e⁻ around As; three nonbonded pairs on each F have been omitted

(e)   13 e⁻, 6.5 e⁻ pairs, odd electron molecule

$$\left[\; :\ddot{O}\!-\!\ddot{O}: \;\right]^{-} \longleftrightarrow \left[\; :\overset{..}{\underset{..}{O}}\!-\!\overset{..}{\underset{..}{O}}: \;\right]^{-}$$

8.55   (a)   16 e⁻ 8 e⁻ pairs $:\overset{..}{\underset{..}{Cl}}\!-\!Be\!-\!\overset{..}{\underset{..}{Cl}}:$

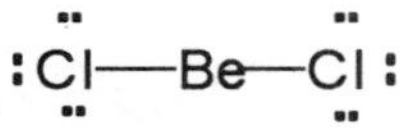

This structure violates the octet rule; Be has only 4 e⁻ around it.

(b) 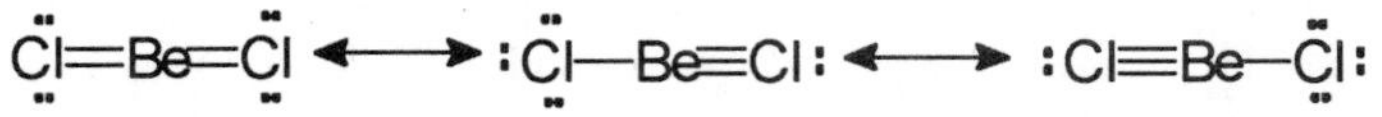 $\ddot{\underset{..}{Cl}}\!\!=\!\!Be\!\!=\!\!\ddot{\underset{..}{Cl}} \longleftrightarrow :\overset{..}{\underset{..}{Cl}}\!-\!Be\!\!\equiv\!\!Cl: \longleftrightarrow :Cl\!\!\equiv\!\!Be\!-\!\overset{..}{\underset{..}{Cl}}:$

(c)   The formal charges on each of the atoms in the four resonance structures are:

$:\overset{..}{\underset{..}{Cl}}\!-\!Be\!-\!\overset{..}{\underset{..}{Cl}}:$ $\ddot{\underset{..}{Cl}}\!\!=\!\!Be\!\!=\!\!\ddot{\underset{..}{Cl}}$ $:\overset{..}{\underset{..}{Cl}}\!-\!Be\!\!\equiv\!\!Cl:$ $:Cl\!\!\equiv\!\!Be\!-\!\overset{..}{\underset{..}{Cl}}:$

  0   0   0   +1 -2 +1   0   -2 +2   +2 -2   0

Since formal charges are minimized on the structure that violates the octet rule, this form is probably most important.

Bond Enthalpies

8.57   (a) $\Delta H = 4D(C\text{-}H) + D(Cl\text{-}Cl) - 2D(C\text{-}H) - 2D(C\text{-}Cl) - D(H\text{-}H)$
        $= 2D(C\text{-}H) + D(Cl\text{-}Cl) - 2D(C\text{-}Cl) - D(H\text{-}H)$
        $= 2(413) + 242 - 2(328) - 436 = -24 \text{ kJ}$

(b) $\Delta H = 6D(C\text{-}H) + 2D(C\text{-}O) + 2D(O\text{-}H) - 6D(C\text{-}H) - 2D(C\text{-}O) - 2D(O\text{-}H)$
        $= 0 \text{ kJ}$

(c) $\Delta H = 2D(C\text{-}H) + D(C\!\!=\!\!N) + D(N\text{-}H) + 2D(O\text{-}H) - 2D(C\text{-}H) - D(C\!\!=\!\!O) - 3D(N\text{-}H)$
        $= D(C\!\!=\!\!N) + 2D(O\text{-}H) - D(C\!\!=\!\!O) - 2D(N\text{-}H)$
        $= 615 + 2(463) - 799 - 2(391) = -40 \text{ kJ}$

8.59   Draw structural formulas so bonds can be visualized.

(a)   2H-O-O-H → 2H-O-H + O=O
      $\Delta H = D(O\text{-}O) - D(O\!\!=\!\!O) = 2(146) - (495) = -203 \text{ kJ}$

(b) 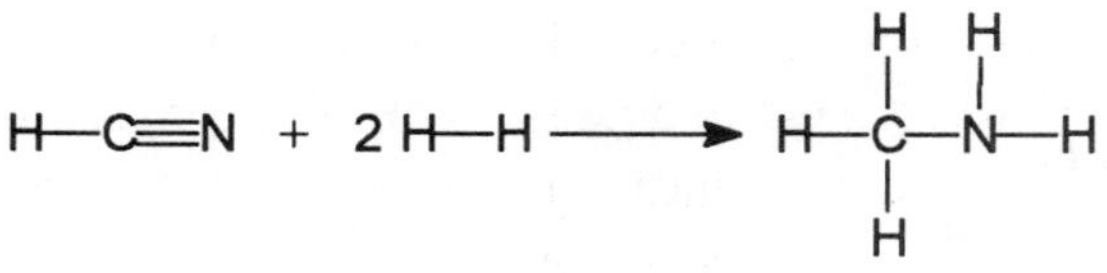 $H\!-\!C\!\!\equiv\!\!N + 2\,H\!-\!H \longrightarrow H\!-\!\underset{\underset{H}{|}}{\overset{\overset{H}{|}}{C}}\!-\!\underset{}{\overset{\overset{H}{|}}{N}}\!-\!H$

      $\Delta H = D(C \equiv N) + 2D(H\text{-}H) - 2D(C\text{-}H) - D(C\text{-}N) - 2D(N\text{-}H)$
        $= 891 + 2(436) - 2(413) - 293 - 2(391) = -138 \text{ kJ}$

(c)

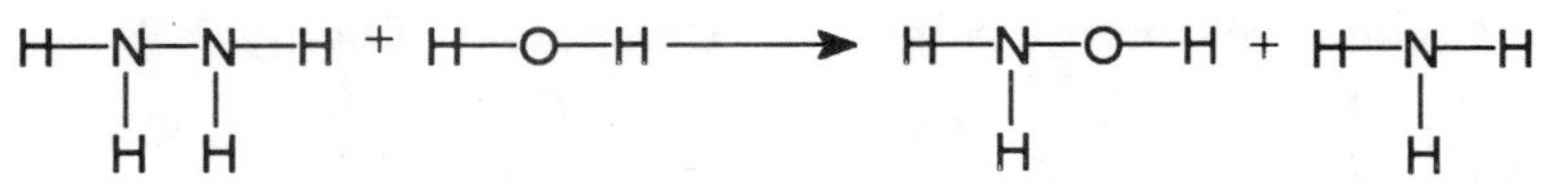

H—N—N—H + H—O—H ⟶ H—N—O—H + H—N—H
　|　|　　　　　　　　　　　　|　　　　　　|
　H　H　　　　　　　　　　　H　　　　　　H

$$\Delta H = D(N\text{-}N) + D(O\text{-}H) - D(N\text{-}O) - D(N\text{-}H)$$
$$= 163 + 463 - 201 - 391 = +34 \text{ kJ}$$

8.61　The average Ti-Cl bond dissociation energy is just the average of the four values listed, 430 kJ/mol.

Oxidation Numbers

8.63　(a)　N, 0　　(b) C, -4　　(c) P, +5　　(d) Mn, +7　　(e) B, +3
　　　(f)　Cl, +1　(g) Fe, +3　(h) S, +6

8.65　(a)　P +5, -3　　(b) C +4, -4　　(c) I +7, -1　　(d) Cr +6, 0

8.67　(a)　MnO_2　　(b) Ga_2S_3　　(c) SeF_6　　(d) copper(I) oxide or cuprous oxide
　　　(e)　chlorine trifluoride
　　　(f)　tellurium trioxide or tellurium(VI) oxide

8.69　(a)　xenon is oxidized from 0 to +4; fluorine is reduced from 0 to -1
　　　(b)　copper is reduced from +2 to +1; iodine is oxidized from -1 to 0
　　　(c)　nitrogen is oxidized from -3 to +3; chlorine is reduced from 0 to -1
　　　(d)　iodine is reduced from 0 to -1; sulfur is oxidized from +4 to +6
　　　(e)　sulfur is oxidized from -2 to +4; oxygen is reduced from 0 to -2

Additional Exercises

8.71　(a)　Group 4A or 4B　　(b) Group 2A　　(c) Group 5A or 5B

8.75　(a)

Compound	Lattice Energy (kJ)		Compound	Lattice Energy (kJ)	
NaCl	788	106 kJ ⌈ ⌉ 56 kJ	LiCl	834	104 kJ ⌈ ⌉ 55 kJ
NaBr	732		**LiBr**	**779**	
NaI	682		LiI	730	

The difference in lattice energy between LiCl and LiI is 104 kJ. The difference between NaCl and NaI is 106 kJ; the difference between NaCl and NaBr is 56 kJ, or 53% of the difference between NaCl and NaI. Applying this relationship to the Li salts, 0.53(104 kJ) = 55 kJ difference between LiCl and LiBr. The approximate lattice energy of LiBr is (834 - 55) kJ = 779 kJ.

(b)

Compound	Lattice Energy (kJ)		Compound	Lattice Energy (kJ)	
NaCl	788		CsCl	657	
NaBr	732	56 kJ	CsBr	**627**	30 k
NaI	682		CsI	600	

106 kJ (bracket for Na column), 57 kJ (bracket for Cs column)

By analogy to the Na salts, the difference between lattice energies of CsCl and CsBr should be approximately 53% of the difference between CsCl and CsI. The lattice energy of CsBr is approximately **627 kJ**.

(c)

Compound	Lattice Energy (kJ)		Compound	Lattice Energy (kJ)	
MgO	3795		$MgCl_2$	2326	
CaO	3414	381 kJ	$CaCl_2$	**2195**	131 kJ
SrO	3217		$SrCl_2$	2127	

578 kJ (bracket for oxide column), 199 kJ (bracket for chloride column)

By analogy to the oxides, the difference between the lattice energies of $MgCl_2$ and $CaCl_2$ should be approximately 66% of the difference between $MgCl_2$ and $SrCl_2$. That is, 0.66(199 kJ) = 131 kJ. The lattice energy of $CaCl_2$ is approximately (2326 - 131) kJ = 2195 kJ.

8.78   Formal charge (FC) = # valence e^- - (# nonbonding e^- + 1/2 # bonding e^-)

(a)   18 e^-, 9 e^- pairs

$$:\overset{..}{\underset{..}{O}}-\overset{..}{O}=\overset{..}{\underset{..}{O}} \longleftrightarrow \overset{..}{\underset{..}{O}}=\overset{..}{O}-\overset{..}{\underset{..}{O}}:$$

FC for the central O = 6 - [2 + 1/2 (6)] = +1

(b)   48 e^-, 24 e^- pairs

FC for P = 5 - [0 + 1/2 (12)] = -1

The three nonbonded pairs on each F have been omitted.

(c)   17 e^- ; 8 e^- pairs, 1 odd e^-

$$\overset{..}{\underset{..}{O}}=\overset{.}{N}-\overset{..}{\underset{..}{O}}: \longleftrightarrow :\overset{..}{\underset{..}{O}}-\overset{.}{N}=\overset{..}{\underset{..}{O}}$$

The odd electron is probably on N because it is less electronegative than O. Assuming the odd electron is on N, FC for N = 5 - [1+ 1/2 (6)] = +1. If the odd electron is on O, FC for N = 5 - [2 + 1/2 (6)] = 0.

(d)   28 e⁻, 14 e⁻ pairs   (e)   32 e⁻, 16 e⁻ pairs

$$:\ddot{C}l—I—\ddot{C}l:$$
$$|$$
$$:\ddot{C}l:$$

$$:\ddot{O}:$$
$$|$$
$$:\ddot{O}—Cl—\ddot{O}—H$$
$$|$$
$$:\ddot{O}:$$

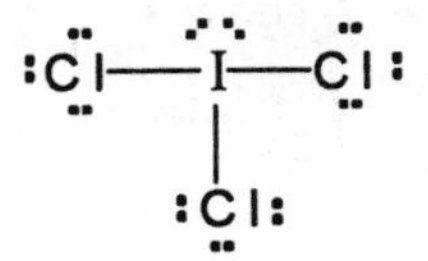

FC for I = 7 - [4 + 1/2 (6)] = 0   FC for Cl = 7 - [0 + 1/2 (8)] = +3

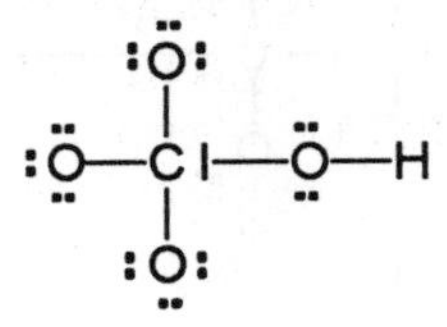

(b)   Resonance predicts that the true picture of bonding is a hybrid of the two resonance forms. In this description, there are no isolated single and double bonds, but rather 6 equivalent C-C bonds, each with some double bond character (see Exercise 8.49).

(c)
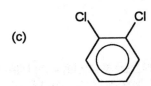

8.81 $$0 \quad :\ddot{F}—B—\ddot{F}: \quad 0 \qquad 0 \quad :\ddot{F}—B=\ddot{F}: \quad +1$$
$$\qquad\quad |\qquad\qquad\qquad\qquad |$$
$$\qquad\quad :\ddot{F}:\qquad\qquad\qquad :\ddot{F}:$$
$$\qquad\qquad 0 \qquad\qquad\qquad\qquad 0$$

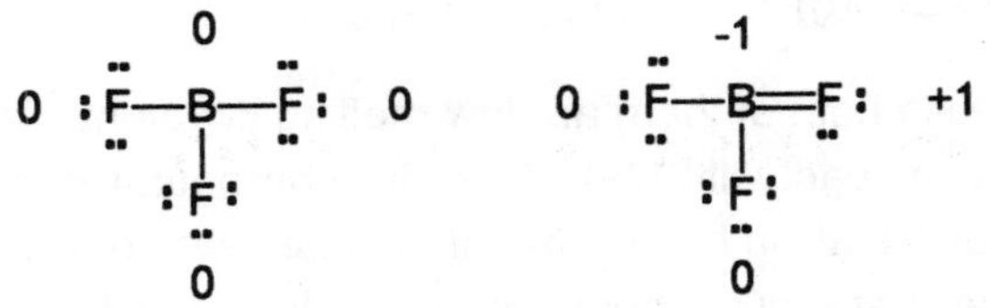

Clearly the structure with less than 8 electrons around B minimizes formal charges. Also, considering the electronegativities, a formal charge of +1 on F seems highly unlikely.

8.83   (a) $:N\equiv N: + 3\,H—H \longrightarrow 2\,H—\overset{..}{N}—H$
                                                       $|$
                                                       H

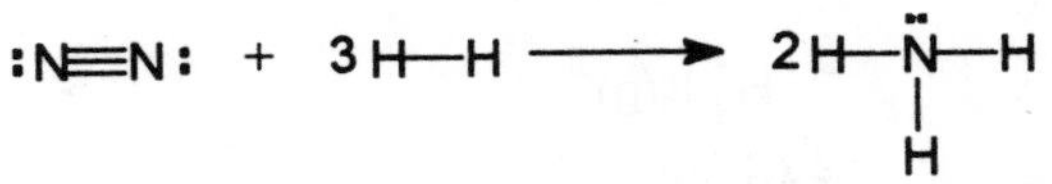

ΔH = D(N ≡ N) + 3D(H-H) - 6(N-H) = 941 kJ + 3(436 kJ) - 6(391 kJ)

= -97 kJ / 2 mol NH₃ ; **exothermic**

(b) ΔH_f° NH₃(g) = -46.19 kJ; ΔH = 2(-46.19) = -92.4 kJ

The ΔH calculated from bond energies is slightly more exothermic (more negative) than that obtained using ΔH_f° values.

8.86 (a)

nitroglycerine

$$\Delta H = 20D(\text{C-H}) + 8D(\text{C-C}) + 12(\text{C-O}) + 24D(\text{O-N}) + 12D(\text{N=O})$$

$$= [6D(\text{N}\equiv\text{N}) + 24D(\text{C=O}) + 20D(\text{H-O}) + D(\text{O=O})]$$

$$= 20(413) + 8(348) + 12(358) + 24(201) + 12(607)$$

$$- [6(941) + 24(799) + 20(463) + 495]$$

$$= -7129 \text{ kJ}$$

$$1.00 \text{ g C}_3\text{H}_5\text{N}_3\text{O}_9 \times \frac{1 \text{ mol C}_3\text{H}_5\text{N}_3\text{O}_9}{227.1 \text{ g C}_3\text{H}_5\text{N}_3\text{O}_9} \times \frac{-7129 \text{ kJ}}{4 \text{ mol C}_3\text{H}_5\text{N}_3\text{O}_9} = 7.85 \text{ kJ/g C}_3\text{H}_5\text{N}_3\text{O}_9$$

(b) $4\text{C}_7\text{H}_5\text{N}_3\text{O}_6(s) \rightarrow 6\text{N}_2(g) + 7\text{CO}_2(g) + 10\text{H}_2\text{O}(g) + 21\text{C}(s)$

8.89 SeO_2 and N_2O_3 can be ruled out because both of them are low melting covalent compounds. SrO is a basic oxide and not likely to react with NaOH, so the compound in question is probably GeO_2. Recall that oxides of metals in higher oxidation states are somewhat acidic in character. Thus, it is reasonable that GeO_2 should dissolve slightly in basic solution. Information that may be of interest: SrO, m.p. 2665°C, GeO_2, m.p. 1115°C.

Integrative Exercises

8.92 The pathway to the formation of K_2O can be written:

$2\text{K}(s) \rightarrow 2\text{K}(g)$	$2\Delta H^\circ_f \text{ K}(g)$
$2\text{K}(g) \rightarrow 2\text{K}^+(g) + 2 e^-$	$2 I_1(\text{K})$
$1/2 \text{ O}_2(g) \rightarrow \text{O}(g)$	$\Delta H^\circ_f \text{ O}(g)$
$\text{O}(g) + 1 e^- \rightarrow \text{O}^-(g)$	$E_1(\text{O})$
$\text{O}^-(g) + 1 e^- \rightarrow \text{O}^{2-}(g)$	$E_2(\text{O})$
$2\text{K}^+(g) + \text{O}^{2-}(g) \rightarrow \text{K}_2\text{O}(s)$	$-\Delta H_{latt} \text{ K}_2\text{O}(s)$

$2\text{K}(s) + 1/2 \text{ O}_2(g) \rightarrow \text{K}_2\text{O}(s)$ $\Delta H^\circ_f \text{ K}_2\text{O}$

$$\Delta H^\circ_f \text{ K}_2\text{O}(s) = 2\Delta H^\circ_f \text{ K}(g) + 2 I_1(\text{K}) + \Delta H^\circ_f \text{ O}(g) + E_1(\text{O}) + E_2(\text{O}) - \Delta H_{latt} \text{ K}_2\text{O}(s)$$

$$E_2(\text{O}) = \Delta H^\circ_f \text{ K}_2\text{O}(s) + \Delta H_{latt} \text{ K}_2\text{O}(s) - 2\Delta H^\circ_f \text{ K}(g) - 2 I_1(\text{K}) - \Delta H^\circ_f \text{ O}(g) - E_1(\text{O})$$

$$E_2(\text{O}) = -363.2 \text{ kJ} + 2238 \text{ kJ} - 2(89.99) \text{ kJ} - 2(419) \text{ kJ} - 247.5 \text{ kJ} - (-141) \text{ kJ}$$

$$= +750 \text{ kJ}$$

8.95 (a) $Br_2(l) \rightarrow 2Br(g)$ $\Delta H° = 2\Delta H_f° \; Br(g) = 2(111.8)$ kJ $= 223.6$ kJ

(b) $CCl_4(l) \rightarrow C(g) + 4Cl(g)$

$\Delta H° = \Delta H_f° \; C(g) + 4\Delta H_f° \; Cl(g) - \Delta H_f° \; CCl_4(l)$
$= 718.4$ kJ $+ 4(121.7)$ kJ $- (-139.3)$ kJ $= 1344.5$

$$\frac{1344.5 \text{ kJ}}{4\,C{-}Cl \text{ bonds}} = 336.1 \text{ kJ}$$

(c) $H_2O_2(l) \rightarrow 2H(g) + 2O(g)$

$2H(g) + 2O(g) \rightarrow 2OH(g)$

_____

$H_2O_2(l) \rightarrow 2OH(g)$

$D(O{-}O)(l) = 2\Delta H_f° \; H(g) + 2\Delta H_f° \; O(g) - \Delta H_f° \; H_2O_2(l) - 2D(O{-}H)(g)$
$= 2(217.94)$ kJ $+ 2(247.5)$ kJ $- (-187.8)$ kJ $- 2(463)$ kJ
$= 193$ kJ

(d) The data are listed below.

bond	D gas kJ/mol	D liquid kJ/mol
Br-Br	193	223.6
C-Cl	328	336.1
O-O	146	192.7

Breaking bonds in the liquid requires more energy than breaking bonds in the gas phase. For simple molecules, bond dissociation from the liquid phase can be thought of in two steps:

molecule (l) → molecule (g)
molecule (g) → atoms (g)

The first step is evaporation or vaporization of the liquid and the second is bond dissociation in the gas phase. Average bond dissociation energy from the liquid phase is then the sum of the enthalpy of vaporization for the molecule and the gas phase bond dissociation enthalpies, divided by the number of bonds dissociated. This is greater than the gas phase bond dissociation enthalpy owing to the contribution from the enthalpy of vaporization.

9 Molecular Geometry and Bonding Theories

Molecular Geometry; the VSEPR Model

9.1 (a) No. A set of bonds with particular lengths can be placed in many different relative orientations. Bond lengths alone do not define the size and shape of a molecule.

(b) Yes. This description means that the three terminal atoms point toward the corners of an equilateral triangle and the central atom is in the plane of this triangle. Only 120° bond angles are possible in this arrangement.

9.3 (a) trigonal planar   (b) tetrahedral   (c) trigonal bipyramidal   (d) octahedral

9.5 The electron pair geometry indicated by VSEPR takes into account the total number of bonding (double and triple bonds count as one effective pair) and nonbonding pairs of electrons. The molecular geometry describes just the atomic positions. NH_3 has the Lewis structure given below; there are four pairs of electrons around nitrogen so the electron pair geometry is tetrahedral, but the molecular geometry of the four atoms is trigonal pyramidal.

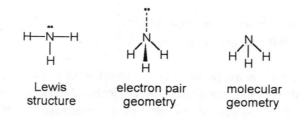

Lewis structure

electron pair geometry

molecular geometry

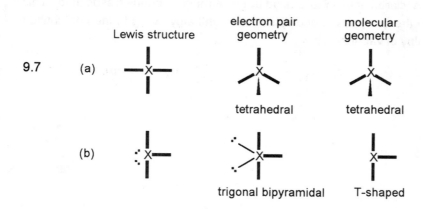

9.7

(c)

octahedral   square pyramidal

9.9   bent (b), linear (l), octahedral (oh), seesaw(ss), tetrahedral (td), trigonal bipyramidal (tbp), trigonal planar (tr), trigonal pyramidal (tp), T-shaped (T)

	Molecule or ion	Valence electrons	Lewis structure		e⁻ pair geometry	Molecular geometry
(a)	ClO_2^-	20	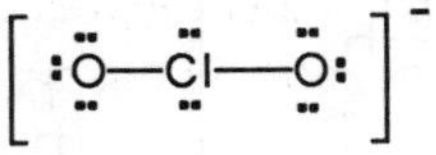	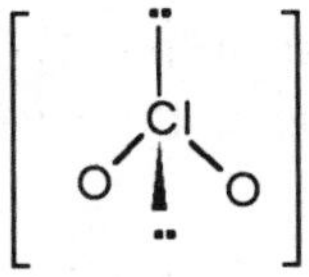	td	b
(b)	SO_3	24	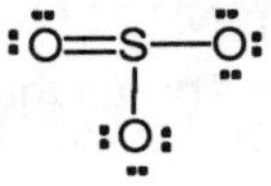	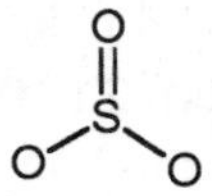	tr	tr
(c)	PCl_3	26	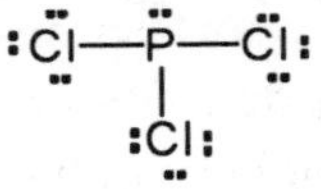	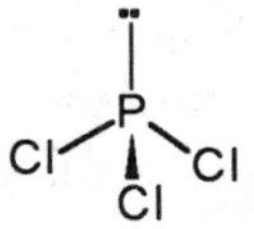	td	tp
(d)	BH_4^-	8	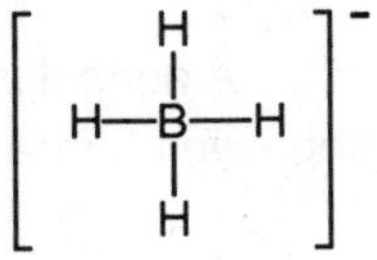	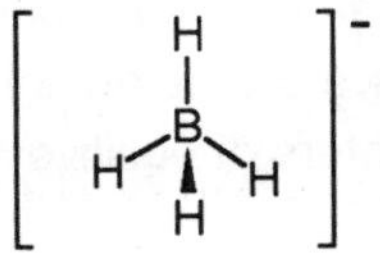	td	td
(e)	SO_3^{2-}	26	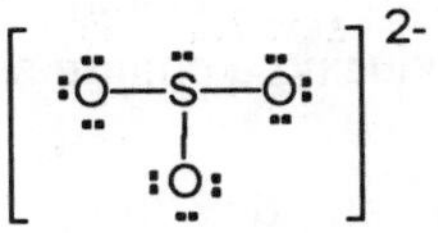	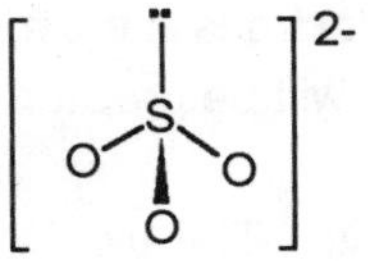	td	tp
(f)	ICl_3	28	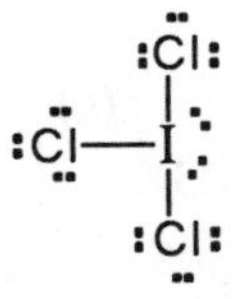	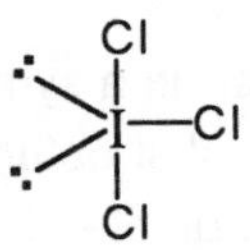	tbp	T

9.11

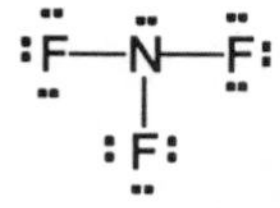

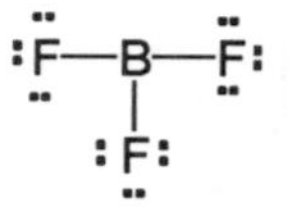

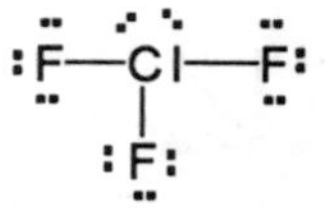

e⁻ pair geometry	td	tr	tbp
molecular geometry	tp	tr	T

The molecular geometries or shapes differ because there are a different number of nonbonding electron pairs around the central atom in each case.

9.13

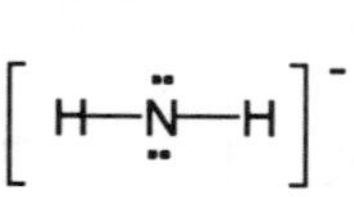

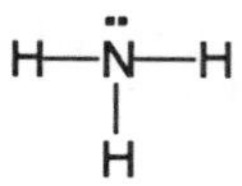

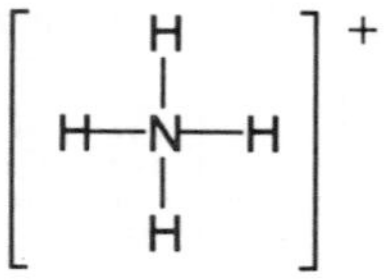

Each molecule has 4 pairs of electrons around the N atom, but the number of nonbonded pairs decreases from 2 to 0 going from NH_2^- to NH_4^+. Since lone pairs occupy more space than bonded pairs, the bond angles expand as the number of lone pairs decreases.

9.15   (a)   1 - 109°, 2 - 109°   (b) 3 - 109°, 4 - 109°
       (c)   5 - 180°   (d) 6 - 120°, 7 - 109°, 8 - 109°

Polarity of Molecules

9.17   (a)   A polar molecule has a measurable dipole moment; its centers of positive and negative charge do not coincide. A nonpolar molecule has a zero net dipole moment; its centers of positive and negative charge do coincide.

       (b)   Yes. If X and Y have different electronegativities, they have different attractions for the electrons in the molecule. The electron density around the more electronegative atom will be greater, producing a charge separation or dipole in the molecule.

       (c) $\mu = Qr$. The dipole moment, μ, is the product of the magnitude of the separated charges, Q, and the distance between them, r.

9.19   (a)   **Nonpolar**, in a symmetrical tetrahedral structure (Figure 9.1), the bond dipoles cancel.   (b)   **Nonpolar**, the molecule is linear and the bond dipoles cancel.

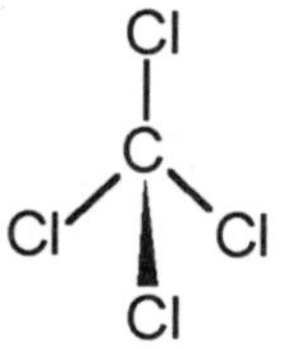

(c)   **Nonpolar,** in a symmetrical trigonal planar structure (Exercise 9.9 (b)), the bond dipoles cancel.

(d)   **Polar,** in the see-saw molecular geometry (Exercise 9.12), the dipoles don't cancel and there is an unequal charge distribution due to the nonbonded electron pair on S.

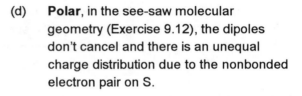

(e)   **Polar,** there is an unequal charge distribution due to the nonbonded electron pair on N.

(f)   **Nonpolar,** in a symmetrical trigonal bipyramid, the bond dipoles cancel.

9.21   In $BeCl_2$ there are just two electron pairs about Be. The structure is linear, so the individual bond dipoles cancel. In SCl_2, there are four electron pairs about S. The two unshared pairs don't cancel with the two S-Cl bonding pairs.

9.23   Q is the charge at either end of the dipole.

$$Q = \frac{\mu}{r} = \frac{1.82\,D}{0.92\,\text{Å}} \times \frac{1\,\text{Å}}{1 \times 10^{-10}\,m} \times \frac{3.34 \times 10^{-30}\,C \cdot m}{1\,D} \times \frac{1\,e^-}{1.60 \times 10^{-19}\,C} = 0.41\,e^-$$

The calculated charge on H and F is $0.41\,e^-$. This can be thought of as the amount of charge "transferred" from H to F.

9.25

polar   nonpolar   polar

All three isomers are planar. The molecules on the left and right are polar because the C-Cl bond dipoles do not point in opposite directions. In the middle isomer, the C-Cl bonds and dipoles are pointing in opposite directions (as are the C-H bonds), the molecule is nonpolar and has a measured dipole moment of zero.

Orbital Overlap; Hybrid Orbitals

9.27   (a)   "Orbital overlap" occurs when a valence atomic orbital on one atom shares the same region of space with a valence atomic orbital on an adjacent atom.

(b)   In valence bond theory, overlap of orbitals allows the two electrons in a chemical bond to mutually occupy the space between the bonded nuclei.

(c)   Valence bond theory is a combination of the atomic orbital concept with the Lewis model of electron pair bonding.

9.29   (a) sp^2 -- 120° angles in a plane
       (b) sp^3d -- 90°, 120° and 180° bond angles (trigonal bipyramid)
       (c) sp^3d^2 -- 90° and 180° bond angles (octahedron)

9.31   (a)   B - $[He]2s^22p^1$

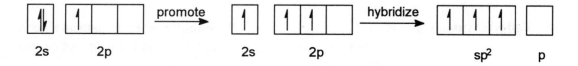

(b)   The hybrid orbitals are called sp^2.   (c)

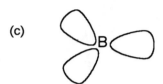

(d)   A single 2p orbital is unhybridized. It lies perpendicular to the trigonal plane of the sp^2 hybrid orbitals.

9.33   (a)   8 e⁻, 4 e⁻ pairs   (b)   8 e⁻, 4 e⁻ pairs

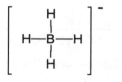

4 e⁻ pairs around B   4 e⁻ pairs around N
tetrahedral e⁻ pair geometry   tetrahedral e⁻ pair geometry
sp^3 hybrid orbitals sp^3 hybrid orbitals

(c)   22 e⁻, 11 e⁻ pairs   (d)   24 e⁻, 12 e⁻ pairs

5 e⁻ pairs around Xe   3 e⁻ pairs around B
trigonal bipyramidal e⁻ pair geometry   trigonal planar e⁻ pair geometry
sp^3d^2 hybrid orbitals sp^2 hybrid orbitals

(e)   48 e⁻, 24 e⁻ pairs   Note that the B-I bond is probably too long for π overlap and formal charge considerations argue against structures with double bonds. However, resonance structures that contain a B = I double bond would use the same hybrid orbitals as the one given above.

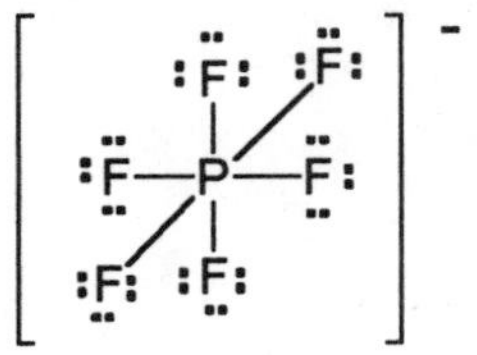

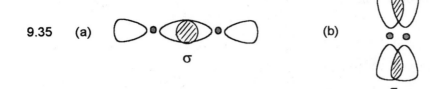

6 e⁻ pairs around P
octahedral e⁻ pair geometry, sp^3d^2

Multiple Bonds

9.35   (a) 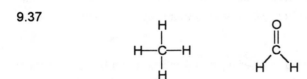   (b)

(c)   A σ bond is generally stronger than a π bond, because there is more extensive orbital overlap.

9.37

$$\begin{array}{c} H \\ | \\ H-C-H \\ | \\ H \end{array} \qquad\qquad \begin{array}{c} O \\ || \\ C \\ H \quad\quad H \end{array}$$

The C atom in CH_4 is sp^3 hybridized; there are no unhybridized p orbitals available for the π overlap required by multiple bonds. In CH_2O, the C atom is sp^2 hybridized, with 1 p atomic orbital available to form the π overlap in the C=O double bond.

9.39   (a)   ~109° about the left most C, sp^3; ~120° about the right-hand C, sp^2

       (b)   The doubly bonded O can be viewed as sp^2, the other as sp^3; the nitrogen is sp^3 with approximately 109° bond angles.

       (c)   nine σ bonds, one π bond

9.41   (a)   In a localized π bond, the electron density is concentrated strictly between the two atoms forming the bond. In a delocalized π bond, parallel p orbitals on more than two adjacent atoms overlap and the electron density is spread over all the atoms that contribute p orbitals to the network. There are still two regions of overlap, above and below the σ framework of the molecule.

(b)   The existence of more than one resonance form is a good indication that a molecule will have delocalized π bonding.

(c)

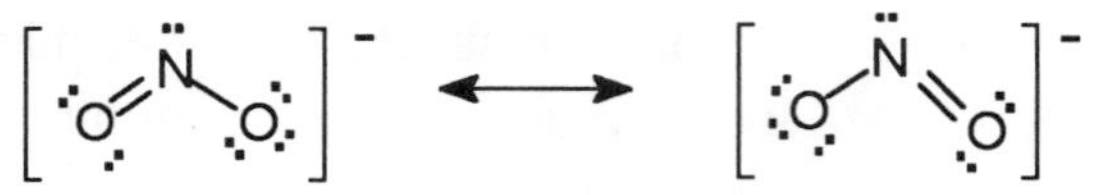

The existence of more than one resonance form for NO_2^- indicates that the π bond is delocalized. From an orbital perspective, the electron pair geometry around N is trigonal planar, so the hybridization at N is sp^2. This leaves a p orbital on N and one on each O atom perpendicular to the trigonal plane of the molecule, in the correct orientation for delocalized π overlap. Physically, the two N-O bond lengths are equal, indicating that the two N-O bonds are equivalent, rather than one longer single bond and one shorter double bond.

Molecular Orbitals

9.43   (a)   Both atomic and molecular orbitals have a characteristic energy and shape (region where there is a high probability of finding an electron). Each atomic or molecular orbital can hold a maximum of two electrons. Atomic orbitals are localized on single atoms and their energies are the result of interactions between the subatomic particles in a single atom. Molecular orbitals can be delocalized over several or even all the atoms in a molecule and their energies are influenced by interactions between electrons on several atoms.

(b)   There is a net stabilization (lowering in energy) that accompanies bond formation because the bonding electrons in H_2 are strongly attracted to both H nuclei.

(c)   2

9.45   (a)

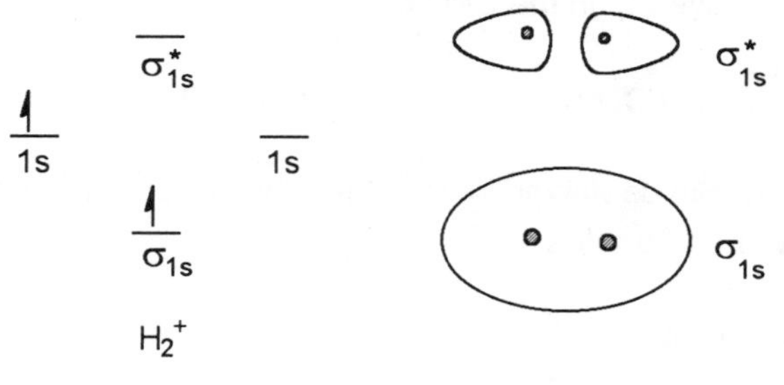

(b)   There is 1 electron in H_2^+. The Lewis model of bonding indicates that a bond is a pair of electrons. Since there is only 1 electron in H_2^+, there is no bond by the Lewis definition, so it is not possible to write a Lewis structure.

(c) 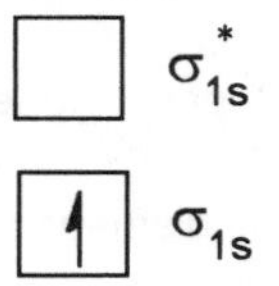   ☐ σ_{1s}^*

         ↑ σ_{1s}

(d) Bond order = 1/2 (1-0) = 1/2

(e)   Yes. The stability of H_2^+ is due to the lower energy state of the σ bonding molecular orbital relative to the energy of a H 1s atomic orbital. If the single electron in H_2^+ is excited to the σ_{1s}^* orbital, its energy is higher than the energy of a H 1s atomic orbital and H_2^+ will decompose into a hydrogen atom and a hydrogen ion,

$$H_2^+ \overset{h\nu}{\rightarrow} H + H^+.$$

9.47   In a σ molecular orbital, the electron density is spherically symmetric about the internuclear axis and is concentrated along this axis. In a π molecular orbital, the electron density is concentrated above and below the internuclear axis and zero along it.

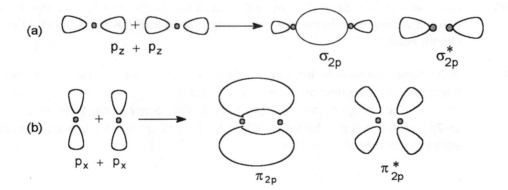

(a) $p_z + p_z$ σ_{2p} σ_{2p}^*

(b) $p_x + p_x$ π_{2p} π_{2p}^*

(c) σ_{2p} is lower in energy than π_{2p} due to greater extent of orbital overlap in the σ molecular orbital. $\sigma_{2p} < \pi_{2p} < \pi_{2p}^* < \sigma_{2p}$

9.49   (a)   Substances with no unpaired electrons are weakly repelled by a magnetic field. This property is called *diamagnetism*.

        (b) H_2 , N_2

        (c) O_2^{2-}, Be_2^{2+} (see Figure 9.39)

9.51   **(a)**

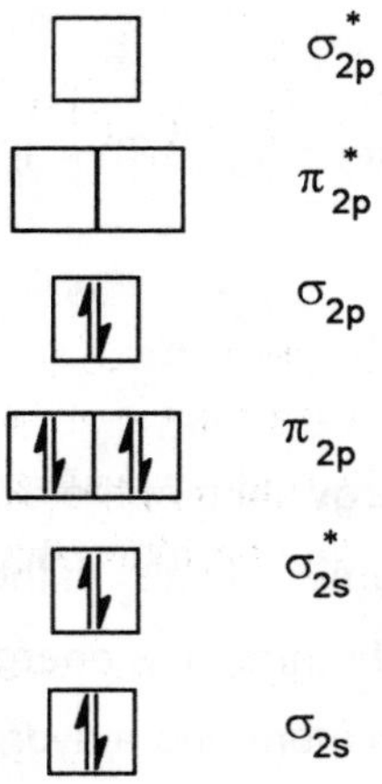

□	σ_{2p}^{*}
□ □	π_{2p}^{*}
⇅	σ_{2p}
⇅ ⇅	π_{2p}
⇅	σ_{2s}^{*}
⇅	σ_{2s}

(a) CO :   B.O. = (8-2) / 2 = 3.0, diamagnetic

(b) NO^{-} :   B.O. = (8-4) / 2 = 2, paramagnetic

(c) CN^{-} :   B.O. = (8-2) / 2 = 3, diamagnetic

(d) OF :   B.O. = (8-5) / 2 = 1.5 paramagnetic

9.53   **(a)**   The separation between related bonding and antibonding orbitals is directly related to the extent of overlap of the atomic orbitals. The Li 2s orbitals are larger than the 1s, the overlap is greater and the energy separation of the resulting σ_{2s} and σ_{2s}^{*} orbitals is larger than that between σ_{1s} and σ_{1s}^{*}.

     (b) O_2^{2-} has a bond order of 1.0, while O_2^{-} has a bond order of 1.5. For the same bonded atoms, the greater the bond order the shorter the bond, so O_2^{-} has the shorter bond.

     (c)   Interactions between electrons in 2s atomic orbitals on one atom and 2p orbitals on the other has an effect on the energies of the $\sigma_{2s}, \sigma_{2s}^{*}, \sigma_{2p}$ and σ_{2p}^{*} molecular orbitals. The energy of the σ_{2s} is lowered and the energy of the σ_{2p} is raised. If the 2s-2p interaction is substantial, as it is in B_2, the energy of the σ_{2p} is actually raised above the energy of the π_{2p}.

Additional Exercises

9.56   **(a)** AsF_3; 26 valence e^{-} **(b)** OCN^{-}; 16 valence e^{-}

$$:\ddot{F}-As-\ddot{F}:$$
$$:\ddot{F}:$$

$$\left[\ddot{O}=C=\ddot{N}\right]^{-} \longleftrightarrow \left[:\ddot{O}-C\equiv N:\right]^{-}$$

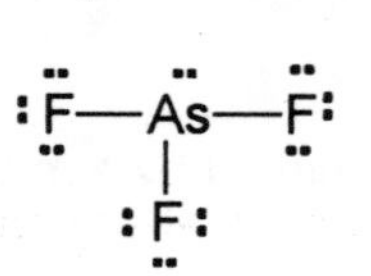

tetrahedral e^{-} pair geometry
trigonal pyramidal molecular geometry

$$\left[:O\equiv C-\ddot{N}:\right]^{-}$$

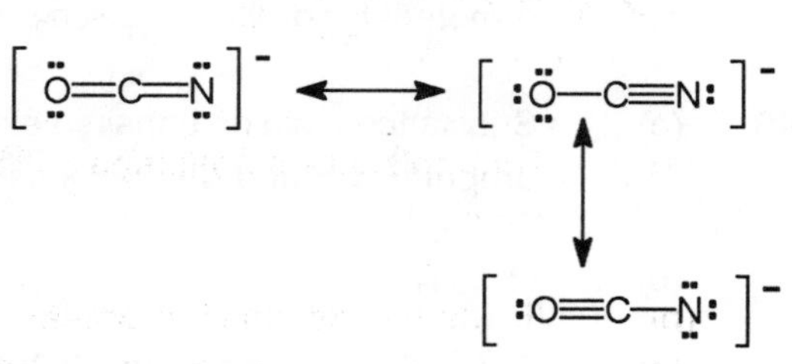

linear e^{-} pair geometry
linear "molecular" geometry

(c) H_2CO; 12 valence e⁻   (d) ICl_2^- ; 22 valence e⁻

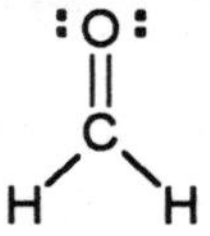

trigonal planar e⁻ pair geometry
trigonal planar molecular geometry

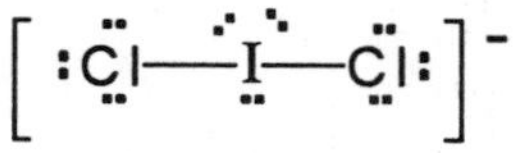

trigonal bipyramidal e⁻ pair geometry
linear molecular geometry
(In a trigonal bipyramid, nonbonded
pairs lie in the trigonal plane.)

9.59   (a) CO_2, 16 valence e⁻   (b) NCS^-, 16 valence e⁻

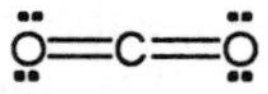

$$\ddot{O}=C=\ddot{O}$$

2σ, 2π

$$[\ddot{N}=C=\ddot{S}]$$

2σ, 2π

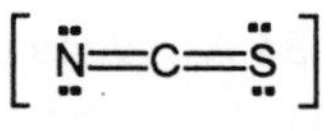

+   two other
resonance
structures

(for any of
the resonance
structures)

(c) SO_4^{2-}, 32 valence e⁻   (d) $HCO(OH)$, 18 valence e⁻

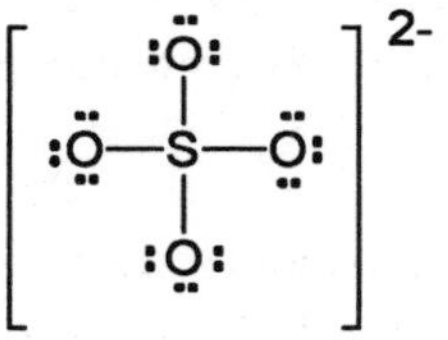

4 σ

(assuming that the best Lewis structure is
the one that does not violate the octet rule)

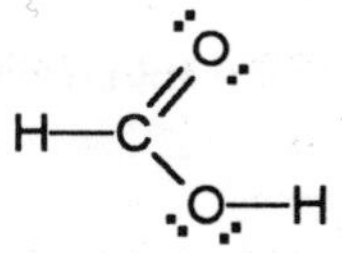

4 σ, 1 π

9.63

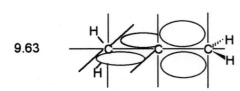

(a)   The molecule is nonplanar. The CH_2 planes at each end are twisted 90° from one
another.

(b)   Allene has no dipole moment.

(c)   The bonding in allene would not be described as delocalized. The π electron
clouds of the two adjacent C=C are mutually perpendicular. The mechanism for
delocalization of π electrons is mutual overlap of parallel p atomic orbitals on
adjacent atoms. If adjacent π electron clouds are mutually perpendicular, there is
no overlap and no delocalization of π electrons.

9.67   (a)   Each C atom is surrounded by 3 VSEPR electron pairs (2 single and 1 double bonds), so bond angles at each C atom will be approximately 120°.

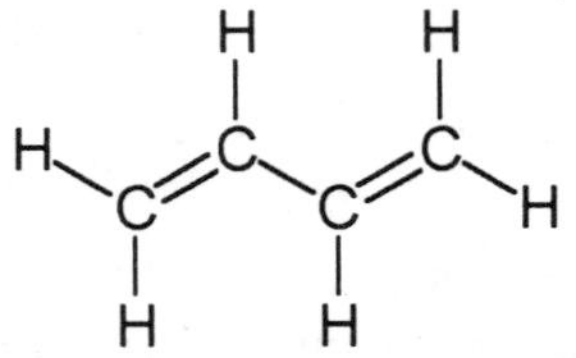

Since there is free rotation around the central C-C single bond, other conformations are possible.

(b)   According to Table 8.4, the average C-C length is 1.54 Å, and the average C=C length is 1.34 Å. While the C=C bonds in butadiene appear "normal," the central C-C is significantly shorter than average. Examination of the bonding in butadiene reveals that each C atom is sp^2 hybridized and the π bonds are formed by the remaining unhybridized 2p orbital on each atom. If the central C-C bond is rotated so that all four C atoms are coplanar, the four 2p orbitals are parallel and some delocalization of the π electrons occurs.

9.70   The two possible orbital energy level diagrams are:

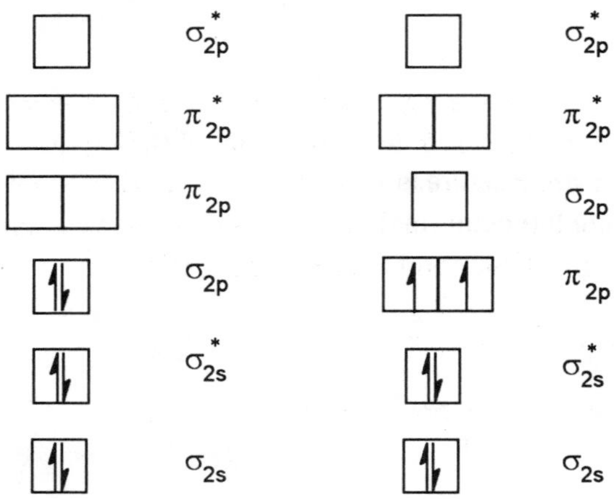

If the σ_{2p} molecular orbital is lower in energy than the π_{2p} orbitals, there are no unpaired electrons, and the molecule is diamagnetic. Switching the order of σ_{2p} and π_{2p} gives 1 unpaired electron in each degenerate π_{2p} orbital and the observed paramagnetism.

Integrative Exercises

9.73   (a)   Assume 100 g of compound

$$2.1 \text{ g H} \times \frac{1 \text{ mol H}}{1.008 \text{ g H}} = 2.1 \text{ mol H}; \ 2.1 / 2.1 = 1$$

$$29.8 \text{ g N} \times \frac{1 \text{ mol N}}{14.01 \text{ g N}} = 2.13 \text{ mol N}; \ 2.13 / 2.1 \approx 1$$

$$68.1 \text{ g O} \times \frac{1 \text{ mol O}}{16.00 \text{ g O}} = 4.26 \text{ mol O}; \ 4.26 / 2.1 \approx 2$$

The empirical formula is HNO_2;   formula weight = 47.   Since the approximate molecular weight is 50, the **molecular formula is HNO_2.**

(b)   Assume N is central, since it is unusual for O to be central, and part (d) indicates as much. HNO_2: 18 valence e^-

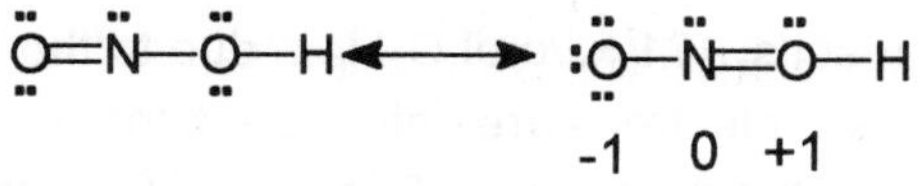

$$\ddot{O}=\ddot{N}-\ddot{O}-H \longleftrightarrow :\ddot{O}-\ddot{N}=\ddot{O}-H$$
$$\qquad\qquad\qquad\qquad -1 \quad 0 \ +1$$

The second resonance form is a minor contributor due to unfavorable formal charges.

(c)   The electron pair geometry around N is trigonal planar; if the resonance structure on the right makes a significant contribution to the molecular structure, all **4 atoms would lie in a plane.** If only the left structure contributes, the H could rotate in and out of the molecular plane.   The contributions of the two resonance structures could be determined by measuring the O-N-O and N-O-H bond angles.

(d)   3 VSEPR e^- pairs around N, sp^2 hybridization

(e)   3 σ, 1 π for both structures (or for H bound to N).

9.76

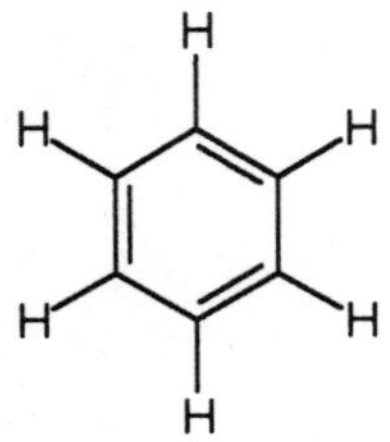

$$\Delta H = 6D(C\text{-}H) + 3D(C\text{-}C) + 3D(C\text{=}C) - 0$$
$$= 6(413 \text{ kJ}) + 3(348 \text{ kJ}) + 3(614 \text{ kJ})$$
$$= 5364 \text{ kJ}$$

(The products are isolated atoms; there is no bond making.)

According to Hess's Law:

$$\Delta H^\circ = 6\Delta H^\circ_f \; C(g) + 6\Delta H^\circ_f \; H(g) - \Delta H^\circ_f \; C_6H_6(g)$$
$$= 6(718.4 \text{ kJ}) + 6(217.94 \text{ kJ}) - (+82.9 \text{ kJ})$$
$$= 5535 \text{ kJ}$$

The difference in the two results, 171 kJ/mol C_6H_6 is due to the resonance stabilization in benzene. That is, because the π electrons are delocalized, the molecule has a lower overall energy than that predicted for the presence of 3 localized C-C and C=C bonds. Thus, the amount of energy actually required to decompose 1 mole of C_6H_6(g), represented by the Hess's Law calculation, is greater than the sum of the localized bond energies (not taking resonance into account) from the first calculation above.

10 Gases

Gas Characteristics; Pressure

10.1   In the gas phase, molecules are far apart, while in the liquid they are touching.

   (a)   A gas is much less dense than a liquid because most of the volume of a gas is empty space.

   (b)   A gas is much more compressible because of the distance between molecules.

   (c)   Gaseous molecules are so far apart that there is no barrier to mixing, regardless of the identity of the molecule. All mixtures of gases are homogeneous. Liquid molecules are touching. In order to mix, they must displace one another. Similar molecules displace each other and form homogeneous mixtures. Very dissimilar molecules form heterogeneous mixtures.

10.3   (a) $F = m \times a$. Since both people have the same mass and both experience the acceleration of gravity, the forces they exert on the floor are exactly equal.

   (b) $P = F / A$. The two forces are equal, but the person standing on one foot exerts this force over a smaller area. Thus, the person standing on one foot exerts a greater pressure on the floor.

10.5   (a) $P_{Hg} = P_{H_2O}$; Using the relationship derived in 10.4: $(d \times h \times a)_{H_2O} = (d \times h \times a)_{Hg}$

   Since a, the acceleration due to gravity, is equal in both liquids,

   $(d \times h)_{H_2O} = (d \times h)_{Hg}$

   $1.00 \text{ g/mL} \times h_{H_2O} = 13.6 \text{ g/mL} \times 760 \text{ mm}$

   $h_{H_2O} = \dfrac{13.6 \text{ g/mL} \times 760 \text{ mm}}{1.00 \text{ g/mL}} = 1.034 \times 10^4 = 1.03 \times 10^4 \text{ mm} = 10.3 \text{ m}$

   (b)   Pressure due to H_2O:

   $1 \text{ atm} = 1.034 \times 10^4 \text{ mm } H_2O$ (from part (a))

   $28 \text{ ft } H_2O \times \dfrac{12 \text{ in}}{1 \text{ ft}} \times \dfrac{2.54 \text{ cm}}{1 \text{ in}} \times \dfrac{10 \text{ mm}}{1 \text{ cm}} \times \dfrac{1 \text{ atm}}{1.034 \times 10^4 \text{ mm}} = 0.826 = 0.83 \text{ atm}$

   $P_{total} = P_{atm} + P_{H_2O} = 0.98 \text{ atm} + 0.826 \text{ atm} = 1.81 \text{ atm}$

10.7 (a) $0.860 \text{ atm} \times \dfrac{101.325 \text{ kPa}}{1 \text{ atm}} = 87.1 \text{ kPa}$

(b) $457 \text{ torr} \times \dfrac{1 \text{ atm}}{760 \text{ torr}} = 0.601 \text{ atm}$

(c) $802 \text{ mm Hg} \times \dfrac{1 \text{ atm}}{760 \text{ mm Hg}} = 1.06 \text{ atm}$

(d) $0.897 \text{ atm} \times \dfrac{760 \text{ torr}}{1 \text{ atm}} = 682 \text{ torr}$

10.9 (a) $30.45 \text{ in Hg} \times \dfrac{25.4 \text{ mm}}{1 \text{ in}} \times \dfrac{1 \text{ torr}}{1 \text{ mm Hg}} = 773.4 \text{ torr}$

[The result has 4 sig figs because 25.4 mm/in is considered to be an exact number. (section 1.5)]

(b) The pressure in Chicago is greater than **standard atmospheric pressure**, 760 torr, so it makes sense to classify this weather system as a "high pressure system."

10.11 $P = \dfrac{m \times a}{A} = \dfrac{135 \text{ lb}}{0.50 \text{ in}^2} \times \dfrac{9.81 \text{ m}}{1 \text{ s}^2} \times \dfrac{0.454 \text{ kg}}{1 \text{ lb}} \times \dfrac{39.4^2 \text{ in}^2}{1 \text{ m}^2} = 1.9 \times 10^3 \text{ kPa}$

10.13 (a) It **is not** necessary to know atmospheric pressure to use a closed-end manometer. The gas on one side of the Hg is working against a vacuum on the other side of the Hg. The pressure of the gas is just the difference in the heights of the two arms.

It **is** necessary to know atmospheric pressure to use an open-end manometer. In an open-end manometer the pressure of the sample is balanced against atmospheric pressure. If the Hg level is higher in the open end than in the end exposed to the sample, the pressure of the sample is atmospheric pressure plus the difference in heights of the two sides. If the Hg level is lower in the open end than in the other end, the pressure of the sample is atmospheric pressure minus the difference in heights of the two sides.

(b) (i) The Hg level is lower in the open end than the closed end, so the gas pressure is less than atmospheric pressure.

$P_{gas} = 0.975 \text{ atm} - \left(52 \text{ cm} \times \dfrac{1 \text{ atm}}{76 \text{ cm}} \right) = 0.29 \text{ atm}$

(ii) The Hg level is higher in the open end, so the gas pressure is greater than atmospheric pressure.

$P_{gas} = 0.975 \text{ atm} + \left(67 \text{ mm Hg} \times \dfrac{1 \text{ atm}}{760 \text{ mm Hg}} \right) = 1.063 \text{ atm}$

(iii)   This is a closed-end manometer so $P_{gas} = h$.

$$P_{gas} = 10.3 \text{ cm} \times \frac{1 \text{ atm}}{76 \text{ cm}} = 0.136 \text{ atm}$$

The Gas Laws; The Ideal-Gas Equation

(In *Solutions to Exercises*, the symbol for molar mass is MM.)

10.15 (a)   If two quantities, X and Y, are **directly proportional**, a change in X causes a proportional change in Y in the **same** direction. Mathematically, if X and Y are directly proportional and C is a constant, Y = CX or Y/X = C.

If X and Y are **inversely proportional**, a change in X causes a change in Y in the **opposite** direction. Mathematically X • Y = C or Y = C/X.

(b)   Volume (V) is inversely proportional to pressure (P); volume is directly proportional to absolute temperature (T); volume is directly proportional to quantity of gas (n).

10.17 $P_1V_1 = P_2V_2$; the proportionality holds true for any pressure or volume units.

$P_1 = 737$ torr, $V_1 = 20.5$ L

(a) $P_2 = 1.80 \text{ atm} \times \dfrac{760 \text{ torr}}{1 \text{ atm}} = 1.368 \times 10^3 = 1.37 \times 10^3 \text{ torr}$

$V_2 = \dfrac{P_1V_1}{P_2} = \dfrac{737 \text{ torr} \times 20.5 \text{ L}}{1.368 \times 10^3 \text{ torr}} = 11.0 \text{ L}$

(b) $V_2 = 16.0 \text{ L}$; $P_2 = \dfrac{P_1V_1}{V_2} = \dfrac{737 \text{ torr} \times 20.5 \text{ L}}{16.0 \text{ L}} = 944 \text{ torr}$

10.19 (a)   Avogadro's hypothesis states that equal volumes of gases at the same temperature and pressure contain equal numbers of molecules. Since molecules react in the ratios of small whole numbers, it follows that the volumes of reacting gases (at the same temperature and pressure) are in the ratios of small whole numbers.

(b)   Since the two gases are at the same temperature and pressure, the ratio of the numbers of atoms is the same as the ratio of volumes. There are 1.5 times as many Xe atoms as Ne atoms.

10.21 (a)   An ideal gas exhibits pressure, volume and temperature relationships which are described by the equation PV = nRT. (An ideal gas obeys the Ideal-Gas Equation.)

(b)   PV = nRT;   P in atmospheres
                        V in liters
                        n in moles
                        T in kelvins

10.23   (a) $n = 8.25 \times 10^{-2}$ mol, $V = 174$ mL, $T = -15°C = 258$ K

$$P = \frac{nRT}{V} = 0.0825 \text{ mol} \times \frac{0.08206 \text{ L} \cdot \text{atm}}{\text{K} \cdot \text{mol}} \times \frac{258 \text{ K}}{0.174 \text{ L}} = 10.0 \text{ atm}$$

(b) $V = 6.38$ L, $T = 35°C = 308$ K, $P = 955$ torr $\times \frac{1 \text{ atm}}{760 \text{ torr}} = 1.257 = 1.26$ atm

$$n = \frac{PV}{RT} = 1.257 \text{ atm} \times \frac{\text{K} \cdot \text{mol}}{0.08206 \text{ L} \cdot \text{atm}} \times \frac{6.38 \text{ L}}{308 \text{ K}} = 0.317 \text{ mol}$$

(c) $n = 2.95$ mol, $P = 0.76$ atm, $T = 52°C = 325$ K

$$V = \frac{nRT}{P} = 2.95 \text{ mol} \times \frac{0.08206 \text{ L} \cdot \text{atm}}{\text{K} \cdot \text{mol}} \times \frac{325 \text{ K}}{0.76 \text{ atm}} = 1.0 \times 10^2 \text{ L}$$

(d) $n = 9.87 \times 10^{-2}$ mol, $V = 164$ mL $= 0.164$ L

$$P = 682 \text{ torr} \times \frac{1 \text{ atm}}{760 \text{ torr}} = 0.8974 = 0.897 \text{ atm}$$

$$T = \frac{PV}{nR} = 0.8974 \text{ atm} \times \frac{0.164 \text{ L}}{0.0987 \text{ mol}} \times \frac{1 \text{ K} \cdot \text{mol}}{0.08206 \text{ L} \cdot \text{atm}} = 18.2 \text{ K}$$

10.25 $V = 0.325$ L, $T = 28°C = 301$ K, $n = 3.0$ g $C_3H_8 \times \frac{1 \text{ mol } C_3H_8}{44.1 \text{ g } C_3H_8} = 0.0680 = 0.068$ mol

$$P = \frac{nRT}{V} = 0.0680 \text{ mol} \times \frac{0.08206 \text{ L} \cdot \text{atm}}{\text{K} \cdot \text{mol}} \times \frac{301 \text{ K}}{0.325 \text{ L}} = 5.2 \text{ atm}$$

10.27   Air is a mixture of N_2 and O_2, but for the purpose of calculating pressure, only the total number of gas molecules is important, not the identity of these molecules.

$$V = 1.05 \text{ L}, \ T = 37°C = 310 \text{ K}, \ P = 740 \text{ mm Hg} \times \frac{1 \text{ atm}}{760 \text{ mm}} = 0.9737 = 0.974 \text{ atm}$$

$$n = \frac{PV}{RT} = 0.9737 \text{ atm} \times \frac{\text{K} \cdot \text{mol}}{0.08206 \text{ L} \cdot \text{atm}} \times \frac{1.05 \text{ L}}{310 \text{ K}} = 0.04019 = 0.0402 \text{ mol of gas}$$

$$0.04019 \text{ mol} \times \frac{6.022 \times 10^{23} \text{ molecules}}{1 \text{ mol}} = 2.42 \times 10^{22} \text{ gas molecules}$$

10.29   (a) $V_2 = \frac{P_1 V_1 T_2}{P_2 T_1} = \frac{0.880 \text{ atm} \times 0.600 \text{ L} \times 273 \text{ K}}{0.205 \text{ atm} \times 319 \text{ K}} = 2.20 \text{ L}$

(b) $V_2 = \frac{0.880 \text{ atm} \times 0.600 \text{ L} \times 273 \text{ K}}{1.00 \text{ atm} \times 319 \text{ K}} = 0.452 \text{ L}$

10.31   (a) $g = \frac{MM \times PV}{RT}$ (MM = molar mass)

$$\frac{32.0 \text{ g } O_2}{1 \text{ mol } O_2} \times \frac{\text{K} \cdot \text{mol}}{0.08206 \text{ L} \cdot \text{atm}} \times 18,000 \text{ kPa} \times \frac{1 \text{ atm}}{101.3 \text{ kPa}} \times \frac{42.0 \text{ L}}{296 \text{ K}}$$

$$= 9.83 \times 10^3 \text{ g } O_2$$

(b) $V_2 = \dfrac{P_1V_1T_2}{T_1P_2} = \dfrac{18{,}000 \text{ kPa} \times 42.0 \text{ L} \times 273 \text{ K}}{296 \text{ K} \times 101.3 \text{ kPa}} = 6.88 \times 10^3$ L

10.33 (a) $5.2 \text{ g} \times 1 \text{ h} \times \dfrac{0.8 \text{ mL O}_2}{1 \text{ g} \cdot \text{hr}} = 4.16 = 4$ mL O_2 consumed

$n = \dfrac{PV}{RT} = 1 \text{ atm} \times \dfrac{\text{K} \cdot \text{mol}}{0.08206 \text{ L} \cdot \text{atm}} \times \dfrac{0.00416 \text{ L}}{297 \text{ K}} = 1.71 \times 10^{-4} = 2 \times 10^{-4}$ mol O_2

(b) $1 \text{ qt air} \times \dfrac{0.946 \text{ L}}{1 \text{ qt}} \times 0.21\% \text{ O}_2 \text{ in air} = 0.199$ L O_2 available

$n = 1 \text{ atm} \times \dfrac{\text{K} \cdot \text{mol}}{0.08206 \text{ L} \cdot \text{atm}} \times \dfrac{0.199 \text{ L}}{297 \text{ K}} = 8.16 \times 10^{-3} = 8 \times 10^{-3}$ mol O_2 available

roach uses $\dfrac{1.71 \times 10^{-4} \text{ mol}}{1 \text{ hr}} \times 48 \text{ hr} = 8.19 \times 10^{-3} = 8 \times 10^{-3}$ mol O_2 consumed

Not only does the roach use 20% of the available O_2, it needs all the O_2 in the jar.

Further Applications of the Ideal-Gas Equation

(In *Solutions to Exercises*, the symbol for molar mass is MM.)

10.35 (a)   Density of a gas $= \dfrac{g}{L}$; $PV = \dfrac{g \text{ RT}}{MM}$; $\dfrac{g}{V} = \dfrac{MM \times P}{RT} = d$

P = 0.96 atm, T = 35°C = 308 K, MM of SO_3 = 80.07 g/mol

$d = \dfrac{80.07 \text{ g}}{\text{mol}} \times \dfrac{\text{K} \cdot \text{mol}}{0.08206 \text{ L} \cdot \text{atm}} \times \dfrac{0.96 \text{ atm}}{308 \text{ K}} = 3.0$ g/L

(b) $MM = \dfrac{gRT}{PV} = \dfrac{4.40 \text{ g}}{3.50 \text{ L}} \times \dfrac{0.08206 \text{ L} \cdot \text{atm}}{\text{K} \cdot \text{mol}} \times \dfrac{314 \text{ K}}{560 \text{ torr}} \times \dfrac{760 \text{ torr}}{1 \text{ atm}} = 44.0$ g/mol

10.37 $MM = \dfrac{gRT}{PV} = \dfrac{1.012 \text{ g}}{0.354 \text{ L}} \times \dfrac{0.08206 \text{ L} \cdot \text{atm}}{\text{K} \cdot \text{atm}} \times \dfrac{372 \text{ K}}{742 \text{ torr}} \times \dfrac{760 \text{ torr}}{1 \text{ atm}} = 89.4$ g/mol

10.39 mol $O_2 = \dfrac{PV}{RT} = 3.5 \times 10^{-6} \text{ torr} \times \dfrac{1 \text{ atm}}{760 \text{ torr}} \times \dfrac{\text{K} \cdot \text{mol}}{0.08206 \text{ L} \cdot \text{atm}} \times \dfrac{0.382 \text{ L}}{300 \text{ K}} = 7.146 \times 10^{-11}$

$= 7.1 \times 10^{-11}$ mol O_2

$7.146 \times 10^{-11} \text{ mol O}_2 \times \dfrac{2 \text{ mol Mg}}{1 \text{ mol O}_2} \times \dfrac{24.3 \text{ g Mg}}{1 \text{ mol Mg}} = 3.5 \times 10^{-9}$ g Mg

10.41   kg H_2SO_4 → g H_2SO_4 → mol H_2SO_4 → mol NH_3 → V NH_3

$$150 \text{ kg} \times \frac{1000 \text{ g}}{1 \text{ kg}} = 1.50 \times 10^5 \text{ g } H_2SO_4 \times \frac{1 \text{ mol}}{98.08 \text{ g}} = 1.529 \times 10^3 = 1.53 \times 10^3 \text{ mol } H_2SO_4$$

$$1.529 \times 10^3 \text{ mol } H_2SO_4 \times \frac{2 \text{ mol } NH_3}{1 \text{ mol } H_2SO_4} = 3.059 \times 10^3 = 3.06 \times 10^3 \text{ mol } NH_3$$

$$V_{NH_3} = \frac{nRT}{P} = 3.059 \times 10^3 \text{ mol} \times \frac{0.08206 \text{ L} \cdot \text{atm}}{K \cdot \text{mol}} \times \frac{293 \text{ K}}{25.0 \text{ atm}} = 2.94 \times 10^3 \text{ L } NH_3$$

Partial Pressures

10.43   (a)   *Partial pressure* is the pressure exerted by a single component of a gaseous mixture at the same temperature and volume as the mixture.

       (b)   The total pressure of a mixture of gases is equal to the sum of the individual pressures the gases would exert if present in the same container alone.

10.45   (a) $P_{He} = \dfrac{nRT}{V} = 0.538 \text{ mol} \times \dfrac{0.08206 \text{ L} \cdot \text{atm}}{K \cdot \text{atm}} \times \dfrac{298 \text{ K}}{7.00 \text{ L}} = 1.88 \text{ atm}$

               $P_{Ne} = \dfrac{nRT}{V} = 0.315 \text{ mol} \times \dfrac{0.08206 \text{ L} \cdot \text{atm}}{K \cdot \text{atm}} \times \dfrac{298 \text{ K}}{7.00 \text{ L}} = 1.10 \text{ atm}$

               $P_{Ar} = \dfrac{nRT}{V} = 0.103 \text{ mol} \times \dfrac{0.08206 \text{ L} \cdot \text{atm}}{K \cdot \text{atm}} \times \dfrac{298 \text{ K}}{7.00 \text{ L}} = 0.360 \text{ atm}$

       (b) P_t = 1.88 atm + 1.10 atm + 0.360 atm = 3.34 atm

10.47   The partial pressure of each component is equal to the mole fraction of that gas times the total pressure of the mixture. Find the mole fraction of each component and then its partial pressure.

n_t = 0.55 mol N_2 + 0.20 mol O_2 + 0.10 mol CO_2 = 0.85 mol

$$\chi_{N_2} = \frac{0.55}{0.85} = 0.647 = 0.65; \quad P_{N_2} = 0.647 \times 1.32 \text{ atm} = 0.85 \text{ atm}$$

$$\chi_{O_2} = \frac{0.20}{0.85} = 0.235 = 0.24; \quad P_{O_2} = 0.235 \times 1.32 \text{ atm} = 0.31 \text{ atm}$$

$$\chi_{CO_2} = \frac{0.10}{0.85} = 0.118 = 0.12; \quad P_{CO_2} = 0.118 \times 1.32 \text{ atm} = 0.16 \text{ atm}$$

10.49 $\chi_{O_2} = \dfrac{P_{O_2}}{P_t} = \dfrac{0.21 \text{ atm}}{8.38 \text{ atm}} = 0.025$; mole % = 0.025 × 100 = 2.5%

10.51 (a) The partial pressure of gas A is **not affected** by the addition of gas C. The partial pressure of A depends only on moles of A, volume of container and conditions; none of these factors change when gas C is added.

(b) The total pressure in the vessel **increases** when gas C is added, because the total number of moles of gas increases.

(c) The mole fraction of gas B **decreases** when gas C is added. The moles of gas B stay the same, but the total moles increase, so the mole fraction of B (n_B/n_t) decreases.

10.53 $P_{N_2} = \dfrac{P_1 V_1 T_2}{V_2 T_1} = \dfrac{3.80 \text{ atm} \times 1.00 \text{ L} \times 293 \text{ K}}{10.0 \text{ L} \times 299 \text{ K}} = 0.372$ atm

$P_{O_2} = \dfrac{P_1 V_1 T_2}{V_2 T_1} = \dfrac{4.75 \text{ atm} \times 5.00 \text{ L} \times 293 \text{ K}}{10.0 \text{ L} \times 299 \text{ K}} = 2.33$ atm

$P_t = 0.372$ atm $+ 2.33$ atm $= 2.70$ atm

10.55 The gas sample is a mixture of $H_2(g)$ and $H_2O(g)$. Find the partial pressure of $H_2(g)$ and then the moles of $H_2(g)$ and $Zn(s)$.

$P_t = 738 \text{ torr} = P_{H_2} + P_{H_2O}$

From Appendix B, the vapor pressure of H_2O at $24°C = 22.38$ torr

$P_{H_2} = (738 \text{ torr} - 22.38 \text{ torr}) \times \dfrac{1 \text{ atm}}{760 \text{ torr}} = 0.9416 = 0.942$ atm

$n_{H_2} = \dfrac{P_{H_2} V}{RT} = 0.9416 \text{ atm} \times \dfrac{K \cdot mol}{0.08206 \text{ L} \cdot atm} \times \dfrac{0.159 \text{ L}}{297 \text{ K}} = 0.006143 = 0.00614 \text{ mol } H_2$

$0.006143 \text{ mol } H_2 \times \dfrac{1 \text{ mol Zn}}{1 \text{ mol } H_2} \times \dfrac{65.39 \text{ g Zn}}{1 \text{ mol Zn}} = 0.402 \text{ g Zn}$

Kinetic - Molecular Theory; Graham's Law

(In *Solutions to Exercises*, the symbol for molar mass is MM.)

10.57 (a) They have the same number of molecules (equal volumes of gases at the same temperature and pressure contain equal numbers of molecules).

(b) N_2 is more dense because it has the larger molar mass. Since the volumes of the samples and the number of molecules are equal, the gas with the larger molar mass will have the greater density.

(c) The average kinetic energies are equal (statement 5, section 10.7).

(d) CH_4 will effuse faster. The lighter the gas molecules, the faster they will effuse (Graham's Law).

10.59 (a) increase in temperature at constant volume, decrease in volume, increase in pressure
       (b) decrease in temperature   (c) increase in volume   (d) increase in temperature

10.61 (a)   In order of increasing speed (and decreasing molar mass):

$$CO_2 \approx N_2O < F_2 < HF < H_2$$

(b) $u_{H_2} = \sqrt{\dfrac{3RT}{MM}} = \left(\dfrac{3 \times 8.314 \, kg \cdot m^2/s^2 \cdot K \cdot mol \times 300 \, K}{2.02 \times 10^{-3} \, kg/mol} \right)^{1/2} = 1.92 \times 10^3 \, m/s$

$$u_{CO_2} = \left(\dfrac{3 \times 8.314 \, kg \cdot m^2/s^2 \cdot K \cdot mol \times 300 \, K}{44.0 \times 10^{-3} \, kg/mol} \right)^{1/2} = 4.12 \times 10^2 \, m/s$$

As expected, the lighter molecule moves at the greater speed.

10.63 The heavier the molecule, the slower the rate of effusion. Thus, the order for increasing rate of effusion is in the order of decreasing mass.

rate $^2H^{37}Cl$ < rate $^1H^{37}Cl$ < rate $^2H^{35}Cl$ < rate $^1H^{35}Cl$

10.65 $\dfrac{rate \, (sulfide)}{rate \, (Ar)} = \left[\dfrac{39.9}{MM \, (sulfide)} \right]^{1/2} = 0.28$

MM (sulfide) = $(39.9 / 0.28)^2$ = 510 g/mol (two significant figures)

The empirical formula of arsenic(III) sulfide is As_2S_3, which has a formula mass of 246.1. Twice this is 490 g/mol, close to the value estimated from the effusion experiment. Thus, the formula of the vapor phase molecule is As_4S_6.

Nonideal-Gas Behavior

10.67 (a)   Nonideal gas behavior is observed at very high pressures and/or low temperatures.

       (b)   The real volumes of gas molecules and attractive intermolecular forces between molecules cause gases to behave nonideally.

10.69 The ratio PV/RT is equal to the number of moles of molecules in an ideal-gas sample; this number should be a constant for all pressure, volume and temperature conditions. If the value of this ratio changes with increasing pressure, the gas sample is not behaving ideally (according to the Ideal-Gas Equation).

10.71 The constants *a* and *b* are part of the correction terms in the van der Waals equation. The smaller the values of *a* and *b*, the smaller the corrections and the more ideal the gas. Ar (*a* = 1.34, *b* = 0.0322) will behave more like an ideal gas than CO_2 (*a* = 3.59, *b* = 0.0427) at high pressures.

10.73   (a) $P = 1.00 \text{ mol} \times \dfrac{0.08206 \text{ L} \cdot \text{atm}}{\text{K} \cdot \text{mol}} \times \dfrac{313 \text{ K}}{28.0 \text{ L}} = 0.917 \text{ atm}$

       (b) $P = \dfrac{nRT}{V \cdot nb} - \dfrac{an^2}{V^2} = \dfrac{1.00 \times 0.08206 \times 313}{28.0 - (1.00 \times 0.1383)} - \dfrac{20.4(1.00)^2}{(28.0)^2} = 0.896 \text{ atm}$

Additional Exercises

(In *Solutions to Exercises*, the symbol for molar mass is MM.)

10.76   (a) $P = \dfrac{nRT}{V}$; $n = 0.29 \text{ kg O}_2 \times \dfrac{1000 \text{ g}}{1 \text{ kg}} \times \dfrac{1 \text{ mol O}_2}{32.00 \text{ g O}_2} = 9.0625 = 9.1 \text{ mol}$; $V = 2.3 \text{ L}$;

              $T = 273 + 9°\text{C} = 282 \text{ K}$

              $P = \dfrac{9.0625 \text{ mol}}{2.3 \text{ L}} \times \dfrac{0.08206 \text{ L} \cdot \text{atm}}{\text{K} \cdot \text{mol}} \times 282 \text{ K} = 91 \text{ atm}$

       (b) $V = \dfrac{nRT}{P}$; $= \dfrac{9.0625 \text{ mol}}{0.95 \text{ atm}} \times \dfrac{0.08206 \text{ L} \cdot \text{atm}}{\text{K} \cdot \text{mol}} \times 299 \text{ K} = 2.3 \times 10^2 \text{ L}$

10.78 $\dfrac{P_1}{T_1} = \dfrac{P_2}{T_2}$; $P_1 = 2.2 \text{ atm}$, $T_1 = 297 \text{ K}$, $P_2 = 3.0 \text{ atm}$, $T_2 = ?$

       (T must be in kelvins for the P, T relationship to hold true.)

       $T_2 = \dfrac{P_2 T_1}{P_1} = \dfrac{3.0 \text{ atm} \times 297 \text{ K}}{2.2 \text{ atm}} = 405 \text{ K or } 132 °\text{C}$

10.81   Volume of laboratory $= 110 \text{ m}^2 \times 2.7 \text{ m} \times \dfrac{1000 \text{ L}}{1 \text{ m}^3} = 2.97 \times 10^5 = 3.0 \times 10^5 \text{ L}$

       Calculate the **total** moles of gas in the laboratory at the conditions given.

       $n_t = \dfrac{PV}{RT} = 1.00 \text{ atm} \times \dfrac{\text{K} \cdot \text{mol}}{0.08206 \text{ L} \cdot \text{atm}} \times \dfrac{2.97 \times 10^5 \text{ L}}{297 \text{ K}} = 1.22 \times 10^4 = 1.2 \times 10^4 \text{ mol gas}$

       An $Ni(CO)_4$ concentration of 1 part in 10^9 means 1 mol $Ni(CO)_4$ in 1×10^9 total moles of gas.

       $\dfrac{x \text{ mol Ni(CO)}_4}{1.22 \times 10^4 \text{ mol gas}} = \dfrac{1}{10^9} = 1.22 \times 10^{-5} \text{ mol Ni(CO)}_4$

       $1.22 \times 10^{-5} \text{ mol Ni(CO)}_4 \times \dfrac{170.74 \text{ g Ni(CO)}_4}{1 \text{ mol Ni(CO)}_4} = 2.1 \times 10^{-3} \text{ g Ni(CO)}_4$

10.84 $MM_{avg} = \dfrac{dRT}{P} = \dfrac{1.104 \text{ g}}{1 \text{ L}} \times \dfrac{0.08206 \text{ L} \cdot \text{atm}}{\text{K} \cdot \text{mol}} \times \dfrac{300 \text{ K}}{435 \text{ mm Hg}} \times \dfrac{760 \text{ mm Hg}}{1 \text{ atm}}$

                                  $= 47.48 = 47.5 \text{ g/mol}$

χ = mole fraction O_2; $1 - \chi$ = mole fraction Kr

47.48 g $= \chi(32.00) + (1-\chi)(83.80)$

$36.3 = 51.8\,\chi$; $\chi = 0.701$; 70.1% O_2

10.86   The balloon will expand; H_2 (MM = 2 g/mol) will effuse in through the walls of the balloon faster than He (MM = 4 g/mol) will effuse out, because the gas with the smaller molar mass effuses more rapidly.

10.90   (a) 80.00 kg $N_2(g) \times \dfrac{1000\,g}{1\,kg} \times \dfrac{1\,mol\,N_2}{28.02\,g\,N} = 2855$ mol N_2

        $P = \dfrac{nRT}{V} = 2855$ mol $\times \dfrac{0.08206\,L\bullet atm}{K\bullet mol} \times \dfrac{573\,K}{1000.0\,L} = 134.2$ atm

        (b)   According to Equation 10.26,

        $P = \dfrac{nRT}{V-nb} - \dfrac{n^2a}{V^2}$

        $P = \dfrac{(2855\,mol)(0.08206\,L\bullet atm/K\bullet mol)(573\,K)}{1000.0\,L - (2855\,mol)(0.0391\,L/mol)} - \dfrac{(2855\,mol)^2\,(1.39\,L^2\bullet atm/mol^2)}{(1000.0\,L)^2}$

        $P = \dfrac{134,243\,L\bullet atm}{1000.0\,L - 111.6\,L} - 11.3$ atm $= 151.1$ atm $- 11.3$ atm $= 139.8$ atm

        (c)   The pressure corrected for the real volume of the N_2 molecules is 151.1 atm, 16.9 atm higher than the ideal pressure of 134.2 atm. The 11.3 atm correction for intermolecular forces reduces the calculated pressure somewhat, but the "real" pressure is still higher than the ideal pressure. The correction for the real volume of molecules dominates. Even though the value of b is small, the number of moles of N_2 is large enough so that the molecular volume correction is larger than the attractive forces correction.

Integrative Exercises

(In *Solutions to Exercises*, the symbol for molar mass is MM.)

10.91   MM $= \dfrac{gRT}{VP} = \dfrac{1.05\,g}{0.500\,L} \times \dfrac{0.08206\,L\bullet atm}{K\bullet mol} \times \dfrac{298\,K}{750\,torr} \times \dfrac{760\,torr}{1\,atm} = 52.0$ g/mol

        0.462 g C $\times \dfrac{1\,mol\,C}{12.01\,g\,C} = 0.0385$ mol C

        0.538 g N $\times \dfrac{1\,mol}{14.01\,g\,N} = 0.0384$ mol N

The mole ratio is 1C:1N and the empirical formula is CN; the formula weight of CN = 12 + 14 = 26 g. Since molar mass of 52.0 g is twice the empirical formula weight, the molecular formula is C_2N_2.

10.93 Strategy: Calculate ΔH°_{rxn} using Hess's Law and data from Appendix C. Using the Ideal-Gas Equation, calculate moles CO_2 produced. Use stoichiometry to calculate the overall enthalpy change, ΔH.

$\Delta H^\circ_{rxn} = \Delta H^\circ_f\ Na^+(aq) + \Delta H^\circ_f\ CO_2(g) + \Delta H^\circ_f\ H_2O(l) - \Delta H^\circ_f\ NaHCO_3(s) - \Delta H^\circ_f\ H^+(aq)$

$= (-240.1\ kJ - 393.5\ kJ - 285.83\ kJ) - (-947.7\ kJ + 0)$

$= 28.27 = 28.3\ kJ$

$n = \dfrac{PV}{RT} = 715\ torr \times \dfrac{1\ atm}{760\ torr} \times \dfrac{K \bullet mol}{0.08206\ L \bullet atm} \times \dfrac{10.0\ L}{292\ K} = 0.3926 = 0.393\ mol\ CO_2$

$\Delta H = \dfrac{28.27\ kJ}{1\ mol\ CO_2} \times 0.3926\ mol\ CO_2 = 11.1\ kJ$

10.95 After reaction, the flask contains $IF_5(g)$ and whichever reactant is in excess. Determine the limiting reactant, which regulates the moles of IF_5 produced and moles of excess reactant.

$$I_2(s) + 5F_2(g) \rightarrow 2\ IF_5(g)$$

$10.0\ g\ I_2 \times \dfrac{1\ mol\ I_2}{253.8\ g\ I_2} \times \dfrac{5\ mol\ F_2}{1\ mol\ I_2} = 0.1970 = 0.197\ mol\ F_2$

$10.0\ g\ F_2 \times \dfrac{1\ mol\ F_2}{38.00\ g\ F_2} = 0.2632 = 0.263\ mol\ F_2$ available

I_2 is the limiting reactant; F_2 is in excess.

0.263 mol F_2 available - 0.197 mol F_2 reacted = 0.066 mol F_2 remain.

$10.0\ g\ I_2 \times \dfrac{1\ mol\ I_2}{253.8\ g\ I_2} \times \dfrac{2\ mol\ IF_5}{1\ mol\ I_2} = 0.0788\ mol\ IF_5$ produced

(a) $P_{IF_5} = \dfrac{nRT}{V} = 0.0788\ mol \times \dfrac{0.08206\ L \bullet atm}{K \bullet mol} \times \dfrac{398\ K}{5.00\ L} = 0.515\ atm$

(b) $\chi_{IF_5} = \dfrac{mol\ IF_5}{mol\ IF_5 + mol\ F_2} = \dfrac{0.0788}{0.0788 + 0.066} = 0.544$

11 Intermolecular Forces, Liquids and Solids

Kinetic-Molecular Theory

11.1   (a)   solid < liquid < gas   (b)   gas < liquid < solid

11.3   (a)   liquid   (b)   gas

11.5   As the temperature of a substance is increased, the average kinetic energy of the particles increases. In a collection of particles (molecules), the state is determined by the strength of interparticle forces relative to the average kinetic energy of the particles. As the average kinetic energy increases, more particles are able to overcome intermolecular attractive forces and move to a less ordered state, from solid to liquid to gas.

Intermolecular Forces

11.7   (a)   London dispersion forces   (b) dipole-dipole forces
         (c)   dipole-dipole or in certain cases hydrogen bonding

11.9   CO is a polar covalent molecule (ΔEN = 1.0) and N_2 is nonpolar. Dipole-dipole forces between CO molecules are stronger than London dispersion forces between N_2 molecules. A higher temperature and greater average kinetic energy is required to overcome the dipole-dipole forces in CO and separate (vaporize) the molecules.

11.11   (a)   *Polarizability* is the ease with which the charge distribution (electron cloud) in a molecule can be distorted to produce a transient dipole.

         (b)   Te is most polarizable because its valence electrons are farthest from the nucleus and least tightly held.

         (c)   Polarizability increases as molecular size (and thus molecular weight) increases. In order of increasing polarizability: CH_4 < SiH_4 < $SiCl_4$ < $GeCl_4$ < $GeBr_4$

11.13   (a) C_6H_{14} - dispersion; C_8H_{18} - dispersion. C_8H_{18} has the higher boiling point due to greater molar mass and similar strength of forces.

         (b) C_3H_8 - dispersion; CH_3OCH_3 - dipole-dipole and dispersion. CH_3OCH_3 has the higher boiling point due to stronger intermolecular forces and similar molar mass.

(c) CH_3OH - hydrogen bonding, dipole-dipole and dispersion; CH_3SH - dipole-dipole and dispersion. CH_3OH has the higher boiling point due to the influence of hydrogen bonding (Figure 11.7).

(d) NH_2NH_2 - hydrogen bonding, dipole-dipole and dispersion; CH_3CH_3 - dispersion. NH_2NH_2 has the higher boiling point due to much stronger intermolecular forces.

11.15   Both hydrocarbons experience dispersion forces. Rodlike butane molecules can contact each other over the length of the molecule, while spherical 2-methylpropane molecules can only touch tangentially. The larger contact surface of butane produces greater polarizability and a higher boiling point.

11.17   Surface tension (Section 11.3), high boiling point (relative to H_2S, H_2Se, H_2Te, Figure 11.7), high heat capacity per gram, high enthalpy of vaporization; the solid is less dense than the liquid; it is a liquid at room temperature despite its low molar mass.

Viscosity and Surface Tension

11.19   Viscosities and surface tensions of liquids both increase as intermolecular forces become stronger.

11.21   (a)   Ethanol molecules experience hydrogen bonding, a stronger intermolecular force than the weak dipole-dipole forces between ether molecules. The stronger forces between ethanol molecules make the liquid more resistant to flow and thus more viscous.

        (b)   The shape of a meniscus depends on the strength of the cohesive forces within a liquid relative to the adhesive forces between the walls of the capillary and the liquid. Because polar water molecules have strong cohesive intermolecular interactions relative to the weak adhesive forces between the polar water molecules and nonpolar polyethylene, the meniscus is concave-downward.

Changes of State

11.23   Endothermic: melting (s → l), vaporization (l → g), sublimation (s → g)
        Exothermic: condensation (g → l), freezing (l → s), deposition (g → s)

11.25   (a)   Ice, $H_2O(s)$, sublimes to water vapor, $H_2O(g)$.

        (b)   The heat energy required to increase the kinetic energy of molecules enough to melt the solid does not produce a large separation of molecules. The specific order is disrupted, but the molecules remain close together. On the other hand, when a liquid is vaporized, the intermolecular forces which maintain close molecular contacts must be overcome. Because molecules are being separated, the energy requirement is higher than for melting.

11.27 Evaporation of 10 g of water requires:

$$10.0 \text{ g H}_2\text{O} \times \frac{2.4 \text{ kJ}}{1 \text{ g H}_2\text{O}} = 24.0 \text{ kJ or } 2.4 \times 10^4 \text{ J}$$

Cooling a certain amount of water by 13°C:

$$2.40 \times 10^4 \text{ J} \times \frac{1 \text{ g} \cdot \text{K}}{4.18 \text{ J}} \times \frac{1}{13°\text{C}} = 442 = 4.4 \times 10^2 \text{ g H}_2\text{O}$$

11.29 Consider the process in steps, using the appropriate thermochemical constant.

Heat the liquid from -50°C to 23.8°C (223 K to 296.8 K), using the specific heat of the liquid.

$$10.0 \text{ g CCl}_3\text{F} \times \frac{0.87 \text{ J}}{\text{g} \cdot \text{K}} \times 73.8 \text{ K} \times \frac{1 \text{ kJ}}{1000 \text{ J}} = 0.642 = 0.64 \text{ kJ}$$

Boil the liquid at 23.8°C (296.8 K), using the enthalpy of vaporization.

$$10.0 \text{ g CCl}_3\text{F} \times \frac{1 \text{ mol CCl}_3\text{F}}{137.4 \text{ g CCl}_3\text{F}} \times \frac{24.75 \text{ kJ}}{\text{mol}} = 1.801 = 1.80 \text{ kJ}$$

Heat the gas from 23.8°C to 50°C (296.8 K to 323 K), using the specific heat of the gas.

$$10.0 \text{ g CCl}_3\text{F} \times \frac{0.59 \text{ J}}{\text{g} \cdot \text{K}} \times 26.2 \text{ K} \times \frac{1 \text{ kJ}}{1000 \text{ J}} = 0.1546 = 0.15 \text{ kJ}$$

The total energy required is 0.642 kJ + 1.801 kJ + 0.1546 kJ = 2.60 kJ.

11.31 (a) The critical temperature is the highest temperature at which a gas can be liquefied, regardless of pressure.

   (b) As the force of attraction between molecules increases, the critical temperature of the compound increases.

   (c) The critical pressure is the pressure required to cause liquefaction at the critical temperature.

Vapor Pressure and Boiling Point

11.33 The boiling point is the temperature at which the vapor pressure of a liquid equals the external pressure acting on the surface of the liquid. If the external pressure is changed, then the temperature required to produce that vapor pressure will also change.

Melting of a solid occurs when the vapor pressures of the solid and liquid phases are equal. Both of these vapor pressures will vary to a degree with the external pressure acting on the solid or liquid, but the variation is not great, and both solid and liquid vapor pressures tend to change in the same way with applied pressure.

11.35 $CBr_4 < CHBr_3 < CH_2Br_2 < CH_2Cl_2 < CH_3Cl < CH_4$

The weaker the intermolecular forces, the higher the vapor pressure, the more volatile the compound. The order of increasing volatility is the order of decreasing strength of intermolecular forces. By analogy to the boiling points of HCl and HBr (Section 11.2), the trend will be dominated by dispersion forces, even though four of the molecules ($CHBr_3$, CH_2Br_2, CH_2Cl_2 and CH_3Cl) are polar. Thus, the order of increasing volatility is the order of decreasing molar mass and decreasing strength of dispersion forces.

11.37   The water in the two pans is at the same temperature, the boiling point of water at the atmospheric pressure of the room. During a phase change, the temperature of a system is constant. All energy gained from the surroundings is used to accomplish the transition, in this case to vaporize the liquid water. The pan of water that is boiling vigorously is gaining more energy and the liquid is being vaporized more quickly than in the other pan, but the temperature of the phase change is the same.

11.39   The boiling point is the temperature at which the vapor pressure of a liquid equals atmospheric pressure.

(a)   The boiling point of ethanol at 300 torr is ~56°C, or, at 56°C, the vapor pressure of ethanol is 300 torr.

(b)   At a pressure of 15 torr, water would boil at ~17.5°C, or, the vapor pressure of water at 17.5°C is 15 torr.

11.41   From Appendix B, the temperature at which the vapor pressure of water is 350 torr is approximately 80°C.

Phase Diagrams

11.43   The liquid/gas line of a phase diagram ends at the critical point, the temperature and pressure beyond which the gas and liquid phases are indistinguishable. At temperatures higher than the critical temperature, a gas cannot be liquefied, regardless of pressure.

11.45   (a)   The water vapor would condense to form a solid at a pressure of around 4 torr. At higher pressure, perhaps 5 atm or so, the solid would melt to form liquid water. This occurs because the melting point of ice, which is 0°C at 1 atm, decreases with increasing pressure.

(b)   In thinking about this exercise, keep in mind that **total** pressure is being maintained constant at 0.3 atm. That pressure is made up of water vapor pressure and some other pressure, which could come from an inert gas. The water at -1.0°C and 0.30 atm is in the solid form. Upon heating, it melts to form liquid water slightly above 0°C. The liquid converts to a vapor when the temperature reaches the point at which the vapor pressure of water reaches 0.3 atm (228 torr). From Appendix B we see that this occurs just below 70°C.

11.47   (a)

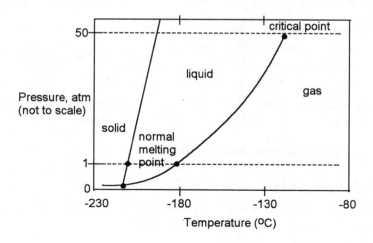

(b) $O_2(s)$ is denser than $O_2(l)$ because the solid-liquid line on the phase diagram is normal. That is, as pressure increases, the melting temperature increases. [Note that the solid-liquid line for O_2 is nearly vertical, indicating a small difference in the densities of $O_2(s)$ and $O_2(l)$].

(c) $O_2(s)$ will melt when heated at a pressure of 1 atm, since this is a much greater pressure than the pressure at the triple point.

Structures of Solids

11.49   In a crystalline solid, the component particles (ions or molecules) are arranged in an ordered repeating pattern. In an amorphous solid, there is no orderly structure.

11.51   The unit cell is the building block of the crystal lattice. When repeated in three dimensions, it produces the crystalline solid. It is a parallelepiped with characteristic distances and angles. Unit cells can be primitive (lattice points only at the corners of the parallelepiped) or centered (lattice points at the corners and at the middle of faces or the middle of the parallelepiped).

11.53   (a)   4 [See Exercise 11.52(c).]

(b)   Each sphere is in contact with 12 nearest neighbors; its coordination number is thus 12.

(c)   The length of the face diagonal of a face-centered cubic unit cell is four times the radius of the metal and $\sqrt{2}$ times the unit cell dimension (usually designated a for cubic cells).

$$4 \times 1.24 \text{ Å} = \sqrt{2} \times a$$

$$a = \frac{4 \times 1.24 \text{Å}}{\sqrt{2}} = 3.507 = 3.51 \times 10^{-8} \text{ cm}$$

(d)   The density of the metal is the mass of the unit cell contents divided by the volume of the unit cell.

$$\text{density} = \frac{4 \text{ Ni atoms}}{(3.507 \times 10^{-8} \text{ cm})^3} \times \frac{58.69 \text{ g Ni}}{6.022 \times 10^{23} \text{ Ni atoms}} = 9.04 \text{ g/cm}^3$$

11.55   (a)   Each sphere is in contact with 12 nearest neighbors; its coordination number is thus 12.

      (b)   Each sphere has a coordination number of six.

      (c)   Each sphere has a coordination number of eight.

11.57   In the face-centered cubic structure, there are four NiO units in the unit cell. Density is the mass of the unit cell contents divided by the unit cell volume (a^3).

$$\text{density} = \frac{4 \text{ NiO units}}{(4.18 \text{ Å})^3} \times \frac{74.7 \text{ g NiO}}{6.022 \times 10^{23} \text{ NiO units}} \times \left(\frac{1 \text{ Å}}{1 \times 10^{-8} \text{ cm}}\right)^3 = \frac{6.79 \text{ g}}{\text{cm}^3}$$

11.59   The volume of the unit cell is $(2.86 \times 10^{-8} \text{ cm})^3$. The mass of the unit cell is:

$$\frac{7.92 \text{ g}}{\text{cm}^3} \times \frac{(2.86 \times 10^{-8})^3 \text{ cm}^3}{\text{unit cell}} = 1.853 \times 10^{-22} \text{ g/unit cell}$$

There are two atoms of the element present in the body-centered cubic unit cell. Thus the atomic weight is:

$$\frac{1.853 \times 10^{-22} \text{ g}}{\text{unit cell}} \times \frac{1 \text{ unit cell}}{2 \text{ atoms}} \times \frac{6.022 \times 10^{23} \text{ atoms}}{1 \text{ mol}} = 55.8 \text{ g/mol}$$

Bonding in Solids

11.61   (a)   Hydrogen bonding, dipole-dipole forces, London dispersion forces

      (b)   covalent chemical bonds (mainly)

      (c)   ionic bonds (mainly)

      (d)   metallic bonds

11.63   Carbon (graphite and diamond) and SiO_2 (quartz)

11.65   (a)   KBr - strong ionic versus weak dispersion forces

      (b) SiO_2 - covalent bonds establish the network structure of the lattice versus weak dispersion forces holding CO_2 molecules together

      (c)   Se - network covalent lattice versus weak dispersion forces

      (d) MgF_2 - due to higher charge on Mg^{2+} than Na^+

11.67   According to Table 11.6, the solid could be either ionic with low water solubility or network covalent. Due to the extremely high sublimation temperature, it is probably network covalent.

11.69 (a)   According to Figure 11.44(b), there are 4 Ag I units in the "zinc blende" unit cell
              [4 complete Ag^+ spheres, $6(\frac{1}{2}) + 8(1/8)$ I^- sphere's.]

$$5.69 \frac{g}{cm^3} = \frac{4 \text{ AgI units}}{a^3} \times \frac{234.8 \text{ g}}{6.022 \times 10^{23} \text{ AgI units}} \times \left(\frac{1 \text{ Å}}{1 \times 10^{-8} \text{ cm}}\right)^3$$

$a^3 = 274.10$ Å^3, $a = 6.50$ Å

       (b)   In an orthonormal coordinate system, the distance between two points (x_1, y_1, z_1) and
              (x_2, y_2, z_2) is $\sqrt{(x_1 - x_2)^2 + (y_1 - y_2)^2 + (z_1 - z_2)^2}$.

On Figure 11.44(b), select a
right-handed coordinate system
and select a bonded pair of ions.
One possibility is shown below.

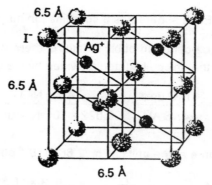
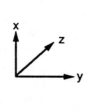

In a cubic unit cell, the lengths of all three cell edges are the same, in this case 6.50 Å.

For Ag^+: x = 0.75(6.50 Å), y = 0.25(6.50 Å), z = 0.25(6.50 Å)

       x = 4.875, y = 1.625, z = 1.625

I^-: x = 6.50, y = 0, z = 0

The Ag-I distance is then $\sqrt{(4.875 - 6.50)^2 + (1.625 - 0)^2 + (1.625 - 0)^2}$.

Ag-I = $\sqrt{(1.625)^2 + (1.625)^2 + (1.625)^2}$ = 2.81 Å

Additional Exercises

11.71 (a)   An *instantaneous dipole* is the temporary, small charge separation that occurs when
              negatively charged electron clouds bump into each other. It is sometimes called an
              induced dipole.

       (b)   *Polarizability* is the tendency of a normally symmetrical electron cloud to adopt an
              instantaneous dipole upon contact with another molecule or outside force.

       (c)   *Intermolecular attractive forces* are electrostatic attractions between neutral mole-
              cules owing to either permanent or induced dipole moments. Portions of molecules
              with partial charges of opposite sign are attracted to each, resulting in an overall
              lowering in energy of the system.

11.74 (a) The *cis* isomer has stronger dipole-dipole forces; the *trans* isomer is nonpolar. The higher boiling point of the *cis* isomer supports this conclusion.

(b) While boiling points are primarily a measure of strength of intermolecular forces, melting points are influenced by crystal packing efficiency as well as intermolecular forces. Since the nonpolar *trans* isomer with weaker intermolecular forces has the higher melting point, it must pack more efficiently.

11.76 When a halogen atom (Cl or Br) is substituted for H in benzene, the molecule becomes polar. These molecules experience dispersion forces similar to those in benzene plus dipole-dipole forces, so they have higher boiling points than benzene. C_6H_5Br has a higher molar mass and is more polarizable than C_6H_5Cl so it has the higher boiling point. C_6H_5OH experiences hydrogen bonding, the strongest force between neutral molecules, so it has the highest boiling point.

11.79 Propylamine experiences hydrogen bonding interactions while trimethylamine, with no N-H bonds, does not. Also, the rod-like shape of propylamine (see Exercise 11.15) leads to stronger dispersion forces than in trimethylamine. The stronger intermolecular forces in propylamine lead to the lower vapor pressure.

11.82 (a) The Clausius-Clapeyron equation is $InP = \dfrac{-\Delta H_{vap}}{RT} + C$.

For two vapor pressures, P_1 and P_2, measured at corresponding temperatures T_1 and T_2, the relationship is

$$InP_1 - InP_2 = \left(\frac{-\Delta H_{vap}}{RT_1} + C \right) - \left(\frac{-\Delta H_{vap}}{RT_2} + C \right)$$

$$InP_1 - InP_2 = \frac{-\Delta H_{vap}}{R}\left(\frac{1}{T_1} - \frac{1}{T_2} \right) + C - C; \quad In\frac{P_1}{P_2} = \frac{-\Delta H_{vap}}{R}\left(\frac{1}{T_1} - \frac{1}{T_2} \right)$$

(b) $P_1 = 10.00$ torr, $T_1 = 716$ K; $P_2 = 400.0$ torr, $T_2 = 981$ K

$$In\frac{10.00}{400.0} = \frac{-\Delta H_{vap}}{8.314 \text{ J/K} \cdot \text{mol}}\left(\frac{1}{716} - \frac{1}{981} \right)$$

$-3.6889 (8.314 \text{ J/K} \cdot \text{mol}) = -\Delta H_{vap}(3.773 \times 10^{-4}/\text{K})$

$\Delta H_{vap} = 8.129 \times 10^4 = 8.13 \times 10^4$ J/mol = 81.3 kJ/mol

(c) The normal boiling point of a liquid is the temperature at which the vapor pressure of the liquid is 760 torr.

$P_1 = 400.0$ torr, $T_1 = 981$ K; $P_2 = 760$ torr, $T_2 = $ b.p. of potassium

$$In\left(\frac{400.0}{760.0} \right) = \frac{-8.129 \times 10^4 \text{ J/mol}}{8.314 \text{ J/K} \cdot \text{mol}}\left(\frac{1}{981 \text{ K}} \times \frac{1}{T_2} \right)$$

$$\frac{-0.64185}{-9.7775 \times 10^3} = 1.0194 \times 10^{-3} - \frac{1}{T_2}; \quad \frac{1}{T_2} = 1.0194 \times 10^{-3} - 6.565 \times 10^{-5}$$

$$\frac{1}{T_2} = 9.5375 \times 10^{-4}; \quad T_2 = 1048 \text{ K } (775°C)$$

(d) P_1 = VP of K(l) at 100°C, T_1 = 373 K; P_2 = 10.00 torr, T_2 = 716 K

$$\ln\frac{P_1}{10.00 \text{ torr}} = \frac{-8.129 \times 10^4 \text{ J/mol}}{8.314 \text{ J/K} \cdot \text{mol}}\left(\frac{1}{373} - \frac{1}{716}\right)$$

$$\ln\frac{P_1}{10.00 \text{ torr}} = \frac{-8.129 \times 10^4 \text{ J/mol}}{8.314 \text{ J/K} \cdot \text{mol}} \times 1.284 \times 10^{-3} = -12.5543$$

$$\frac{P_1}{10.00 \text{ torr}} = e^{-12.5543} = 3.530 \times 10^{-6}; \quad P_1 = 3.53 \times 10^{-5} \text{ torr}$$

11.84   The length of the body diagonal in a body-centered cubic cell is 4r, where r is the radius of a Cr atom. To deduce the value of r, look at the triangle formed by the cube side (length = a), a face diagonal (length = $\sqrt{2}\ a$) and the body diagonal (length = 4r).

The angle θ is arctan $1/\sqrt{2}$ = 35.264°.

Sin θ = a/4r;   r = 2.884 Å/4 sin 35.264° = 1.249 Å.

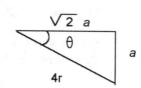

Integrative Exercises

11.87   (a)   The greater dipole moment of HCl (1.08 D vs 0.82 D for HBr) indicates that it will experience greater dipole-dipole forces.

       (b)   The longer bond length in HBr (1.41 Å vs 1.27 Å for HCl) indicates a more diffuse and therefore more polarizable electron cloud in HBr. It experiences stronger London dispersion forces.

       (c)   For molecules with similar structures, the compound with the higher boiling point experiences stronger intermolecular forces. Since HBr has the higher boiling point, weaker dipole-dipole forces but stronger dispersion forces, dispersion forces must determine the boiling point.

       (d)   HF experiences hydrogen bonding, a much stronger force than the dipole-dipole and dispersion forces operating in the other hydrogen halides, which results in the high boiling point.

             HI, with the longest bond length and greatest polarizability in the series, has the strongest dispersion forces and highest boiling point of the compounds that do not experience hydrogen bonding.

11.89 Strategy:

  i) Using thermochemical data from Appendix B, calculate the energy (enthalpy) required to melt and heat the H_2O.

  ii) Using Hess's Law, calculate the enthalpy of combustion, ΔH_{comb}, for C_3H_8.

  iii) Solve the stoichiometry problem.

  i) Heat H_2O(s) from -10.0°C to 0.0°C; $1000 \text{ g } H_2O \times \dfrac{2.092 \text{ J}}{\text{g} \cdot °\text{C}} \times 10.0\,°\text{C} = 20.92 = 20.9 \text{ kJ}$

  Melt H_2O(s); $1000 \text{ g } H_2O \times \dfrac{6.008 \text{ kJ}}{\text{mol } H_2O} \times \dfrac{1 \text{ mol } H_2O}{18.02 \text{ g } H_2O} = 333.41 = 333.4 \text{ kJ}$

  Heat H_2O(l) from 0.0°C to 75.0°C; $1000 \text{ g } H_2O \times \dfrac{4.184 \text{ J}}{\text{g} \cdot °\text{C}} \times 75.0\,°\text{C} = 313.8 = 314 \text{ kJ}$

  Total energy = 20.92 kJ + 333.4 kJ + 313.8 kJ = 668.1 = 668 kJ

  (The result has zero decimal places because 314 kJ has zero decimal places.)

  ii) C_3H_8(g) + $5O_2$(g) → $3CO_2$(g) + $4H_2O$(l)

  Assume that one product is H_2O(l), since this leads to a more negative ΔH_{comb} and fewer grams of C_3H_8(g) required.

  $\Delta H_{comb} = 3\Delta H^{\circ}_f \, CO_2(g) + 4\Delta H^{\circ}_f \, H_2O(l) - \Delta H^{\circ}_f \, C_3H_8(g) - \Delta H^{\circ}_f \, O_2(g)$

  $= 3(-393.5 \text{ kJ}) + 4(-285.83 \text{ kJ}) - (-103.85 \text{ kJ}) - 0 = -2219.97 = -2220 \text{ kJ}$

  iii) $668.1 \text{ kJ required} \times \dfrac{1 \text{ mol } C_3H_8}{2219.97 \text{ kJ}} \times \dfrac{44.094 \text{ g } C_3H_8}{1 \text{ mol } C_3H_8} = 13.3 \text{ g } C_3H_8$

  (668 kJ required has 3 sig figs and so does the result)

11.91 $PV = \dfrac{g}{MM} \times RT$; $g = \dfrac{PV \cdot MM}{RT}$; T = 313 K; MM = 18.02 g/mol

  P (the vapor pressure of H_2O at 40°C) = $55.3 \text{ torr} \times \dfrac{1 \text{ atm}}{760 \text{ torr}} = 0.07276 = 0.0728 \text{ atm}$

  $V = 4.0 \text{ m} \times 4.0 \text{ m} \times 3.0 \text{ m} \times \dfrac{10^3 \text{ dm}^3}{1 \text{ m}^3} = 4.8 \times 10^4 \text{ L}$

  $g = \dfrac{0.07276 \text{ atm} \times 4.8 \times 10^4 \text{ L} \times 18.02 \text{ g}}{313 \text{ K} \times \text{mol}} \times \dfrac{\text{K} \cdot \text{mol}}{0.08206 \text{ L} \cdot \text{atm}} = 2.450 \times 10^3 \text{ g} = 2.5 \text{ kg } H_2O$

12 Modern Materials

Liquid Crystals

12.1   Both an ordinary liquid and a nematic liquid crystal phase are fluids; they are converted directly to the solid phase upon cooling. The nematic phase is cloudy and more viscous than an ordinary liquid. Upon heating, the nematic phase is converted to an ordinary liquid.

12.3   Reinitzer observed that cholesteryl benzoate has a phase that exhibits properties intermediate between those of the solid and liquid phases. This "liquid-crystalline" phase, formed by melting at 145°C, is opaque, changes color from red to blue as the temperature is increased, and becomes clear at 179°C.

12.5   Because order is maintained in at least one dimension, the molecules in a liquid crystalline phase are not totally free to change orientation. This makes the liquid crystalline phase more resistant to flow, more viscous, than the isotropic liquid.

12.7   In the nematic phase, the long axes of the molecules are aligned. Translational motion is allowed, but rotational motion is restricted. In the smectic phase, both the long axes and the ends of the molecules are aligned; the molecules are organized into sheets. Both translational and rotational motion are restricted.

12.9   nematic - one dimension; smectic - at least two dimensions.

Polymers

12.11 *n*-decane does not have a sufficiently high chain length or molecular mass to be considered a polymer.

12.13

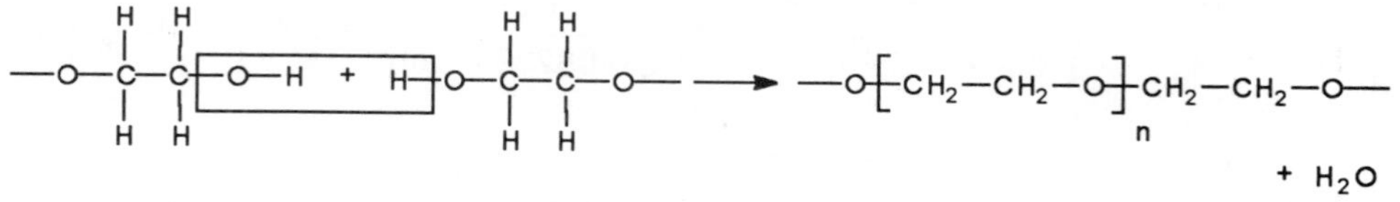

12.15 (a)

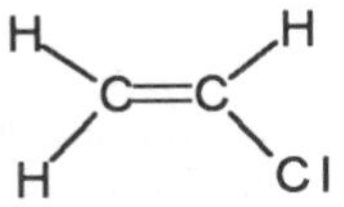

vinyl chloride (chloroethylene or chloroethene)

(b)

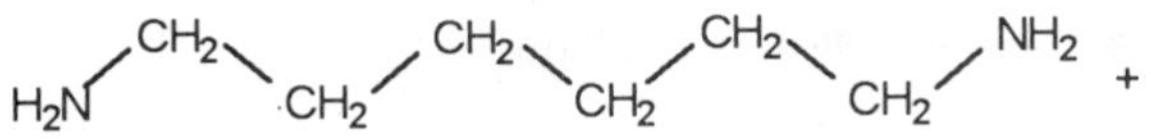

hexanediamine

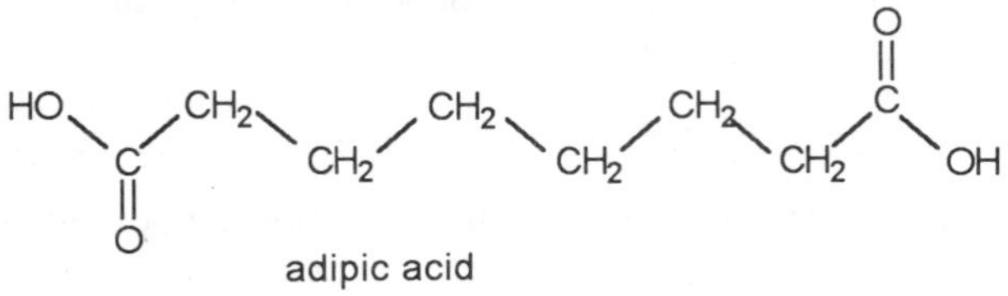

adipic acid

(Formulas given in Equation 12.3.)

(c)

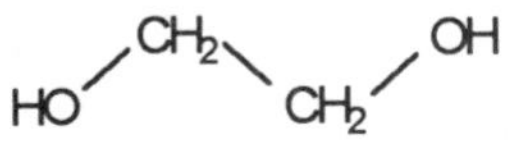

ethylene glycol 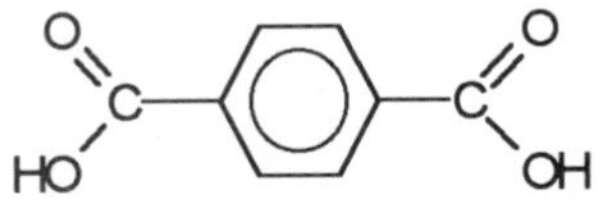   terephthalic acid

12.17

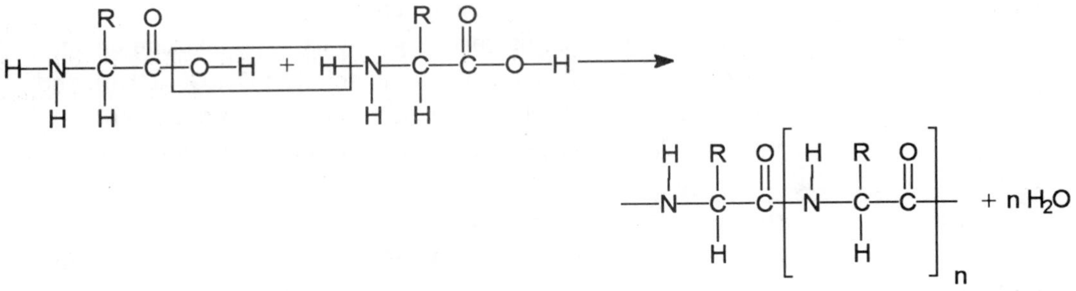

12.19 High density polyethylene (HDPE) has a higher molecular mass, melting point, density and mechanical strength than low density polyethylene (LDPE). At the molecular level, the longer, unbranched chains of HDPE fit closer together and have more crystalline (ordered, aligned) regions than the shorter, branched chains of LDPE. Closer packing leads to higher density. The stronger dispersion forces that result from tighter packing and more alignment lead to higher melting point and greater mechanical strength.

12.21 The function of the material (polymer) determines whether high molecular mass and high degree of crystallinity are desirable properties. If the material will be formed into containers or pipes, rigidity and structural strength are required. If the polymer will be used as a flexible wrapping or as a garment material, rigidity is an undesirable property.

Ceramics

12.23   Structurally, polymers are formed from organic monomers held together by covalent bonds, whereas ceramics are formed from inorganic materials linked by ionic or highly polar covalent bonds. Ceramics are often stabilized by a three-dimensional bonding network, whereas in polymers, covalent bonds link atoms into a long, chain-like molecule with only weak interactions between molecules. Polymers are nearly always amorphous, and though ceramics can be amorphous, they are often crystalline.

In terms of physical properties, ceramics are generally much harder, more heat-stable and higher melting than polymers. (These properties are true in general for network solids relative to molecular solids, as described in Table 11.6.)

12.25   Since Zr and Ti are in the same family, assume that the stoichiometry of the compounds in a sol-gel process will be the same for the two metals.

i.   Alkoxide formation - oxidation-reduction reaction
$$Zr(s) + 4CH_3CH_2OH(l) \rightarrow Zr(OCH_2CH_3)_4(s) + 2H_2(g)$$
                                   alkoxide

ii.   Sol formation - metathesis reaction
$$Zr(OCH_2CH_3)_4(soln) + 4H_2O(l) \rightarrow Zr(OH)_4(s) + 4CH_3CH_2OH(l)$$
                                   "precipitate"   nonelectrolyte
                                        sol

$Zr(OCH_2CH_3)_4(s)$ is dissolved in an alcohol solvent and then reacted with water. In general, reaction with water is called *hydrolysis*. The alkoxide anions ($CH_3CH_2O^-$) combine with H^+ from H_2O to form the nonelectrolyte $CH_3CH_2OH(l)$, and Zr^{2+} cations combine with OH^- to form the $Zr(OH)_4$ solid. The product $Zr(OH)_4(s)$ is not a traditional coagulated precipitate, but finely divided evenly dispersed particles called a sol.

iii.   Gel formation - condensation reaction
$$(OH)_3Zr-O-H(s) + H-O-Zr(OH)_3(s) \rightarrow (HO)_3Zr-O-Zr(OH)_3(s) + H_2O(l)$$
                                   gel

Adjusting the acidity of the $Zr(OH)_4$ sol initiates condensation, the splitting-out of $H_2O(l)$ and formation of a zirconium-oxide network solid. The solid remains suspended in the solvent mixture and is called a gel.

iv.   Processing - physical changes
The gel is heated to drive off solvent and the resulting solid consists of dry, uniform and finely divided ZrO_2 particles.

12.27   Concrete is a typically brittle ceramic and susceptible to catastrophic fracture. Steel reinforcing rods are added to resist stress applied along the long direction of the rod. By analogy, the shape of the reinforcing material in the ceramic composite should be rod-like, with a length much greater than its diameter. This is the optimal shape because rods have great strength when the load or stress is applied parallel to the long direction of the rod. Rods can be oriented in many directions, so that the material (concrete or ceramic composite) is strengthened in all directions.

12.29   By analogy to the ZnS structure, the C atoms form a face-centered cubic array with Si atoms occupying **alternate** tetrahedral holes in the lattice. This means that the coordination numbers of both Si and C are 4; each Si is bound to 4 C atoms in a tetrahedral arrangement, and each C is bound to 4 Si atoms in a tetrahedral arrangement, producing an extended three-dimensional network. ZnS, an ionic solid, sublimes at 1185° and 1 atm pressure and melts at 1850° and 150 atm pressure. The considerably higher melting point of SiC, 2800° at 1 atm, indicates that SiC is probably not a purely ionic solid and that the Si-C bonding network has significant covalent character. This is reasonable, since the electronegativities of Si and C are similar (Figure 8.7). SiC is high-melting because a great deal of chemical energy is stored in the covalent Si-C bonds, and it is hard because the three-dimensional lattice resists any change that would weaken the Si-C bonding network.

12.31   A superconducting material offers no resistance to the flow of electrical current; it is the frictionless flow of electrons. Superconductive materials could transmit electricity with no heat loss and therefore much greater efficiency than current carriers. Because of the Meisner effect, they are also potential materials for magnetically levitated trains.

Thin Films

12.33   In general, a useful thin film should:

      (a)   be chemically stable in its working environment
      (b)   adhere to its substrate
      (c)   have a uniform thickness
      (d)   have an easily controllable composition
      (e)   be nearly free of imperfections

12.35   There are three major methods of producing thin films.

      (i)   In *vacuum deposition*, a substance is vaporized or evaporated by heating under vacuum and then deposited on the desired substrate. No chemical change occurs.

      (ii)   In *sputtering*, ions accelerated to high energies by applying a high voltage are allowed to strike the target material, knocking atoms from its surface. These target material atoms are further accelerated toward the substrate, forming a thin film. No net chemical change occurs, because the material in the film is the target material.

      (iii)   In *chemical vapor deposition*, two gas phase substances react at the substrate surface to form a stable product which is deposited as a thin film. This involves a net chemical change.

Additional Exercises

12.37 A dipole moment (permanent, partial charge separation) roughly parallel to the long dimension of the molecule would cause the molecules to reorient when an electric field is applied perpendicular to the usual direction of molecular orientation.

12.40 At the temperature where a substance changes from the solid to the liquid-crystalline phase, kinetic energy sufficient to overcome most of the long range order in the solid has been supplied. A few van der Waals forces have sufficient attractive energy to impose the one-dimensional order characteristic of the liquid-crystalline state. Very little additional kinetic energy (and thus a relatively small increase in temperature) is required to overcome these aligning forces and produce an isotropic liquid.

12.43 Ceramics are usually three-dimensional network solids, whereas plastics most often consist of large, chain-like molecules (the chain may be branched) held loosely together by relatively weak van der Waals forces. Ceramics are rigid precisely because of the many strong bonding interactions intrinsic to the network. Once a crack forms, atoms near the defect are subject to great stress, and the crack is propagated. They are stable to high temperatures because tremendous kinetic energy (temperature) is required for an atom to break free from the bonding network. On the other hand, plastics are flexible because the molecules themselves are flexible (free rotation around the sigma bonds in the polymer chain), and it is easy for the molecules to move relative to one another (weak intermolecular forces). (However, recall that rigidity of the plastic increases as crosslinking of the polymer chain increases. The melamine-formaldehyde polymer in Figure 12.20 is a very rigid, brittle polymer.) Plastics are not thermally stable because their largely organic molecules are subject to oxidation and/or bond breaking at high temperatures.

12.46 (a)

$$\left[\begin{array}{c} CH_3 \\ | \\ -Si- \\ | \\ CH_3 \end{array}\right]_n \xrightarrow{400^\circ C} \left[\begin{array}{c} H \\ | \\ -Si-CH_2- \\ | \\ CH_3 \end{array}\right]_n$$

$$\left[\begin{array}{c} H \\ | \\ -Si-CH_2- \\ | \\ CH_3 \end{array}\right]_n \xrightarrow{1200^\circ C} [SiC]_n + CH_4(g) + H_2(g)$$

(b) $2NbBr_5(g) + 5H_2(g) \rightarrow 2Nb(s) + 10HBr(g)$

(c) $SiCl_4(l) + C_2H_5OH(l) \rightarrow Si(OC_2H_5)_4(s) + 4HCl(g)$

12.50   The formula of the compound deposited as a thin film is indicated by boldface type.

(a) $SiH_4(g) + 2H_2(g) + 2CO_2(g) \rightarrow$ **$SiO_2(s)$** $+ 2H_2(g) + 2CO(g)$

(b) $TiCl_4(g) + 2H_2O(g) \rightarrow$ **$TiO_2(s)$** $+ 4HCl(g)$

(c) $GeCl_4(g) \rightarrow$ **$Ge(s)$** $+ 2Cl_2(g)$ The H_2 carrier gas dilutes the $GeCl_4(g)$ so the reaction occurs more evenly and at a controlled rate; it does not participate in the reaction.

12.53   Application of an electric field in the direction perpendicular to the planes of the plates would cause the molecules to orient along the direction of the applied field, vertical in the diagram below.

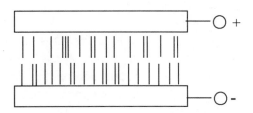

Because the molecules have thermal energy which causes them to move randomly, alignment parallel to the applied field is not complete, but some net reorientation does take place. This happens because the molecules in a nematic substance generally have a dipole moment (partial charge separation) oriented roughly parallel to the long direction of the molecule. Also, the molecules are more polarizable in their long direction, so London dispersion forces between molecules are stronger when the molecules are aligned in the direction of the applied field.

Integrative Exercises

12.57   (a)   The data (14.99%) has 4 sig figs, so use molar masses to 5 sig figs.

$$\text{mass \% O} = 14.99 = \frac{(8+x)\,15.999}{746.04 + (8+x)\,15.999} \times 100$$

rounded (to show sig figs)   **unrounded**

(8+x)15.999   (8+x)15.999

    = 0.1499 [746.04 + (8+x) 15.999]   = 0.1499 [746.04 + (8+x) 15.999]

127.99 + 15.999x   127.992 + 15.999x

    = 0.1499(874.04 + 15.999x)   = 0.1499 (874.036 + 15.999x

15.999x - 2.398x   15.999x - 2.3983x

    = 131.0 - 127.99   = 131.018 - 127.992

13.601x = 3.0;   x = 0.22   13.6007x = 3.026;   x = 0.2225

(b)   **Hg** and **Cu** both have more than one stable oxidation state. If different Cu ions (or Hg ions) in the solid lattice have different charges, then the average charge is a noninteger value. Ca and Ba are stable only in the +2 oxidation state; they are unlikely to have noninteger average charge.

(c) Ba^{2+} is largest; Cu^{2+} is smallest. For ions with the same charge, size decreases going up or across the periodic table. In the +2 state, Hg is smaller than Ba. If Hg has an average charge greater than 2+, it will be smaller yet. The same argument is true for Cu and Ca.

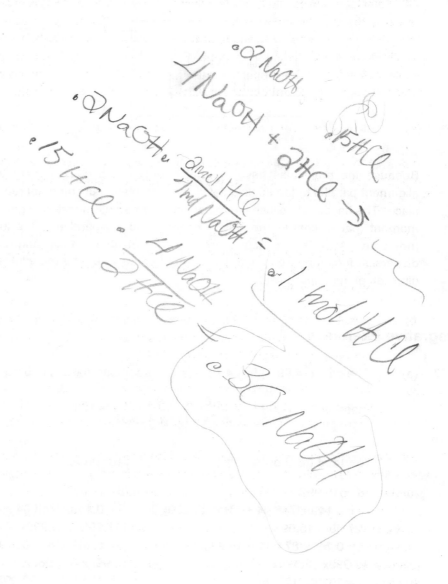

13 Properties of Solutions

The Solution Process

13.1   If the enthalpy released due to solute-solvent attractive forces (ΔH_3) is at least as large as the enthalpy required to separate the solute particles (ΔH_1), the overall enthalpy of solution (ΔH_{soln}) will be either slightly endothermic (owing to $+\Delta H_2$) or exothermic. Even if ΔH_{soln} is slightly endothermic, the increase in disorder due to mixing will cause a significant amount of solute to dissolve. If the magnitude of ΔH_3 is small relative to the magnitude of ΔH_1, ΔH_{soln} will be large and endothermic (energetically unfavorable) and not much solute will dissolve.

13.3   (a)   dispersion   (b) ion-dipole   (c) hydrogen bonding   (d) dipole-dipole

13.5   Water, H_2O, is a polar solvent that forms hydrogen bonds with other H_2O molecules. The more soluble solute in each case will have intermolecular interactions that are most similar to the hydrogen bonding in H_2O.

   (a) CH_3CH_2OH is more soluble because it has a shorter nonpolar hydrocarbon chain. The longer nonpolar hydrocarbon chain of $CH_3CH_2CH_2CH_2OH$ is capable only of dispersion forces and interferes with hydrogen bonding interactions between the -OH group and water.

   (b)   Ionic $CaCl_2$ is more soluble because ion-dipole solute-solvent interactions are more similar to ionic solute-solute and hydrogen bonding solvent-solvent interactions than the weak dispersion forces between CCl_4 and H_2O.

   (c) C_6H_5OH is more soluble because it is capable of hydrogen bonding. Nonpolar C_6H_6 is capable only of dispersion force interactions and does not have strong intermolecular interactions with polar (hydrogen bonding) H_2O.

13.7   The overall energy change associated with dissolution depends on the relative magnitudes of the solute-solute, solvent-solvent, and solute-solvent interactions. If disrupting the "old" solute-solute and solvent-solvent interactions requires more energy than is released by the "new" solute-solvent interactions, the process is endothermic. When $NH_4NO_3(s)$ dissolves in water, disrupting the ion-ion interactions among NH_4^+ and NO_3^- ions of the solute (ΔH_1) and the hydrogen bonding among H_2O molecules of the solvent (ΔH_2) requires more energy than is released by subsequent ion-dipole and hydrogen bonding interactions among NH_4^+ and H_2O and NO_3^- and H_2O (ΔH_3).

13.9 (a) ΔH_{soln} is determined by the relative magnitudes of the "old" solute-solute (ΔH_1) and solvent-solvent (ΔH_2) interactions and the new solute-solvent interactions (ΔH_3); $\Delta H_{soln} = \Delta H_1 + \Delta H_2 + \Delta H_3$. Since the solute and solvent in this case experience very similar London dispersion forces, the energy required to separate them individually and the energy released when they are mixed are approximately equal. $\Delta H_1 + \Delta H_2 \approx -\Delta H_3$. Thus, ΔH_{soln} is nearly zero.

(b) Mixing hexane and heptane produces a homogeneous solution from two pure substances, and the randomness of the system increases. Since no strong intermolecular forces prevent the molecules from mixing, they do so spontaneously due to the increase in disorder.

Concentrations of Solutions

13.11 (a) $\text{mass \%} = \dfrac{\text{mass solute}}{\text{total mass solution}} \times 100 = \dfrac{16.5 \text{ g CaCl}_2}{16.5 \text{ g CaCl}_2 + 456 \text{ g H}_2\text{O}} \times 100 = 3.49\%$

(b) $\text{ppm} = \dfrac{\text{mass solute}}{\text{total mass solution}} \times 10^6; \; \dfrac{83.5 \text{ g Ag}}{1 \text{ ton ore}} \times \dfrac{1 \text{ ton}}{2000 \text{ lb}} \times \dfrac{1 \text{ lb}}{453.6 \text{ g}} \times 10^6$

$= 92.0 \text{ ppm}$

13.13 (a) $8.5 \text{ g CH}_3\text{OH} \times \dfrac{1 \text{ mol CH}_3\text{OH}}{32.04 \text{ g CH}_3\text{OH}} = 0.265 = 0.27 \text{ mol CH}_3\text{OH}$

$224 \text{ g H}_2\text{O} \times \dfrac{1 \text{ mol H}_2\text{O}}{18.02 \text{ g H}_2\text{O}} = 12.43 = 12.4 \text{ mol H}_2\text{O}$

$\chi_{\text{CH}_3\text{OH}} = \dfrac{0.265}{0.265 + 12.43} = 0.021$

(b) $\dfrac{65.2 \text{ g CH}_3\text{OH}}{32.04 \text{ g/mol}} = 2.035 = 2.04 \text{ mol CH}_3\text{OH}; \; \dfrac{144 \text{ g CCl}_4}{153.8 \text{ g/mol}} = 0.9363 = 0.936 \text{ mol CCl}_4$

$\chi_{\text{CH}_3\text{OH}} = \dfrac{2.035}{2.035 + 0.9363} = 0.685$

13.15 (a) $M = \dfrac{\text{mol solute}}{\text{L soln}}; \; \dfrac{10.5 \text{ g NaCl}}{0.350 \text{ L soln}} \times \dfrac{1 \text{ mol NaCl}}{58.44 \text{ NaCl}} = 0.513 \; M$

(b) $\dfrac{40.7 \text{ g LiClO}_4 \cdot 3\text{H}_2\text{O}}{0.125 \text{ L soln}} \times \dfrac{1 \text{ mol LiClO}_4 \cdot 3\text{H}_2\text{O}}{160.4 \text{ g LiClO}_4 \cdot 3\text{H}_2\text{O}} = 2.03 \; M$

(c) $M_c \times L_c = M_d \times L_d; \; 1.50 \; M \text{ HNO}_3 \times 0.0400 \text{ L} = ?M \text{ HNO}_3 \times 0.500 \text{ L}$

500 mL of 0.120 M HNO$_3$

13.17 (a) $m = \dfrac{\text{mol solute}}{\text{kg solvent}}; \; \dfrac{13.0 \text{ g C}_6\text{H}_6}{17.0 \text{ g CCl}_4} \times \dfrac{1 \text{ mol C}_6\text{H}_6}{78.11 \text{ g C}_6\text{H}_6} \times \dfrac{1000 \text{ g CCl}_4}{1 \text{ kg CCl}_4} = 9.79 \; m \; \text{C}_6\text{H}_6$

(b) The density of H_2O = 0.997 g/mL = 997 kg/L.

$$\frac{5.85 \text{ mol NaCl}}{0.250 \text{ L } H_2O} \times \frac{1 \text{ mol NaCl}}{58.44 \text{ g NaCl}} \times \frac{1 \text{ L } H_2O}{0.997 \text{ kg } H_2O} = 0.402 \text{ } M \text{ NaCl}$$

13.19 Given: 100.0 mL of $CH_3CN(l)$, 0.786 g/mL; 20.0 mL CH_3OH, 0.791 g/mL

(a) $\text{mol } CH_3CN = \dfrac{0.786 \text{ g}}{1 \text{ mL}} \times 100.0 \text{ mL} \times \dfrac{1 \text{ mol } CH_3CN}{41.05 \text{ g } CH_3CN} = 1.9147 = 1.91 \text{ mol}$

$\text{mol } CH_3OH = \dfrac{0.791 \text{ g}}{1 \text{ mL}} \times 20.0 \text{ mL} \times \dfrac{1 \text{ mol } CH_3OH}{32.04 \text{ g } CH_3OH} = 0.4938 = 0.494 \text{ mol}$

$\chi_{CH_3OH} = \dfrac{0.4938 \text{ mol } CH_3OH}{1.9147 \text{ mol } CH_3CN + 0.4938 \text{ mol } CH_3OH} = 0.205$

(b) Assuming CH_3OH is the solute and CH_3CN is the solvent,

$100.0 \text{ mL } CH_3CN \times \dfrac{0.786 \text{ g}}{1 \text{ mL}} \times \dfrac{1 \text{ kg}}{1000 \text{ g}} = 0.0786 \text{ kg } CH_3CN$

$m_{CH_3OH} = \dfrac{0.4938 \text{ mol } CH_3OH}{0.0786 \text{ kg } CH_3CN} = 6.28 \text{ } m \text{ } CH_3OH$

(c) The total volume of the solution is 120.0 mL, assuming volumes are additive.

$M = \dfrac{0.4938 \text{ mol } CH_3OH}{0.1200 \text{ L solution}} = 4.12 \text{ } M \text{ } CH_3OH$

13.21 (a) $\text{mass \%} = \dfrac{\text{mass } C_6H_8O_6}{\text{total mass solution}} \times 100;$

$\dfrac{80.5 \text{ g } C_6H_8O_6}{80.5 \text{ g } C_6H_8O_6 + 210 \text{ g } H_2O} \times 100 = 27.71 = 27.7\%$

(b) $\text{mol } C_6H_8O_6 = \dfrac{80.5 \text{ g } C_6H_8O_6}{176.1 \text{ g/mol}} = 0.4571 = 0.457 \text{ mol } C_6H_8O_6$

$\text{mol } H_2O = \dfrac{210 \text{ g } H_2O}{18.02 \text{ g/mol}} = 11.654 = 11.7 \text{ mol } H_2O$

$\chi_{C_6H_8O_6} = \dfrac{0.4571 \text{ mol } C_6H_8O_6}{0.4571 \text{ mol } C_6H_8O_6 + 11.654 \text{ mol } H_2O} = 0.0377$

(c) $m = \dfrac{0.4571 \text{ mol } C_6H_8O_6}{0.210 \text{ kg } H_2O} = 2.18 \text{ } m \text{ } C_6H_8O_6$

(d) $M = \dfrac{\text{mol } C_6H_8O_6}{\text{L solution}};$ $290.5 \text{ g soln} \times \dfrac{1 \text{ mL}}{1.22 \text{ g}} \times \dfrac{1 \text{ L}}{1000 \text{ mL}} = 0.2381 = 0.238 \text{ L}$

$M = \dfrac{0.4571 \text{ mol } C_6H_8O_6}{0.2381 \text{ L soln}} = 1.92 \text{ } M$

13.23   Assume a solution volume of 1.00 L. Calculate the mass of 1.00 L of solution and the mass of HNO_3 in 1.00 L of solution.

$$1.00 \text{ L} \times \frac{1000 \text{ mL}}{1 \text{ L}} \times \frac{1.42 \text{ g soln}}{\text{mL soln}} = 1.42 \times 10^3 \text{ g soln}$$

$$16 \text{ } M = \frac{16 \text{ mol HNO}_3}{1 \text{ L soln}} \times \frac{63.02 \text{ g HNO}_3}{1 \text{ mol HNO}_3} = 1008 = 1.0 \times 10^3 \text{ g HNO}_3$$

$$\text{mass \%} = \frac{1008 \text{ g HNO}_3}{1.42 \times 10^3 \text{ g soln}} \times 100 = 71\%$$

13.25   (a) $\dfrac{0.240 \text{ mol MgBr}_2}{1 \text{ L soln}} \times 0.400 \text{ L} = 9.60 \times 10^{-2} \text{ mol MgBr}_2$

(b) $\dfrac{0.460 \text{ mol glucose}}{1 \text{ L soln}} \times 80.0 \text{ } \mu\text{L} \times \dfrac{1 \times 10^{-6} \text{ L}}{1 \text{ } \mu\text{L}} = 3.68 \times 10^{-5} \text{ mol glucose}$

(c) $\dfrac{0.040 \text{ mol Na}_2\text{CrO}_4}{1 \text{ L soln}} \times 3.00 \text{ L} = 0.12 \text{ mol Na}_2\text{CrO}_4$

13.27   (a) $\dfrac{1.50 \times 10^{-2} \text{ mol KBr}}{1 \text{ L soln}} \times 0.85 \text{ L} \times \dfrac{119.0 \text{ g KBr}}{1 \text{ mol KBr}} = 1.5 \text{ g KBr}$

Weigh out 1.5 g KBr, dissolve in water, dilute with stirring to 0.85 L (850 mL).

(b)   Determine the mass % of KBr:

$$\frac{0.180 \text{ mol KBr}}{1000 \text{ g H}_2\text{O}} \times \frac{119.0 \text{ g KBr}}{1 \text{ mol KBr}} = 21.42 = 21.4 \text{ g KBr/kg H}_2\text{O}$$

Thus, mass fraction $= \dfrac{21.42 \text{ g KBr}}{1000 + 21.42} = 0.02097 = 0.0210$

In 165 g of the 0.180 m solution, there are

$$(165 \text{ g soln}) \times \frac{0.02097 \text{ g KBr}}{1 \text{ g soln}} = 3.460 = 3.46 \text{ g KBr}$$

Weigh out 3.46 g KBr, dissolve it in 165 - 3.46 = 162 g H_2O to make exactly 165 g of 0.180 m solution.

(c)   Calculate the total mass of 1.85 L of solution, and from the mass % of KBr, the mass of KBr required.

$$1.85 \text{ L soln} \times \frac{1000 \text{ mL}}{1 \text{ L}} \times \frac{1.10 \text{ g soln}}{1 \text{ mL}} = 2035 = 2.04 \times 10^3 \text{ g soln}$$

0.120 (2035 g soln) = 244.2 = 244 g KBr

Dissolve 224 g KBr in water, dilute with stirring to 1.85 L.

(d)   Calculate moles KBr needed to precipitate 16.0 g AgBr.

$$16.0 \text{ g AgBr} \times \frac{1 \text{ mol AgBr}}{187.8 \text{ g AgBr}} \times \frac{1 \text{ mol KBr}}{1 \text{ mol AgBr}} = 0.08520 = 0.0852 \text{ mol KBr}$$

$$0.0852 \text{ mol KBr} \times \frac{1 \text{ L soln}}{0.150 \text{ mol KBr}} = 0.568 \text{ L soln}$$

Weigh out 0.0852 mol KBr (10.1 g KBr), dissolve it in a small amount of water and dilute to 0.568 L.

Saturated Solutions: Factors Affecting Solubility

13.29 (a)   Supersaturated

(b)   Add a seed crystal. Supersaturated solutions exist because not enough solute molecules are properly aligned for crystallization to occur. A seed crystal provides a nucleus of already aligned molecules, so that ordering of the dissolved particles is more facile.

13.31 (a)   unsaturated   (b) saturated   (c) saturated   (d) unsaturated

13.33 In order for the dissolving process to be energetically favorable, the energy required to disrupt solute-solute and solvent-solvent intermolecular interactions (ΔH_1 + ΔH_2) must be similar to the energy released when solute-solvent interactions form (ΔH_3). This can only occur if the same types of interactions are being disrupted and formed. That is, if the types of intermolecular interactions in the pure solute and solvent are similar.

13.35 $\dfrac{1.38 \times 10^{-3} M}{1 \text{ atm O}_2 \text{ pressure}} \times 0.21 \text{ atm O}_2 \text{ pressure} = 2.9 \times 10^{-4} M$

Colligative Properties

13.37 Freezing point depression, boiling point elevation, osmotic pressure, vapor pressure lowering

13.39 For a solution with a nonvolatile solute, the vapor pressure above the solution is equal to the mole fraction of the *solvent* times the vapor pressure of the pure solvent. Nonvolatile solute particles at the surface of a solution physically limit the number of chances for a solvent molecule to escape to the vapor phase. This causes the vapor pressure above the solution to be less than the vapor pressure of the pure solvent.

13.41 For this problem, it will be convenient to express Raoult's law in terms of the lowering of the vapor pressure of the solvent, ΔP_A.

$$\Delta P_A = P_A^{\circ} - \chi_A P_A^{\circ} = P_A^{\circ} (1 - \chi_A)$$

$1 - \chi_A = \chi_B$, the mole fraction of the *solute* particles

$\Delta P_A^\circ = \chi_B P_A^\circ$; the vapor pressure of the solvent (A) is lowered according to the mole fraction of solute (B) particles present.

(a)   Calculate χ_B by vapor pressure lowering; $\chi_B = \Delta P_A / P_A^\circ$. Given moles solvent, calculate moles solute from the definition of mole fraction.

$$\chi_{C_2H_6O_2} = \frac{13.2 \text{ torr}}{100 \text{ torr}} = 0.132$$

$$\frac{1.00 \times 10^3 \text{ g } C_2H_5OH}{46.07 \text{ g/mol}} = 21.71 = 21.7 \text{ mol } C_2H_5OH; \text{ let } y = \text{mol } C_2H_6O_2$$

$$\chi_{C_2H_6O_2} = \frac{y \text{ mol } C_2H_6O_2}{y \text{ mol } C_2H_6O_2 + 21.71 \text{ mol } C_2H_5OH} = 0.132 = \frac{y}{y + 21.71}$$

$0.132\,y + 2.866 = y;\ 0.868\,y = 2.866;\ y = 3.302 = 3.30 \text{ mol } C_2H_6O_2$

$3.302 \text{ mol } C_2H_6O_2 \times \dfrac{62.07 \text{ g}}{1 \text{ mol}} = 205 \text{ g } C_2H_6O_2$

(b) $\Delta P_A = \chi_B P_A^\circ$; vapor pressure of the solvent is lowered according to the mole fraction of solute particles present.

$$P_{H_2O} \text{ at } 40^\circ C = 55.3 \text{ torr}; \quad \frac{500 \text{ g } H_2O}{18.02 \text{ g/mol}} = 27.747 = 27.7 \text{ mol } H_2O$$

$$\chi_{ions} = \frac{4.60 \text{ torr}}{55.3 \text{ torr}} = \frac{y \text{ mol ions}}{y \text{ mol ions} + 27.747 \text{ mol } H_2O} = 0.08318 = 0.0832$$

$$0.08318 = \frac{y}{y + 27.747}; 0.08318\,y + 2.308 = y;\ 0.9168\,y = 2.308,$$

$$y = 2.517 = 2.52 \text{ mol of ions}$$

This result has 3 sig figs because $(27.7 \times 0.0832 = 2.31)$ has 3 sig figs.

$2.517 \text{ mol ions} \times \dfrac{1 \text{ mol KBr}}{2 \text{ mol ions}} \times \dfrac{119.0 \text{ g KBr}}{\text{mol KBr}} = 149.8 = 150 \text{ g KBr}$

13.43   At $63.5^\circ C$, $P_{H_2O}^\circ = 175 \text{ torr}$, $P_{Eth}^\circ = 400 \text{ torr}$.

Let G = the mass of H_2O and/or C_2H_5OH.

(a) $$\chi_{Eth} = \frac{\dfrac{G}{46.07 \text{ g } C_2H_5OH}}{\dfrac{G}{46.07 \text{ g } C_2H_5OH} + \dfrac{G}{18.02 \text{ g } H_2O}}$$

Multiplying top and bottom of the right side of the equation by $1/G$ gives:

$$\chi_{Eth} = \frac{1/46.07}{1/46.07 + 1/18.02} = \frac{0.02171}{0.02171 + 0.05549} = 0.2812$$

(b) $P_T = P_{Eth} + P_{H_2O}$; $P_{Eth} = \chi_{Eth} P^{\circ}_{Eth}$; $P_{H_2O} = \chi_{H_2O} P^{\circ}_{H_2O}$

       $\chi_{Eth} = 0.2812$, $P_{Eth} = 0.2812\,(400\text{ torr}) = 112.48 = 112$ torr

       $\chi_{H_2O} = 1 - 0.2812 = 0.7188$; $P_{H_2O} = 0.7188(175\text{ torr}) = 125.8 = 126$ torr

       $P_T = 112.5$ torr $+ 125.8$ torr $= 238.3 = 238$ torr

(c) χ_{Eth} in vapor $= \dfrac{P_{Eth}}{P_{total}} = \dfrac{112.5\text{ torr}}{238.3\text{ torr}} = 0.4721 = 0.472$

13.45 $\Delta T = K\,(m)$; first, calculate the **molality** of each solution

(a) $0.17\ m$ (b) 16.8 mol $CHCl_3 \times \dfrac{119.4\text{ g }CHCl_3}{\text{mol }CHCl_3} = 2.006 = 2.01$ kg;

                                $\dfrac{1.92\text{ mol }C_{10}H_8}{2.006\text{ kg }CHCl_3} = 0.9571 = 0.957\ m$

(c) 5.44 g KBr $\times \dfrac{1\text{ mol KBr}}{119.0\text{ g KBr}} \times \dfrac{2\text{ mol particles}}{1\text{ mol KBr}} = 0.09143 = 0.0914$ mol particles

       6.35 g $C_6H_{12}O_6 \times \dfrac{1\text{ mol }C_6H_{12}O_6}{180.2\text{ g }C_6H_{12}O_6} = 0.03524 = 0.0352$ mol particles

       $m = \dfrac{(0.09143 + 0.03524)\text{ mol particles}}{0.200\text{ kg }H_2O} = 0.63335 = 0.633\ m$

       Then, f.p. $= T_f - K_f(m)$; b.p. $= T_b + K_b(m)$; T in °C

	m	T_f	$-K_f(m)$		f.p.	T_b	$+K_b(m)$		b.p.
(a)	0.17	-114.6	-1.99(0.17)	= -0.34	-114.9	78.4	1.22(0.17)	= 0.21	78.6
(b)	0.957	-63.5	-4.68(0.957)	= -4.48	-68.0	61.2	3.63(0.957)	= 3.47	64.7
(c)	0.633	0.0	-1.86(0.633)	= -1.20	-1.2	100.0	0.52(0.633)	= 0.33	100.3

13.47 $0.030\ m$ phenol $< 0.040\ m$ glycerin $= 0.020\ m$ KBr. Phenol is very slightly ionized in water, but not enough to match the number of particles in a $0.040\ m$ glycerin solution. The KBr solution is $0.040\ m$ in particles, so it has the same boiling point as 0.040 glycerin, which is a nonelectrolyte.

13.49 $\pi = MRT$; T $= 20°C + 273 = 293$ K

      $M = \dfrac{\text{mol urea}}{\text{L soln}} = \dfrac{2.02\text{ g }(NH_2)_2CO}{0.145\text{ L soln}} \times \dfrac{1\text{ mol urea}}{60.06\text{ g }(NH_2)_2CO} = 0.2320 = 0.232\ M$

      $\pi = \dfrac{0.232\text{ mol}}{L} \times \dfrac{0.08206\text{ L} \cdot \text{atm}}{K \cdot \text{mol}} \times 293\text{ K} = 5.557 = 5.58$ atm

13.51 $\Delta T_b = K_b\,m\ ;\quad m = \dfrac{\Delta T_b}{K_b} = \dfrac{+0.49}{5.02} = 0.0976 = 0.098\ m$ adrenaline

$$m = \frac{\text{mol adrenaline}}{\text{kg CCl}_4} = \frac{\text{g adrenaline}}{\text{MM adrenaline} \times \text{kg CCl}_4}$$

$$\text{MM adrenaline} = \frac{\text{g adrenaline}}{m \times \text{kg CCl}_4} = \frac{0.64\ \text{g adrenaline}}{0.0976\ m \times 0.0360\ \text{kg CCl}_4} = 1.8 \times 10^2\ \text{g/mol adrenaline}$$

13.53 $\pi = MRT;\quad M = \dfrac{\pi}{RT}\ ;\quad T = 25\,^\circ C + 273 = 298\ K$

$$M = 0.953\ \text{torr} \times \frac{1\ \text{atm}}{760\ \text{torr}} \times \frac{K \cdot \text{mol}}{0.08206\ L \cdot \text{atm}} \times \frac{1}{298\ K} = 5.128 \times 10^{-5} = 5.13\ M$$

$$\text{mol} = M \times L = 5.128 \times 10^{-5} \times 0.210\ L = 1.077 \times 10^{-5} = 1.08 \times 10^{-5}\ \text{mol lysozyme}$$

$$\text{MM} = \frac{g}{\text{mol}} = \frac{0.150\ g}{1.077 \times 10^{-5}\ \text{mol}} = 1.39 \times 10^4\ \text{g/mol lysozyme}$$

13.55   Attractive forces between ions in electrolyte solutions lead to ion-pairing, which reduces the effective number of particles in solution. Fewer effective particles in solution leads to smaller ΔT_f and ΔT_b values, higher freezing points and lower boiling points than predicted for ideal solutions.

Colloids

13.57   The outline of a light beam passing through a colloid is visible, whereas light passing through a true solution is invisible unless collected on a screen. This is the Tyndall effect. To determine whether Faraday's (or anyone's) apparently homogeneous dispersion is a true solution or a colloid, shine a beam of light on it and see if the light is scattered.

13.59   (a)   hydrophobic   (b) hydrophilic   (c) hydrophobic

13.61   Colloid particles are stabilized by attractive intermolecular forces with the dispersing medium (solvent) and do not coalesce because of electrostatic repulsions between groups at the surface of the dispersed particles. Colloids can be coagulated by heating (more collisions, greater chance that particles will coalesce); hydrophilic colloids can be coagulated by adding electrolytes, which neutralize surface charges allowing the colloid particles to collide more freely.

Additional Exercises

13.64   (a) $\dfrac{1.80\ \text{mol CH}_3\text{CN}}{1\ \text{L soln}} \times \dfrac{86.85\ \text{g LiBr}}{1\ \text{mol LiBr}} = 156.3 = 156\ \text{g LiBr}$

1 L soln = 826 g soln; g CH_3CN = 826 - 156.3 = 669.7 = 670 g CH_3CN

m LiBr $= \dfrac{1.80\ \text{mol LiBr}}{0.6697\ \text{kg CH}_3\text{CN}} = 2.69\ m$

(b) $\dfrac{669.7 \text{ g CH}_3\text{CN}}{41.05 \text{ g/mol}} = 16.31 = 16.3 \text{ mol CH}_3\text{CN}$; $\chi_{\text{LiBr}} = \dfrac{1.80}{1.80 + 16.31} = 0.0994$

(c) mass % $= \dfrac{669.7 \text{ g CH}_3\text{CN}}{826 \text{ g soln}} \times 100 = 81.1\% \text{ CH}_3\text{CN}$

13.68 $P_{\text{Rn}} = \chi_{\text{Rn}} P_{\text{total}}$; $P_{\text{Rn}} = 3.5 \times 10^{-6} (36 \text{ atm}) = 1.26 \times 10^{-4} = 1.3 \times 10^{-4} \text{ atm}$

$C_{\text{Rn}} = k\, P_{\text{Rn}}$; $C_{\text{Rn}} = \dfrac{7.27 \times 10^{-3}\, M}{1 \text{ atm}} \times 1.26 \times 10^{-4} \text{ atm} = 9.2 \times 10^{-7}\, M$

13.73   Assume the radiator solution is prepared by mixing 1.00 L of each of the liquids. To calculate the freezing and boiling points, we need the molality of this solution. Assume H_2O is the solvent.

$$m = \dfrac{\text{mol C}_2\text{H}_6\text{O}_2}{\text{kg H}_2\text{O}}; \quad \text{kg H}_2\text{O} = 1.00 \text{ L} \times \dfrac{1000 \text{ mL}}{1 \text{ L}} \times \dfrac{1.00 \text{ g}}{\text{mL}} = 1.00 \text{ kg}$$

$$\text{mol C}_2\text{H}_6\text{O}_2 = 1.00 \text{ L} \times \dfrac{1000 \text{ mL}}{1 \text{ L}} \times \dfrac{1.12 \text{ g}}{1 \text{ mL}} \times \dfrac{1 \text{ mol C}_2\text{H}_6\text{O}_2}{62.07 \text{ g C}_2\text{H}_6\text{O}_2} = 18.04 = 18.0 \text{ mol C}_2\text{H}_6\text{O}_2$$

$$m = \dfrac{18.04 \text{ mol C}_2\text{H}_6\text{O}_2}{1.00 \text{ kg H}_2\text{O}} = 18.04\, m$$

$\Delta T_f = K_f m = -1.86(18.04) = -33.6°C$; f.p. $= 0.0 - 33.6 = -33.6°C$

$\Delta T_b = K_b m = 0.52(18.04) = 9.4°C$; b.p. $= 100.0 + 9.4 = +109.4°C$

13.76   The solvent vapor pressure over each solution is determined by the total particle concentrations present in the solutions. When the particle concentrations are equal, the vapor pressures will be equal and equilibrium established. The particle concentration of the nonelectrolyte is just 0.065 M; the ion concentration of the NaCl is $2 \times 0.035\ M = 0.070\ M$. Solvent will diffuse from the less concentrated nonelectrolyte solution. Let x = volume of solvent transferred.

$$\dfrac{0.065\ M \times 20.0 \text{ mL}}{(20.0 - x) \text{ mL}} = \dfrac{0.070\ M \times 20.0 \text{ mL}}{(20.0 + x) \text{ mL}}; \quad 1.3(20.0 + x) = 1.4(20.0 - x)$$

2.7x = 2, x = 0.74 = 0.7 mL transferred

The nonelectrolyte beaker contains 20.0 - 0.74 = 19.3 mL solution; the NaCl beaker contains 20.0 + 0.74 = 20.7 mL solution.

Integrative Exercises

13.80 (a) $\Delta T_f = K_f m = K_f \times \dfrac{\text{mol } C_7H_6O_2}{\text{kg } C_6H_6} = K_f \times \dfrac{\text{g } C_7H_6O_2}{\text{kg } C_6H_6 \times \text{MM } C_7H_6O_2}$

$MM = \dfrac{K_f \times \text{g } C_7H_6O_2}{\Delta T_f \times \text{kg } C_6H_6} = \dfrac{5.12 \times 0.55}{0.360 \times 0.032} = 244.4 = 2.4 \times 10^2$ g/mol

(b)   The formula weight of $C_7H_6O_2$ is 122 g/mol. The experimental molar mass is twice this value, indicating that benzoic acid is associated into dimers in benzene solution. This is reasonable, since the carboxyl group

is capable of strong hydrogen bonding with itself. Many carboxylic acids exist as dimers in solution. The structure of benzoic acid dimer in benzene solution is:

13.84 $\chi_{CHCl_3} = \chi_{C_3H_6O} = 0.500$

(a)   For an ideal solution, Raoult's Law is obeyed.

$P_T = P_{CHCl_3} + P_{C_3H_6O}$; $P_{CHCl_3} = 0.5(300 \text{ torr}) = 150$ torr

$P_{C_3H_6O} = 0.5(360 \text{ torr}) = 180$ torr; $P_T = 150$ torr + 180 torr = 330 torr

(b)   The real solution has a lower vapor pressure, 250 torr, than an ideal solution of the same composition, 330 torr. Thus, fewer molecules escape to the vapor phase from the liquid. This means that fewer molecules have sufficient kinetic energy to overcome intermolecular attractions. Clearly, even weak hydrogen bonds such as this one are stronger attractive forces than dipole-dipole or dispersion forces. These hydrogen bonds prevent molecules from escaping to the vapor phase and result in a lower than ideal vapor pressure for the solution. There is essentially no hydrogen bonding in the individual liquids.

(c)   According to Coulomb's law, electrostatic attractive forces lead to an overall lowering of the energy of the system. Thus, when the two liquids mix and hydrogen bonds are formed, the energy of the system is decreased and $\Delta H_{soln} < 0$; the solution process is exothermic.

14 Chemical Kinetics

Reaction Rates

14.1   (a)   *Reaction rate* is the change in the amount of products or reactants in a given amount of time; it is the speed of a chemical reaction.

(b)   Rates depend on concentration of reactants, surface area of reactants, temperature and presence of catalyst.

(c)   The stoichiometry of the reaction (mole ratios of reactants and products) must be known to relate rate of disappearance of products to rate of appearance of reactants.

14.3

Time(s)	Mol A	(a) Mol B	Δ Mol A	(b) Rate (Δ mol A/s)
0	0.100	0		
30	0.0595	0.0405	-0.0405	13.5×10^{-4}
60	0.0354	0.0646	-0.0241	8.03×10^{-4}
90	0.0210	0.0790	-0.0144	4.80×10^{-4}
120	0.0125	0.0875	-0.0085	2.8×10^{-4}

(c)   The volume of the container must be known to report the rate in units of concentration (mol/L) per time.

14.5

Time (sec)	Time Interval (sec)	Concentration (M)	ΔM	Rate (M/s)
0		0.0165		
2,000	2,000	0.0110	-0.0055	28×10^{-7}
5,000	3,000	0.00591	-0.0051	17×10^{-7}
8,000	3,000	0.00314	-0.00277	9.23×10^{-7}
12,000	4,000	0.00137	-0.00177	4.43×10^{-7}
15,000	3,000	0.00074	-0.00063	2.1×10^{-7}

14.7   From the slopes of the lines in the figure at right, the rates are 1.5×10^{-6} M/s at 4000 s, 4.3×10^{-7} M/s at 10,000 s.

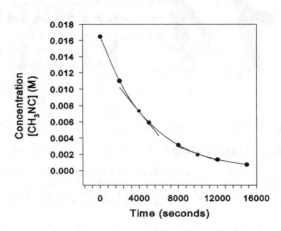

14.9   (a) $-\Delta[H_2O_2]/\Delta t = \Delta[H_2]/\Delta t = \Delta[O_2]/\Delta t$

        (b) $-\Delta[N_2O]/2\Delta t = \Delta[N_2]/2\Delta t = \Delta[O_2]/\Delta t$
               $-\Delta[N_2O]/\Delta t = \Delta[N_2]/\Delta t = 2\Delta[O_2]/\Delta t$

        (c) $-\Delta[N_2]/\Delta t = \Delta[NH_3]/2\Delta t$
               $-\Delta[H_2]/3\Delta t = \Delta[NH_3]/2\Delta t; \ -\Delta[H_2]/\Delta t = 3\Delta[NH_3]/2\Delta t$

14.11   (a) $$\frac{\Delta[H_2O]}{2\Delta t} = \frac{-\Delta[H_2]}{2\Delta t} = \frac{-\Delta[O_2]}{\Delta t}$$

               H_2 is burning, $\dfrac{-\Delta[H_2]}{\Delta t} = 4.6$ mol/s

               O_2 is consumed, $\dfrac{-\Delta[O_2]}{\Delta t} = \dfrac{-\Delta[H_2]}{2\Delta t} = \dfrac{4.6 \text{ mol/s}}{2} = 2.3$ mol/s

               H_2O is produced, $\dfrac{+\Delta[H_2O]}{\Delta t} = \dfrac{-\Delta[H_2]}{\Delta t} = 4.6$ mol/s

        (b)   The change in total pressure is the sum of the changes of each partial pressure. NO and Cl_2 are disappearing and NOCl is appearing.

               $-\Delta P_{NO}/\Delta t = -30$ torr/min

               $-\Delta P_{Cl_2}/\Delta t = \Delta P_{NO}/2\Delta t = -15$ torr/min

               $+\Delta P_{NOCl}/\Delta t = -\Delta P_{NO}/\Delta t = +30$ torr/min

               $\Delta P_T/\Delta t = -30$ torr/min - 15 torr/min + 30 torr/min = -15 torr/min

Rate Laws

14.13   (a)   A *rate law* is an algebraic expression that shows how the rate of a reaction varies with concentrations of reactants. The *rate constant*, k, is a number, the proportionality constant in the rate law; it depends on temperature and the system.

(b)   The *reaction order* for a specific reactant is the exponent of that reactant in the rate law. It indicates how a change in that reactant affects the reaction rate. The *overall reaction order* is the sum of the exponents in the rate law.

(c)   Units of $k = \dfrac{\text{units of rate}}{(\text{units of concentration})^2} = \dfrac{M/s}{M^2} = M^{-1}s^{-1}$

14.15   (a)   rate $= k[N_2O_5] = 6.08 \times 10^{-4}\,s^{-1}\,[N_2O_5]$

           (b)   rate $= 6.08 \times 10^{-4}\,s^{-1}\,(0.100\,M) = 6.08 \times 10^{-5}\,M/s$

           (c)   rate $= 6.08 \times 10^{-4}\,s^{-1}\,(0.200\,M) = 12.16 \times 10^{-5} = 1.22 \times 10^{-4}\,M/s$

When the concentration of N_2O_5 doubles, the rate of the reaction doubles.

14.17   (a,b)   rate $= k[CH_3Br][OH^-]$; $k = \dfrac{\text{rate}}{[CH_3Br]\,[OH^-]}$

                 at 298 K, $k = \dfrac{0.28\,M/s}{(0.010\,M)(0.10\,M)} = 2.8 \times 10^2\,M^{-1}s^{-1}$

           (c)   Since the rate law is first order in $[OH^-]$, if $[OH^-]$ is tripled, the rate triples.

14.19   (a)   rate $= k[A][C]^2$

           (b)   new rate $= (1/2)(1/2)^2 = 1/8$ of old rate

14.21   (a)   Doubling [NO] while holding $[O_2]$ constant increases the rate by a factor of 4 (experiments 1 and 3). Reducing $[O_2]$ by a factor of 2 while holding [NO] constant reduces the rate by a factor of 2 (experiments 2 and 3). The rate is second order in [NO] and first order in $[O_2]$. rate $= k[NO]^2[O_2]$

           (b,c)   From experiment 1: $k = \dfrac{1.41 \times 10^{-2}\,M/s}{(0.0126\,M)^2(0.0125\,M)} = 7.11 \times 10^3\,M^{-2}s^{-1}$

(Any of the three sets of initial concentrations and rates could be used to calculate the rate constant k.)

14.23   (a)   Increasing [NO] by a factor of 2.5 while holding $[Br_2]$ constant (experiments 1 and 2) increases the rate by a factor 6.25 or $(2.5)^2$. Increasing $[Br_2]$ by a factor of 2.5 while holding [NO] constant increases the rate by a factor of 2.5. The rate law for the appearance of NOBr is: rate $= \Delta[NOBr]/\Delta t = k[NO]^2[Br_2]$.

           (b)   From experiment 1: $k = \dfrac{24\,M/s}{(0.10\,M)^2(0.20\,M)} = 1.2 \times 10^4\,M^{-2}s^{-1}$

           (c) $\Delta[NOBr]/2\Delta t = -\Delta[Br_2]/\Delta t$; the rate of disappearance of Br_2 is half the rate of appearance of NOBr.

           (d) $\dfrac{-\Delta[Br_2]}{\Delta t} = \dfrac{k[NO]^2[Br_2]}{2} = \dfrac{1.2 \times 10^4}{2\,M^2 s} \times (0.075\,M)^2 \times (0.185\,M) = 6.2\,M/s$

Change of Concentration with Time

14.25 (a)   A *first-order reaction* depends on the concentration, raised to the first power, of only one reactant; rate = $k[A]^1$.

(b)   A graph of ln[A] vs time yields a straight line for a first-order reaction.

(c)   The half-life of a first-order reaction **does not** depend on initial concentration; it is determined by the value of the rate constant, k.

14.27   For a first order reaction, $t_{1/2} = 0.693/k$, $k = 5.1 \times 10^{-4} s^{-1}$

$$t_{1/2} = \frac{0.693}{5.1 \times 10^{-4} s^{-1}} = 1.4 \times 10^3 \text{ s or } 23 \text{ min}$$

14.29 (a)   Rearranging Equation [14.13] for a first order reaction:

$\ln[A]_t = -kt + \ln[A]_0$

1.5 min = 90 s; $[N_2O_5]_0 = (0.300 \text{ mol}/0.500 \text{ L}) = 0.600 \ M$

$\ln[N_2O_5]_{90} = -(6.82 \times 10^{-3} s^{-1})(90 \text{ s}) + \ln(0.600)$

$\ln[N_2O_5]_{90} = -0.6138 + (-0.5108) = -1.1246 = -1.125$

$[N_2O_5]_{90} = 0.3248 = 0.325 \ M$; mol $N_2O_5 = 0.3248 \ M \times 0.500 \text{ L} = 0.162$ mol

(b) $[N_2O_5]_t = 0.030 \text{ mol}/0.500 \text{ L} = 0.060 \ M$; $[N_2O_5]_0 = 0.600 \ M$

$\ln(0.060) = -(6.82 \times 10^{-3} s^{-1}) (t) + \ln(0.600)$

$$t = \frac{-[\ln(0.060) - \ln(0.600)]}{(6.82 \times 10^{-3} s^{-1})} = 337.6 = 338 \text{ s} \times \frac{1 \text{ min}}{60 \text{ s}} = 5.63 \text{ min}$$

(c) $t_{1/2} = 0.693/k = 0.693/6.82 \times 10^{-3} s^{-1} = 101.6 = 102$ s or 1.69 min

14.31

t(s)	$P_{SO_2Cl_2}$	$\ln P_{SO_2Cl_2}$
0	1.000	0
2500	0.947	-0.0545
5000	0.895	-0.111
7500	0.848	-0.165
10000	0.803	-0.219

Graph ln $P_{SO_2Cl_2}$ vs. time. (Pressure is a satisfactory concentration unit for a gas, since the concentration in moles/liter is proportional to P.) The graph is linear with slope $-2.19 \times 10^{-5} s^{-1}$ as shown on the figure.   The rate constant k = -slope = $2.19 \times 10^{-5} s^{-1}$.

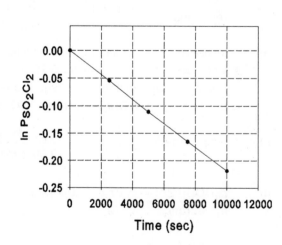

14.33 (a)   Make both first- and second-order plots to see which is linear. Moles is a satisfactory concentration unit, since volume is constant.

time(s)	mol A	ln (mol A)	1/mol A
0	0.1000	-2.303	10.00
30	0.0595	-2.822	16.8
60	0.0354	-3.341	28.2
90	0.0210	-3.863	47.6
120	0.0125	-4.382	80.0

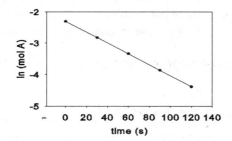   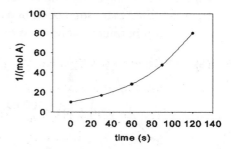

The plot of ln (mol A) vs time is linear, so the reaction is first-order in A.

(b) $k = -\text{slope} = -[-4.382 - (-2.822)]/90\ s = 0.01733 = 0.017\ s^{-1}$

(The best fit to this line yields the same value for the slope.

(c) $t_{1/2} = 0.693/k = 0.693/0.0173\ s^{-1} = 40\ s$

14.35 (a)   Make both first and second order plots to see which is linear.

time(s)	[NO$_2$](M)	ln[NO$_2$]	1/[NO$_2$]
0.0	0.100	-2.303	10.0
5.0	0.017	-4.08	59
10.0	0.0090	-4.71	110
15.0	0.0062	-5.08	160
20.0	0.0047	-5.36	210

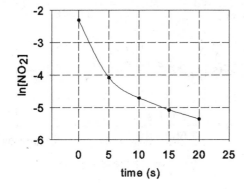   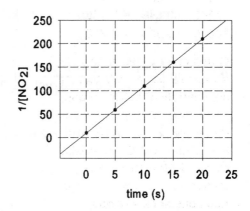

The plot of 1/[NO$_2$] vs time is linear, so the reaction is second order in NO$_2$.

(b)   The slope of the line is (210 - 59) M^{-1} / 15.0 s = 10.07 = 10 $M^{-1}s^{-1}$ = k.
(The slope of the best-fit line is 10.02 = 10 $M^{-1}s^{-1}$.)

Temperature and Rate

14.37   (a)   The central idea of the *collision model* is that molecules must collide to react.

(b)   The energy of the collision and the orientation of the molecules when they collide determine whether a reaction will occur.

(c)   According to the Kinetic Molecular Theory (Chapter 10), the higher the temperature, the greater the speed and kinetic energy of the molecules. Therefore, at a higher temperature, there are more total collisions and each collision is more energetic.

14.39   (a)   (b)   123 kJ/mol

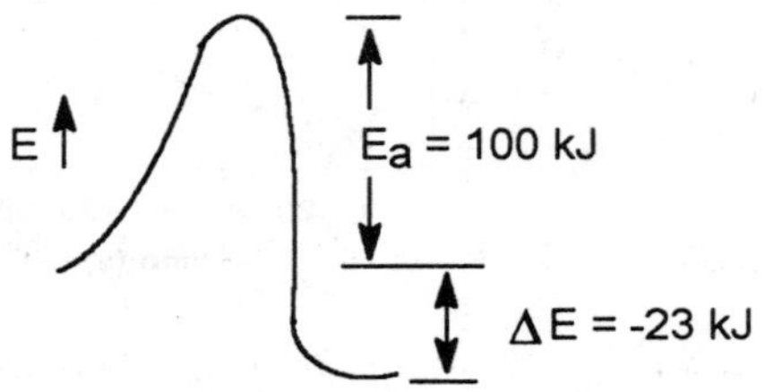

E_a = 100 kJ

ΔE = -23 kJ

14.41   Assuming all collision factors (A) to be the same, reaction rate depends only on E_a; it is independent of ΔE. Based on the magnitude of E_a, reaction (c) is fastest and reaction (b) is slowest.

14.43   No. The value of A, which is related to frequency and effectiveness of collisions, can be different for each reaction and k is proportional to A.

14.45 T_1 = 20°C + 273 = 293 K; T_2 = 60°C + 273 = 333 K; k_1 = 1.75 × $10^{-1}s^{-1}$

(a) $\ln\left(\dfrac{k_1}{k_2}\right) = \dfrac{E_a}{R}\left(\dfrac{1}{333} - \dfrac{1}{293}\right) = \dfrac{55.5 \times 10^3 \text{ J/mol}}{8.314 \text{ J/mol}}(-4.100 \times 10^{-4})$

$\ln(k_1/k_2)$ = -2.7367 = -2.74; k_1/k_2 = 0.0648 = 0.065; $k_2 = \dfrac{0.175 \text{ s}^{-1}}{0.0648} = 2.7 \text{ s}^{-1}$

(b) $\ln\left(\dfrac{k_1}{k_2}\right) = \dfrac{121 \times 10^3 \text{ J/mol}}{8.314 \text{ J/mol}}\left(\dfrac{1}{333} - \dfrac{1}{293}\right) = -5.9666 = -5.97$

k_1/k_2 = 2.563 × 10^{-3} = 2.6 × 10^{-3} ; $k_2 = \dfrac{0.175 \text{ s}^{-1}}{2.563 \times 10^{-3}} = 68 \text{ s}^{-1}$

14.47

k	ln k	T(K)	1/T(× 10^3)
0.0521	-2.995	288	3.47
0.101	-2.293	298	3.36
0.184	-1.693	308	3.25
0.332	-1.103	318	3.14

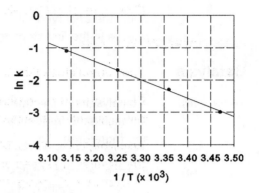

The slope, -5.7×10^3, equals $-E_a/R$. Thus,
$E_a = 5.7 \times 10^3 \times 8.314$ J/mol = 47 kJ/mol.

14.49 $T_1 = 50°C + 273 = 323$ K; $T_2 = 0°C + 273 = 273$ K

$$\ln\left(\frac{k_1}{k_2}\right) = \frac{E_a}{R}\left[\frac{1}{T_2} - \frac{1}{T_1}\right] = \frac{76.7 \text{ kJ/mol}}{8.314 \text{ J/mol}} \times \frac{1000 \text{ J}}{1 \text{ kJ}}\left[\frac{1}{273} - \frac{1}{323}\right]$$

$\ln(k_1/k_2) = 9.225 \times 10^3 (5.670 \times 10^{-4}) = 5.231 = 5.23$; $k_1/k_2 = 187 = 1.9 \times 10^2$

The reaction will occur 190 times faster at 50°C, assuming equal initial concentrations.

Reaction Mechanisms

14.51 (a)   An *elementary step* is a process that occurs in a single event; the order is given by the coefficients in the balanced equation for the step.

(b)   A *unimolecular* elementary step involves only one reactant molecule; the activated complex is derived from a single molecule. A *bimolecular* elementary step involves two reactant molecules in the activated complex and the overall process.

(c)   A *reaction mechanism* is a series of elementary steps that describe how an overall reaction occurs and explain the experimentally determined rate law.

14.53 (a)   bimolecular, rate = $k[N_2O][Cl]$
(b)   unimolecular, rate = $k[Cl_2]$
(c)   bimolecular, rate = $k[NO][Cl_2]$

14.55 (a)
$$NO(g) + NO(g) \rightarrow N_2O_2(g)$$
$$N_2O_2(g) + H_2(g) \rightarrow N_2O(g) + H_2O(g)$$

$$2NO(g) + N_2O_2(g) + H_2(g) \rightarrow N_2O_2(g) + N_2O(g) + H_2O(g)$$
$$2NO(g) + H_2(g) \rightarrow N_2O(g) + H_2O(g)$$

(b)   First step: $-\Delta[NO]/\Delta t = k[NO][NO] = k[NO]^2$
Second step: $-\Delta[H_2]/\Delta t = k[H_2][N_2O_2]$

(c) N_2O_2 is the intermediate; it is produced in the first step and consumed in the second.

(d)   Since $[H_2]$ appears in the rate law, the second step must be slow relative to the first.

14.57   (a)   rate = $k[NO][Cl_2]$

       (b)   Since the observed rate law is second-order in [NO], the second step must be slow relative to the first step; the second step is rate determining.

Catalysis

14.59   (a)   A catalyst increases the rate of reaction by decreasing the activation energy, E_a, or increasing the frequency factor A. Lowering the activation energy is more common and more dramatic.

       (b)   A homogeneous catalyst is in the same phase as the reactants; a hetereogeneous catalyst is in a different phase and is usually a solid.

14.61   (a)

$$2[NO_2(g) + SO_2(g) \rightarrow NO(g) + SO_3(g)]$$
$$2NO(g) + O_2(g) \rightarrow 2NO_2(g)$$
$$\overline{2SO_2(g) + O_2(g) \rightarrow 2SO_3(g)}$$

       (b) $NO_2(g)$ is a catalyst because it is consumed and then reproduced in the reaction sequence. ($NO(g)$ is an intermediate because it is produced and then consumed.)

       (c)   Since NO_2 is in the same state as the other reactants, this is homogeneous catalysis.

14.63   Use of chemically stable supports such as alumina and silica makes it possible to obtain very large surface areas per unit mass of the precious metal catalyst. This is so because the metal can be deposited in a very thin, even monomolecular, layer on the surface of the support.

14.65   As illustrated in Figure 14.23, the two C-H bonds that exist on each carbon of the ethylene molecule before adsorption are retained in the process in which a D atom is added to each C (assuming we use D_2 rather than H_2). To put two deuteriums on a single carbon, it is necessary that one of the already existing C-H bonds in ethylene be broken while the molecule is adsorbed, so the H atom moves off as an adsorbed atom, and is replaced by a D. This requires a larger activation energy than simply adsorbing C_2H_4 and adding one D atom to each carbon.

14.67   (a)   In the laboratory, if a reaction is too slow, we usually heat it to speed it up.

       (b)   Living organisms operate efficiently in a very narrow temperature range; heating to increase reaction rate is not an option. Therefore, the role of enzymes as homogeneous catalysts that speed up desirable reactions without heating and undesirable side-effects is crucial for biological systems.

       (c)   *catalase*: $2H_2O_2 \rightarrow 2H_2O + O_2$; *nitrogenase*: $N_2 \rightarrow 2NH_3$ (nitrogen fixation)

14.69 Let k = the rate constant for the uncatalyzed reaction,

k_c = the rate constant for the catalyzed reaction

According to Equation 14.19, $\ln k = -E_a/RT + \ln A$

Subtracting $\ln k$ from $\ln k_c$,

$$\ln k_c - \ln k = -\left[\frac{55\ \text{kJ/mol}}{RT} + \ln A\right] - \left[-\frac{85\ \text{kJ/mol}}{RT} + \ln A\right]$$

(a) $RT = 8.314\ \text{J/K} \cdot \text{mol} \times 298\ \text{k} \times 1\ \text{kJ/1000 J} = 2.478\ \text{kJ/mol}$; $\ln A$ is the same for both reactions.

$$\ln(k_c/k) = \frac{85\ \text{kJ/mol} - 55\ \text{kJ/mol}}{2.478\ \text{kJ/mol}}; \quad k_c/k = 1.8 \times 10^5$$

The catalyzed reaction is approximately 180,000 times faster at 25°C.

(b) $RT = 8.314\ \text{J/K} \cdot \text{mol} \times 398\ \text{k} \times 1\ \text{kJ/1000 J} = 3.309\ \text{kJ/mol}$

$$\ln(k_c/k) = \frac{30\ \text{kJ/mol}}{3.309\ \text{kJ/mol}}; \quad k_c/k = 8.7 \times 10^3$$

The catalyzed reaction is 8700 times faster at 125°C.

Additional Exercises

14.71 $\text{rate} = \dfrac{-\Delta[H_2S]}{\Delta t} = \dfrac{\Delta[Cl^-]}{2\Delta t} = k[H_2S][Cl_2]$

$$\frac{-\Delta[H_2S]}{\Delta t} = (3.5 \times 10^{-2}\ M^{-1}s^{-1})(1.6 \times 10^{-4}\ M)(0.070\ M) = 3.92 \times 10^{-7} = 3.9 \times 10^{-7}\ M/s$$

$$\frac{\Delta[Cl^-]}{\Delta t} = \frac{2\Delta[H_2S]}{\Delta t} = 2(3.92 \times 10^{-7}\ M/s) = 7.8 \times 10^{-7}\ M/s$$

14.74 (a) Because a plot of $\ln[SO_2Cl_2]$ vs time is linear, the reaction is first order. The rate law is rate = $k[SO_2Cl_2]$. For a first order reaction, Equation [14.13] is appropriate.

$$\ln([A]_t/[A]_o) = -kt; \quad k = -\ln[[A]_t/[A]_o]/t$$

$$k = \frac{-\ln(0.280/0.400)}{240\ s} = \frac{-(-0.3567)}{240\ s} = 1.486 \times 10^{-3} = 1.49 \times 10^{-3}\ s^{-1}$$

(b) For a first order reaction, $t_{1/2} = 0.693/k$ (Equation [14.15]).

$t_{1/2} = 0.693/1.486 \times 10^{-3}\ s^{-1} = 466\ s = 7.77\ \text{min}$

14.77   (a) $T_1 = 37°C = 310$ K; $T_2 = 50°C = 323$ K; $t_{1/2}(50) = 2$ min

               $t_{1/2} = 0.693/k$; $\quad k_2 = 0.693 / t_{1/2} = 0.693 / 2$ min $= 0.347 = 0.3$ min^{-1}

               $\ln(k_1 / k_2) = E_a / R \left(\dfrac{1}{T_2} - \dfrac{1}{T_1} \right)$; $\quad \ln(k_1 / k_2) = \dfrac{400 \times 10^3 \text{ J/mol}}{8.314 \text{ J/mol}} \left(\dfrac{1}{323} - \dfrac{1}{310} \right)$

               $\ln(k_1 / k_2) = -6.2464$; $\quad k_1 / 0.347$ min$^{-1} = 1.937 \times 10^{-3}$;

                                               $k_1 = 6.72 \times 10^{-4} = 7 \times 10^{-4}$ min^{-1}

               $t_{1/2}(37) = 0.693 / 6.72 \times 10^{-4}$ min$^{-1} = 1 \times 10^3$ min ($\sim$ 17 hr)

    (b)   A 13° increase in temperature causes the half-life of coiled DNA to decrease by a factor of 500. Clearly temperature increases (fevers) that are sustained over a period of time can have negative effects on human DNA. To prevent these effects, the human body needs (and has developed) a robust temperature regulation system.

14.80   Ice crystals in the upper atmosphere act as **heterogeneous** catalysts for the destruction of ozone. The system is similar to the one pictured in Figure 14.23, where ice crystals act as the solid support and $O_3(g)$ and $Cl(g)$ are adsorbed onto the surface.

14.84   The fact that the rate doubles with a doubling of the concentration of sugar tells us that the fraction of enzyme tied up in the form of an enzyme-substrate complex is small. A doubling of the substrate concentration leads to a doubling of the concentration of enzyme-substrate complex, because most of the enzyme molecules are available to bind substrates. The behavior of inositol suggests that it acts as a competitor with sucrose for binding at the active sites of the enzyme system. Such a competition results in a lower effective concentration of active sites for binding of sucrose, and thus results in a lower reaction rate.

Integrative Exercises

14.87   Enzymes and proteins are biopolymers, with much of the same structural flexibility as synthetic polymers (Chapter 12). The three dimensional shape of the protein is determined by many relatively weak intermolecular interactions and is sensitive to changes in local environment. Changes in temperature change the kinetic energy of the various groups on the enzyme and their tendency to form intermolecular associations or break free from them. Thus, changing the temperature changes the overall shape of the protein and specifically the shape of the active site. At body temperature, the competition between kinetic energy driving groups apart and intermolecular attraction pulling them together forms an active site that is optimum for a specific substrate. At other temperatures, a different structural equilibrium is reached, the shape of the active site is slightly different and the enzyme is less active.

14.89   (a)   The two figures both have the same basic shape and represent the energies of the reactants, intermediates and products of a chemical reaction.

       (b)   Activation energies are lower than the sum of bond dissociation enthalpies for net bonds broken, because the transition state often involves partially broken old bonds as well as partially formed new ones.

       (c) E_a (max) = $D(C=O) + D(C-C) + D(C-H)$

                          = 799 + 348 +413 =1560 kJ/mol CH_3CHO

       (d)   Assuming that the stretching of the C-C bond and the partial formation of a new C-H bond roughly cancel, the net change for the production of this transition state is

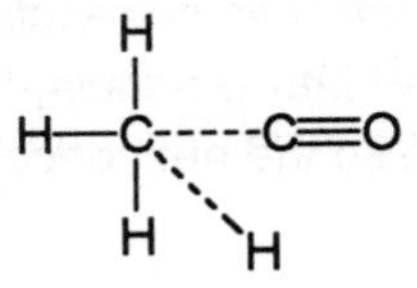

$$
\begin{array}{c}
\overset{\displaystyle H}{|} \\
H-\underset{\displaystyle |}{\underset{\displaystyle H}{C}}\text{-----}C\equiv O \\
\diagdown H
\end{array}
$$

$E_a = D(C=O) - D(C \equiv O) + D(C-H) = 799 - 1072 + 413 = 140$ kJ/mol.

15 Chemical Equilibrium

The Concept of Equilibrium; Equilibrium Expressions

15.1 (a) At equilibrium the forward and reverse reactions proceed at equal rates. Reactants are continually transformed into products, but products are also transformed into reactants at the same rate, so the net concentrations of reactants and products are constant at equilibrium.

(b) At equilibrium, it is the forward and reverse rates, not the rate constants, that are equal. It is true that the ratio of the two rate constants is constant at equilibrium, but this ratio is not constrained to have a value of 1.

(c) At equilibrium, the net concentrations of reactants and products are *constant* (see part (a)), but not necessarily equal. The relative concentrations of reactants and products at equilibrium are determined by their initial concentrations and the value of the equilibrium constant.

15.3 (a) $K = \dfrac{k_f}{k_r}$, Equation [15.3]; $\quad K = \dfrac{9.6 \times 10^2\,s^{-1}}{3.8 \times 10^4\,s^{-1}} = 2.5 \times 10^{-2}$

(b) $rate_f = rate_r;\quad k_f\,[A] = k_r\,[B]$

Since $k_r > k_f$, in order for the two rates to be equal, [A] must be greater than [B].

15.5 (a) The *law of mass action* expresses the relationship between the concentrations of reactants and products at equilibrium for any reaction. The law of mass action is a generic equilibrium expression.

$$K_c = \frac{[H_2O_2]}{[H_2]\,[O_2]}$$

(b) The *equilibrium expression* is an algebraic equation where the variables are the equilibrium concentrations of the reactants and products for a specific chemical reaction. The *equilibrium constant* is a number; it is the ratio calculated from the equilibrium expression for a particular chemical reaction. For any reaction, there are an infinite number of sets of equilibrium concentrations, depending on initial concentrations, but there is only one equilibrium constant.

(c) Introduce a known quantity of $H_2O_2(g)$ into a vessel of known volume at constant (known) temperature. After equilibrium has been established, measure the total pressure in the flask. Using an equilibrium table, such as the one in Sample Exercise 15.7, calculate equilibrium pressures and concentrations of $H_2(g)$, $O_2(g)$ and $H_2O_2(g)$ and calculate K_c.

15.7 (a) $K_c = \dfrac{[NO]^2}{[N_2][O_2]}$ (b) $K_c = \dfrac{[N_2][H_2]^2}{[N_2H_4]}$ (c) $K_c = \dfrac{[C_2H_6]^2[O_2]}{[C_2H_4]^2[H_2O]^2}$

(d) $K_c = \dfrac{[H_2O]}{[H_2]}$ (e) $K_c = \dfrac{1}{[Cl_2]^2}$

homogeneous: (a), (b), (c); heterogeneous: (d), (e)

15.9 (a) $K_c' = (K_c)^{-1} = 1/2.4 \times 10^3 = 4.2 \times 10^{-4}$

(b) Since $K_c > 1$ when N_2 and O_2 are products and $K_c' < 1$ when N_2 and O_2 are reactants, the equilibrium favors N_2 and O_2 at this temperature.

15.11 (a) $K_c = \dfrac{[Na_2O][SO_2]}{[Na_2SO_3]}$

(b) The molar concentration, the ratio of moles of a substance to volume occupied by the substance, is a constant for pure solids and liquids.

(c) constant 1 = $[Na_2O]$; constant 2 = $[Na_2SO_3]$

$K_c = \dfrac{\text{constant 1 }[SO_2]}{\text{constant 2}}$; $K_c' = K_c \dfrac{\text{constant 2}}{\text{constant 1}} = [SO_2]$

Calculating Equilibrium Constants

15.13 $K_c = \dfrac{[H_2][I_2]}{[HI]^2} = \dfrac{(4.79 \times 10^{-4})(4.79 \times 10^{-4})}{(3.53 \times 10^{-3})^2} = 1.84 \times 10^{-2}$

15.15 $2NO(g) + Cl_2(g) \rightleftharpoons 2NOCl$

$K_p = \dfrac{P_{NOCl}^2}{P_{NO}^2 \times P_{Cl_2}} = 52.0$; $P_{NOCl}^2 = 52.0\left(P_{NO}^2 \times P_{Cl_2}\right)$

$P_{NOCl} = \left[52.0\left(P_{NO}^2 \times P_{Cl_2}\right)\right]^{1/2} = [52.0\,((0.095)^2 \times 0.171)]^{1/2}$

$P_{NOCl} = 0.28$ atm

15.17 (a) Since the reaction is carried out in a 1.00 L vessel, the moles of each component are equal to the molarity.

	2NO	+	2H$_2$	$\rightleftharpoons$	N$_2$	+	2H$_2$O
initial	0.100 M		0.050 M		0 M		0.100 M
change	-0.038 M		-0.038 M		+0.019 M		+0.038 M
equil.	0.062 M		0.012 M		0.019 M		0.138 M

First calculate the change in [NO], 0.100 - 0.062 = 0.038 M. From the stoichiometry of the reaction, calculate the change in the other concentrations. Finally, calculate the equilibrium concentrations.

(b) $$K_c = \frac{[N_2][H_2O]^2}{[NO]^2[H_2]^2} = \frac{(0.019)(0.138)^2}{(0.062)^2(0.012)^2} = 6.5 \times 10^2$$

15.19 (a) $K_c' = 1/K_c = 1/20.4 = 0.0490$

(b) $K_c' = K_c^2 = (20.4)^2 = 416.16 = 416$

(c) $K_p = K_c(RT)^{\Delta n}$; $\Delta n = -1$; $T = 700 + 273 = 973$

$K_p = 416.16/(0.08206)(973) = 5.21$

15.21 (a) Since the reaction is carried out in a 1.00 L vessel, the moles of each component are equal to the molarity.

	CO$_2$	+	H$_2$	$\rightleftharpoons$	CO	+	H$_2$O
initial	0.1000 M		0.0500 M		0 M		0.1000 M
change	-0.0046 M		-0.0046 M		+0.0046 M		+0.0046 M
equil	0.0954 M		0.0454 M		0.0046 M		0.1046 M

First calculate the change in [CO$_2$], 0.1000 - 0.0954 = 0.0046 M. From the stoichiometry of the reaction, calculate the change in the other concentrations. Finally, calculate the equilibrium concentrations.

(b) $$K_c = \frac{[CO][H_2O]}{[CO_2][H_2]} = \frac{(0.0046)(0.1046)}{(0.0954)(0.0454)} = 0.1111 = 0.11$$

(c) No. In order to calculate K_p from K_c, the temperature of the reaction must be known.

Applications of Equilibrium Constants

15.23　(a)　A *reaction quotient* is the result of the law of mass action for a general set of concentrations, whereas the equilibrium constant requires equilibrium concentrations.

(b)　In the direction of more products, to the right.

(c)　If Q = K, the system is at equilibrium; the concentrations used to calculate Q must be equilibrium concentrations.

15.25　$K_c = \dfrac{[CO][Cl_2]}{[COCl_2]} = 2.19 \times 10^{-10}$ at 100°C

(a)　$Q = \dfrac{(3.31 \times 10^{-6})(3.31 \times 10^{-6})}{(5.00 \times 10^{-2})} = 2.19 \times 10^{-10}$;　$Q = K_c$

The mixture is at equilibrium.

(b)　$Q = \dfrac{(1.11 \times 10^{-5})(3.25 \times 10^{-6})}{(3.50 \times 10^{-3})} = 1.03 \times 10^{-8}$;　$Q > K_c$

The reaction will proceed to the left to attain equilibrium.

(c)　$Q = \dfrac{(1.56 \times 10^{-6})(1.56 \times 10^{-6})}{(1.45)} = 1.68 \times 10^{-12}$,　$Q < K_c$

The reaction will proceed to the right to attain equilibrium.

15.27　$K_c = \dfrac{[SO_2][Cl_2]}{[SO_2Cl_2]}$;　$[Cl_2] = \dfrac{K_c[SO_2Cl_2]}{[SO_2]} = \dfrac{(0.078)(0.136)}{(0.072)} = 0.15\ M$

15.29　$K_c = 1.04 \times 10^{-3} = \dfrac{[Br]^2}{[Br_2]}$;　$[Br_2] = \dfrac{0.245\ g\ Br_2}{159.8\ g\ Br_2/mol \times 0.200\ L} = 7.666 \times 10^{-3}$

$= 7.67 \times 10^{-3}\ M$

$[Br] = (1.04 \times 10^{-3}\ [Br_2])^{1/2} = (1.04 \times 10^{-3}\ (7.666 \times 10^{-3}))^{1/2} = 2.824 \times 10^{-3} = 2.82 \times 10^{-3}\ M$

g Br = $\dfrac{2.824 \times 10^{-3}\ mol\ Br}{1\ L} \times \dfrac{79.90\ g\ Br}{1\ mol\ Br} \times 0.200\ L = 0.04513 = 0.0451\ g\ Br$

$[Br_2] = 7.67 \times 10^{-3}\ M$;　$[Br] = 2.82 \times 10^{-3}\ M$;　0.0451 g Br

15.31

	2NO(g)	⇌	N₂(g)	+	O₂(g)	$K_c = \dfrac{[N_2][O_2]}{[NO]^2} = 2.4 \times 10^3$
initial	0.500 *M*		0		0	
change	-2x		+x		+x	
equil.	0.500-2x		+x		+x	

$$2.4 \times 10^3 = \frac{x^2}{(0.500-2x)^2}; \quad (2.4 \times 10^3)^{1/2} = \frac{x}{0.500 - 2x}$$

$x = (2.4 \times 10^3)^{1/2} (0.500 - 2x); \quad 97.98x + x = 24.495$

$98.98x = 24.495, \quad x = 0.2474 = 0.25$

$[N_2] = [O_2] = 0.25 \ M; \quad [NO] = 0.500 - 2(0.2474) = 0.0051 = 0.005 \ M$

15.33 $K_p = \dfrac{P_{NO}^2 P_{Br_2}}{P_{NOBr}^2}$

When $P_{NOBr} = P_{NO}$, these terms cancel and $P_{Br_2} = K_p = 0.416$ atm. This is true for all cases where $P_{NOBr} = P_{NO}$.

15.35   **(a)**   Starting with only $PH_3BCl_3(s)$, the equation requires that the equilibrium concentrations of $PH_3(g)$ and $BCl_3(g)$ are equal.

         $K_c = [PH_3][BCl_3]; \quad 1.87 \times 10^{-3} = x^2; \quad x = 0.04324 = 0.0432 \ M \ PH_3$ and BCl_3

       (b)   Since the mole ratios are 1:1:1, mol $PH_3BCl_3(s)$ required = mol PH_3 or BCl_3 produced.

$$\frac{0.04324 \ mol \ PH_3}{1 \ L} \times 0.500 \ L \times \frac{151.2 \ g \ PH_3BCl_3}{1 \ mol \ PH_3BCl_3} = 3.27 \ g \ PH_3BCl_3$$

In fact, some $BH_3BCl_3(s)$ must remain for the system to be in equilibrium, so a bit more than 3.27 g PH_3BCl_3 is needed.

15.37 $K_c = \dfrac{[PCl_3][Cl_2]}{[PCl_5]}; \quad [PCl_5]$ initial $= \dfrac{0.100 \ mol}{5.00 \ L} = 0.0200 \ M$

	PCl_2	$\rightleftharpoons$	PCl_3	+	Cl_2
initial	0.0200 M		0		0
change	-x		+x		+x
equil.	0.0200-x		x		x

$K_c = 1.80 = \dfrac{x^2}{0.0200-x}$ Assume x is small compared to 0.0200.

$1.80 \approx \dfrac{x^2}{0.0200}; \quad x^2 = 0.0360; \quad x = 0.190 \ M$

Clearly, our assumption is not true. Solve the quadratic formula to obtain the value of x.

$x^2 = 1.80(0.0200-x); \quad x^2 + 1.80x - 0.0360 = 0;$ for equations in the form $ax^2 + bx + c = 0,$

$$x = \frac{-b + \sqrt{b^2 - 4ac}}{2} = \frac{-1.80 + \sqrt{(1.80)^2 + 4(0.0360)}}{2}$$

$$x = \frac{-1.80 + \sqrt{3.384}}{2} = \frac{0.03957}{2} = 0.0198 = 0.02 \ M$$

$[PCl_3] = [Cl_2] = 0.0198 \ M; \ [PCl_5] = 0.0200 \ M - 0.0198 \ M = 2 \times 10^{-4} \ M$

(For this initial condition, the reaction essentially goes to completion.)

LeChatelier's Principle

15.39 (a)   Shift equilibrium to the right; more $SO_3(g)$ is formed, the amount of $SO_2(g)$ decreases.

(b)   Heating an exothermic reaction decreases the value of K. More SO_2 and O_2 will form, the amount of SO_3 will decrease.

(c)   Since, $\Delta n = -1$, a change in volume will affect the equilibrium position and favor the side with more moles of gas. The amounts of SO_2 and O_2 increase and the amount of SO_3 decreases.

(d)   No effect. Speeds up the forward and reverse reactions equally.

(e)   No effect. Does not appear in the equilibrium expression.

(f)   Shift equilibrium to the right; amounts of SO_2 and O_2 decrease.

15.41 (a)   No effect   (b) no effect   (c) increase equilibrium constant   (d) no effect

15.43 (a) $\Delta H° = \Delta H°_f \ NO_2(g) + \Delta H°_f \ N_2O(g) - 3\Delta H°_f \ NO(g)$
         $\Delta H° = 33.84 \ kJ + 81.6 \ kJ - 3(90.37 \ kJ) = -155.7 \ kJ$

(b)   The reaction is exothermic $(-\Delta H°)$, so the equilibrium constant will decrease with increasing temperature.

(c) Δn does not equal zero, so a change in volume at constant temperature will affect the fraction of products in the equilibrium mixture. An increase in container volume would favor reactants, while a decrease in volume would favor products.

Additional Exercises

15.45   (a)   Since both the forward and reverse processes are elementary steps, we can write the rate laws directly from the chemical equation.

$$\text{rate}_f = k_f\,[CO][Cl_2] = \text{rate}_r = k_r\,[COCl][Cl]$$

$$\frac{k_f}{k_r} = \frac{[COCl][Cl]}{[CO][Cl_2]} = K$$

$$K = \frac{k_f}{k_r} = \frac{1.4 \times 10^{-28}\,M^{-1}\,s^{-1}}{9.3 \times 10^{10}\,M^{-1}\,s^{-1}} = 1.5 \times 10^{-39}$$

       (b)   Since the K is quite small, reactants are much more plentiful than products at equilibrium.

15.48   First, calculate the number of moles of each component present.

$$\frac{3.22\text{ g NOBr}}{109.9\text{ g/mol}} = 0.02930 = 0.0293\text{ mol NOBr}; \quad \frac{3.08\text{ g NO}}{30.01\text{ g/mol}} = 0.1026 = 0.103\text{ mol NO}$$

$$\frac{4.19\text{ g Br}_2}{159.8\text{ g/mol}} = 0.02622 = 0.0262\text{ mol Br}_2$$

       (a)   In calculating K_c, divide each number of moles by 5.00 L to convert to moles/L, then insert into the expression for K_c to obtain:

$$K_c = \frac{[Br_2][NO]^2}{[NOBr]^2} = \frac{(5.244 \times 10^{-3})(2.053 \times 10^{-2})^2}{(5.860 \times 10^{-3})^2} = 6.436 \times 10^{-2} = 6.44 \times 10^{-2}$$

       (b) $K_p = K_c(RT)^{\Delta n} = (6.436 \times 10^{-2})\,(0.08206 \times 373) = 1.97$

       (c)   The total moles of gas present is $0.02930 + 0.1026 + 0.02622 = 0.1581 = 0.158$

$$P = (0.1581\text{ mol}) \times \frac{0.08206\text{ L}\cdot\text{atm}}{1\text{ mol}\cdot\text{K}} \times \frac{373\text{ K}}{5.00\text{ L}} = 0.968\text{ atm}$$

15.51 $K_c = \dfrac{[I_2][Br_2]}{[IBr]^2}$; initial $[Br] = \dfrac{0.040\text{ mol}}{1.00\text{ L}} = 0.040\ M$

	2IBr	$\rightleftharpoons$	I_2	+	Br_2
initial	0.040 *M*		0		0
change	-2x		x		x
equil.	0.040-2x		x		x

$K_c = 8.5 \times 10^{-3} = \dfrac{x^2}{(0.040-2x)^2}$; Taking the square root of both sides

$$\frac{x}{0.040-2x} = \sqrt{8.5 \times 10^{-3}} = 0.0922; \quad x = 0.0922(0.040-2x)$$

$x + 0.1844x = 0.003688;\ 1.1844x = 0.003688,\ x = 0.003114 = 0.0031$

$[IBr] = 0.040 - 2(0.0031) = 0.034\ M$

15.54 First find the initial moles of SO_3 and then the total moles of gas at equilibrium.

$$\frac{0.831 \text{ g } SO_3}{80.07 \text{ g/mol}} = 0.01038 = 0.0104 \text{ mol } SO_3 \; ; \quad n_t = \frac{1.30 \text{ atm} \times 1.00 \text{ L}}{1100 \text{ K}} \times \frac{\text{K} \cdot \text{mol}}{0.08206 \text{ L} \cdot \text{atm}}$$

$$= 0.01440 = 0.0144 \text{ mol}$$

	$2SO_3$	$\rightleftharpoons$	$2SO_2$	+	O_2
initial	0.01038		0		0
change	-2x		+2x		+x
equil.	0.01038-2x		2x		x
[equil.]	0.00234 M		0.00804 M		0.00402 M

$n_t = 0.0144 = 0.01038 - 2x + 2x + x; \quad x = 0.00402 = 0.0040 \text{ mol } O_2$

Since the volume is 1 L, the equilibrium molar concentrations are equal to the moles of each component.

$$K_c = \frac{[SO_2]^2[O_2]}{[SO_3]^2} = \frac{(0.00804)^2(0.00402)}{(0.00234)^2} = 0.04746 = 0.047$$

$$K_p = K_c(RT)^{\Delta n} = 4.746 \times 10^{-2}(0.08206 \times 1100)^1 = 4.3$$

15.56 $K_c = [CO_2] = 0.0108$

(a) $[CO_2] = 15.0 \text{ g } CO_2 \times \dfrac{1 \text{ mol } CO_2}{44.01 \text{ g } CO_2} = \dfrac{0.3408 \text{ mol}}{10.0 \text{ L}} = 0.0341 \; M$

$Q = 0.0341 > K_c$. The reaction proceeds to the left to achieve equilibrium and the amount of $CaCO_3(s)$ increases.

(b) $[CO_2] = 4.75 \text{ g } CO_2 \times \dfrac{1 \text{ mol } CO_2}{44.01 \text{ g } CO_2} \times \dfrac{1}{10.0 \text{ L}} = 0.0108 \; M$

$Q = 0.0108 = K_c$. The mixture is at equilibrium and the amount of $CaCO_3(s)$ remains constant.

(c) $[CO_2] = 2.50 \text{ g } CO_2 \times \dfrac{1 \text{ mol } CO_2}{44.01 \text{ g } CO_2} \times \dfrac{1}{10.0 \text{ L}} = 0.00568 \; M$

$Q = 0.00568 < K_c$. The reaction proceeds to the right to achieve equilibrium and the amount of $CaCO_3(s)$ decreases.

15.61 First calculate K_c for the equilibrium

$$H_2 + I_2 \rightleftharpoons 2HI$$

$$K_c = \frac{[HI]^2}{[H_2][I_2]} = \frac{(0.155)^2}{(2.24 \times 10^{-2})(2.24 \times 10^{-2})} = 47.88 = 47.9$$

The added HI represents a concentration of $\dfrac{0.100 \text{ mol}}{5.00 \text{ L}} = 0.0200 \; M.$

	H_2	$+$	I_2	$\rightleftharpoons$	$2HI$
initial	$2.24 \times 10^{-2}\ M$		$2.24 \times 10^{-2}\ M$		$0.155 + 0.0200\ M$
change	$+ x\ M$		$+ x\ M$		$-2x\ M$
equil.	$2.24 \times 10^{-2}\ M + x\ M$		$2.24 \times 10^{-2}\ M + x\ M$		$0.175 - 2x\ M$

$$\frac{(0.175 - 2x)^2}{(2.24 \times 10^{-2} + x)^2} = 47.88. \quad \text{Take the square root of both sides:}$$

$$\frac{0.175 - 2x}{2.24 \times 10^{-2} + x} = (47.88)^{1/2} = 6.920 = 6.92$$

$$0.175 - 2x = 0.155 + 6.92x; \quad x = 2.242 \times 10^{-3} = 2.24 \times 10^{-3}$$

$$[I_2] = [H_2] = 2.24 \times 10^{-2} + 2.24 \times 10^{-3} = 2.464 \times 10^{-2} = 2.46 \times 10^{-2}\ M$$

$$[HI] = 0.175 - 2x = 0.1705 = 0.171\ M$$

15.64   The patent claim is false. A catalyst does not alter the position of equilibrium in a system, only the rate of approach to the equilibrium condition.

Integrative Exercises

15.65   (a)   (i) $K_c = [Na^+]/[Ag^+]$ (ii) $K_c = [Hg^{2+}]^3 / [Al^{3+}]^2$

                    (iii) $K_c = [Zn^{2+}][H_2] / [H^+]^2$

        (b)   According to Table 4.5, the activity series of the metals, a metal can be oxidized by any metal cation below it on the table.

                (i) Ag^+ is far below Na, so the reaction will proceed to the right and K_c will be large.

                (ii) Al^{3+} is above Hg, so the reaction will not proceed to the right and K_c will be small.

                (iii) H^+ is below Zn, so the reaction will proceed to the right and K_c will be large.

        (c) $K_c < 1$ for this reaction, so Fe^{2+} (and thus Fe) is above Cd on the table. In other words, Cd is below Fe. The value of K_c, 0.06, is small but not extremely small, so Cd will be only a few rows below Fe.

15.67   (a)   At equilibrium, the forward and reverse reactions occur at <u>equal</u> rates.

        (b)   One expects the reactants to be favored at equilibrium since they are lower in energy.

(c)   A catalyst lowers the activation energy for both the forward and reverse reactions; the "hill" would be lower.

(d)   Since the activation energy is lowered for both processes, the new rates would be equal and the ratio of the rate constants, k_f / k_r, would remain unchanged.

(e)   Since the reaction is endothermic (the energy of the reactants is lower than that of the products, ΔE is positive), the value of K should increase with increasing temperature.

15.71   (a) $H_2O(l) \rightleftharpoons H_2O(g)$; $K_p = P_{H_2O}$

        (b)   At 30°C, the vapor pressure of $H_2O(l)$ is 31.82 torr.

               $K_p = P_{H_2O} = 31.82$ torr $= 0.041868 = 0.04187$ atm

        (c) $K_c = K_p/(RT)^{\Delta n}$; $\Delta n = +1$; $K_c = 0.041868/(0.08206)(303) = 1.684 \times 10^{-3}$

        (d)   From part (b), the value of K_p is the vapor pressure of the liquid at that temperature. By definition, vapor pressure = atmospheric pressure = 1 atm at the normal boiling point. $K_p = 1$ atm

16 Acid-Base Equilibria

Acid-Base Equilibria

16.1   Solutions of HCl and H_2SO_4 taste sour, turn litmus paper red (are acidic), neutralize solutions of bases, react with active metals to form $H_2(g)$ and conduct electricity. The two solutions have these properties in common because both solutes are strong acids. That is, they both dissociate completely in H_2O to form $H^+(aq)$ and an anion. (The first dissociation step for H_2SO_4 is complete, but HSO_4^- is not completely dissociated.) The presence of ions enables the solutions to conduct electricity; the presence of $H^+(aq)$ in excess of 1×10^{-7} M accounts for all the other properties listed.

16.3   (a)   *Autoionization* is the ionization of a neutral molecule (in the absence of any other reactant) into an anion and a cation. The equilibrium expression for the autoionization of water is $H_2O(l) \rightleftharpoons H^+(aq) + OH^-(aq)$.

   (b)   Pure water is a poor conductor of electricity because it contains very few ions. Ions, mobile charged particles, are required for the conduction of electricity in liquids.

   (c)   If a solution is *acidic*, it contains more H^+ than OH^- ($[H^+] > [OH^-]$).

16.5   In pure water at 25°C, $[H^+] = [OH^-] = 1 \times 10^{-7}$ M. If $[H^+] > 1 \times 10^{-7}$ M, the solution is acidic; if $[H^+] < 1 \times 10^{-7}$ M, the solution is basic.

   (a) $[H^+] = \dfrac{K_w}{[OH^-]} = \dfrac{1.0 \times 10^{-14}}{7 \times 10^{-4}\ M} = 1.43 \times 10^{-11} = \mathbf{1 \times 10^{-11}\ \textit{M}} < 1 \times 10^{-7}\ M$; basic

   (b) $[H^+] = \dfrac{K_w}{[OH^-]} = \dfrac{1.0 \times 10^{-14}}{8.2 \times 10^{-10}\ M} = \mathbf{1.2 \times 10^{-5}\ \textit{M}} > 1 \times 10^{-7}\ M$; acidic

   (c) $[OH^-] = 100[H^+]$; $K_w = [H^+] \times 100[H^+] = 100[H^+]^2$;

   $[H^+] = (K_w/100)^{1/2} = \mathbf{1.0 \times 10^{-8}\ \textit{M}} < 1 \times 10^{-7}\ M$; basic

16.7   At 37°C, $K_w = 2.4 \times 10^{-14} = [H^+][OH^-]$.

   In pure water, $[H^+] = [OH^-]$; $2.4 \times 10^{-14} = [H^+]^2$; $[H^+] = (2.4 \times 10^{-14})^{1/2}$

   $[H^+] = [OH^-] = 1.5 \times 10^{-7}\ M$

The pH Scale

16.9   A change of one pH unit (in either direction) is:

$\Delta pH = pH_2 - pH_1 = -(\log[H^+]_2 - \log[H^+]_1) = -\log\dfrac{[H^+]_2}{[H^+]_1} = \pm 1$.   The antilog of +1 is 10; the antilog of -1 is 1×10^{-1}. Thus, a ΔpH of one unit represents an increase or decrease in $[H^+]$ by a factor of 10.

(a) $\Delta pH = \pm 2.00$ is a change of $10^{2.00}$; $[H^+]$ changes by a factor of 100.

(b) $\Delta pH = \pm 0.5$ is a change of $10^{0.50}$; $[H^+]$ changes by a factor of 3.2.

16.11   (a) $K_w = [H^+][OH^-]$. If NaOH is added to water, it dissociates into $Na^+(aq)$ and $OH^-(aq)$. This increases $[OH^-]$ and necessarily decreases $[H^+]$. When $[H^+]$ decreases, pH increases.

(b) $pH = -\log[H^+] = -\log(0.005) = 2.3$ If pH < 7, the solution is acidic.

(c) $pH = 6.3$ $pOH = 14.0 - 6.3 = 7.7$
$[H^+] = 10^{-pH} = 10^{-6.3} = 5 \times 10^{-7}\ M$ $[OH^-] = 10^{-pOH} = 10^{-7.7} = 2 \times 10^{-8}\ M$

16.13

$[H^+]$	$[OH^-]$	pH	pOH	acidic or basic
$2.5 \times 10^{-4}\ M$	$4.0 \times 10^{-11}\ M$	3.60	10.40	acidic
$1.4 \times 10^{-7}\ M$	$6.9 \times 10^{-8}\ M$	6.84	7.16	acidic
$6 \times 10^{-4}\ M$	$2 \times 10^{-11}\ M$	3.2	10.8	acidic
$5 \times 10^{-9}\ M$	$2 \times 10^{-6}\ M$	8.3	5.7	basic

16.15   (a)   According to the Arrhenius definition, an acid when dissolved in water increases $[H^+]$. According to the Brønsted-Lowry definition, an acid is capable of donating H^+, regardless of physical state. The Arrhenius definition of an acid is confined to an aqueous solution; the Brønsted-Lowry definition applies to any physical state.

(b) $HCl(g) + NH_3(g) \rightarrow NH_4^+Cl^-(s)$ HCl is the B-L (Brønsted-Lowry) acid; it donates an H^+ to NH_3 to form NH_4^+. NH_3 is the B-L base; it accepts the H^+ from HCl.

16.17   A conjugate base has one less H^+ than its conjugate acid.

(a) ClO_2^- (b) HS^- (c) SO_4^{2-} (d) NH_3

16.19

B-L acid	+	**B-L base**	$\rightleftharpoons$	**Conjugate acid**	+	**Conjugate base**
(a) $NH_4^+(aq)$		$CN^-(aq)$		$HCN(aq)$		$NH_3(aq)$
(b) $H_2O(l)$		$(CH_3)_3N(aq)$		$(CH_3)_3NH^+(aq)$		$OH^-(aq)$
(c) $HCHO_2(aq)$		$PO_4^{3-}(aq)$		$HPO_4^{2-}(aq)$		$CHO_2^-(aq)$

16.21 Acid: $HC_2O_4^-(aq) + H_2O(l) \rightleftharpoons C_2O_4^{2-}(aq) + H_3O^+(aq)$
         B-L acid   B-L base   conj. base   conj. acid

      Base: $HC_2O_4^-(aq) + H_2O(l) \rightleftharpoons H_2C_2O_4(aq) + OH^-(aq)$
         B-L base   B-L acid   conj. acid   conj. base

16.23 (a) weak, NO_2^- (b) strong, HSO_4^- (c) weak, PO_4^{3-}
      (d) negligible, CH_3^- (e) weak, CH_3NH_2

16.25 (a) HNO_3. It is one of the seven strong acids (Section 16.5); it has the more electronegative central atom.

      (b) H_2O. When NH_3 and H_2O are combined, as in $NH_3(aq)$, H_2O acts as the B-L acid. It has the greater tendency to donate H^+. For binary hydrides in general, acid strength increases going to the right across a row on the periodic chart; the more polar the H-X bond, the stronger the acid (Section 16.10).

16.27 Acid-base equilibria favor formation of the weaker acid and base. Compare the substances acting as acids on opposite sides of the equation. (Bases can also be compared; the conclusion should be the same.)

             Base + **Acid** $\rightleftharpoons$ **Conjugate acid** + **Conjugate base**

      (a) $NH_2^-(aq) + H_2O(l) \rightleftharpoons NH_3(aq) + OH^-(aq)$

      H_2O is a stronger acid than NH_3 (Exercise 16.25), so the equilibrium lies to the right.

      (b) $H_2O(l) + HClO_2(aq) \rightleftharpoons H_3O^+(aq) + ClO_2^-(aq)$

      H_3O^+ is a stronger acid than $HClO_2$, so the equilibrium lies to the left.

      (c) $F^-(aq) + H_3O^+(aq) \rightleftharpoons HF(aq) + H_2O(l)$

      H_3O^+ is a stronger acid than HF, so the equilibrium lies to the right.

Strong Acids and Bases

16.29 (a) A *strong* acid is completely ionized in aqueous solution; a strong acid is a strong electrolyte.

      (b) For a strong acid such as HCl, $[H^+]$ = initial acid concentration. $[H^+]$ = 0.500 M

      (c) HCl, HBr, HI

16.31 For a strong acid, $[H^+]$ = initial acid concentration.

      (a) 1.8×10^{-4} M HBr = 1.8×10^{-4} M H^+; pH = -log (1.8×10^{-4}) = 3.74

(b) $\dfrac{1.02 \text{ g HNO}_3}{0.250 \text{ L soln}} \times \dfrac{1 \text{ mol HNO}_3}{63.02 \text{ g HNO}_3} = 0.06474 = 0.0647 \, M \text{ HNO}_3$

$[H^+] = 0.0647 \, M$; pH = -log (0.06474) = 1.189

(c) $M_c \times L_c = M_d \times L_d$; $0.500 \, M \times 0.00200 \text{ L} = ? \, M \times 0.500 \text{ L}$

$M_d = \dfrac{0.500 \, M \times 0.00200}{0.0500} = 0.0200 \, M \text{ HCl}$

$[H^+] = 0.0200 \, M$; pH = -log (0.0200) = 1.699

(d) $[H^+]_{total} = \dfrac{\text{mol H}^+ \text{ from HBr} + \text{mol H}^+ \text{ from HCl}}{\text{total L solution}}$

$[H^+]_{total} = \dfrac{(0.0100 \, M \text{ HBr} \times 0.0100 \text{ L}) + (2.50 \times 10^{-3} \, M \times 0.0200 \text{ L})}{0.0300 \text{ L}}$

$[H^+]_{total} = \dfrac{1.00 \times 10^{-4} \text{ mol H}^+ + 0.500 \times 10^{-4} \text{ mol H}^+}{0.0300 \text{ L}} = 5.00 \times 10^{-3} \, M$

pH = -log $(5.00 \times 10^{-3} \, M)$ = 2.301

16.33 (a) $[OH^-] = 2[Sr(OH)_2] = 2(3.5 \times 10^{-4} \, M) = 7.0 \times 10^{-4} \, M \text{ OH}^-$ (see Exercise 16.30(b))

pOH = -log (7.0×10^{-4}) = 3.15; pH = 14 - pOH = 10.85

(b) $\dfrac{1.50 \text{ g LiOH}}{0.250 \text{ L soln}} \times \dfrac{1 \text{ mol LiOH}}{23.95 \text{ g LiOH}} = 0.2505 = 0.251 \, M \text{ LiOH} = [OH^-]$

pOH = -log (0.2505) = 0.601; pH = 14 - pOH = 13.399

(c) $M_c \times L_c = M_d \times L_d$; $0.095 \, M \times 0.00100 \text{ L} = ? \, M \times 2.00 \text{ L}$

$M_d = \dfrac{0.095 \, M \times 0.00100 \text{ L}}{2.00 \text{ L}} = 4.75 \times 10^{-5} = 4.8 \times 10^{-5} \, M \text{ NaOH} = [OH^-]$

pOH = -log (4.75×10^{-5}) = 4.32; pH = 14 - pOH = 9.68

(d) $[OH^-]_{total} = \dfrac{\text{mol OH}^- \text{ from KOH} + \text{mol OH}^- \text{ from Ca(OH)}_2}{\text{total L soln}}$

$[OH^-]_{total} = \dfrac{(0.0105 \, M \times 0.00500 \text{ L}) + 2(3.5 \times 10^{-3} \times 0.0150 \text{ L})}{0.0200 \text{ L}}$

$[OH^-]_{total} = \dfrac{5.250 \times 10^{-5} \text{ mol OH}^- + 10.5 \times 10^{-5} \text{ mol OH}^-}{0.0200 \text{ L}} = 7.875 \times 10^{-3}$

$= 7.9 \times 10^{-3} \, M$

$pOH = -\log (7.875 \times 10^{-3}) = 2.104$; $pH = 14 - pOH = 11.90$

(3.5×10^{-3} M has 2 sig figs, so the $[OH^-]$ has 2 sig figs and pH and pOH have 2 decimal places.)

16.35   Upon dissolving, Li_2O dissociates to form Li^+ and O^{2-}. According to Equation 16.18, O^{2-} is completely protonated in aqueous solution.

Thus, initial $[Li_2O] = [O_2^-]$; $[OH^-] = 2[O^{2-}] = 2[Li_2O]$

$$[Li_2O] = \frac{mol\ Li_2O}{L\ solution} = 2.00\ g\ Li_2O \times \frac{1\ mol\ Li_2O}{29.88\ g\ Li_2O} \times \frac{1}{0.600\ L} = 0.1116 = 0.112\ M$$

$[OH^-] = 0.2232 = 0.223\ M$; $pOH = 0.651$ $pH = 14.00 - pOH = 13.349$

Weak Acids

16.37   (a) $HBrO_2(aq) \rightleftharpoons H^+(aq) + BrO_2^-(aq)$; $K_a = \dfrac{[H^+][BrO_2^-]}{[HBrO_2]}$

              $HBrO_2(aq) + H_2O(l) \rightleftharpoons H_3O^+(aq) + BrO_2^-(aq)$; $K_a = \dfrac{[H_3O^+][BrO_2^-]}{[HBrO_2]}$

      (b) $HC_3H_5O_2(aq) \rightleftharpoons H^+(aq) + C_3H_5O_2^-(aq)$; $K_a = \dfrac{[H^+][C_3H_5O_2^-]}{[HC_3H_5O_2]}$

              $HC_3H_5O_2(aq) + H_2O(l) \rightleftharpoons H_3O^+(aq) + C_3H_5O_2^-(aq)$; $K_a = \dfrac{[H_3O^+][C_3H_5O_2^-]}{[HC_3H_5O_2]}$

16.39 $HC_3H_5O_3(aq) \rightleftharpoons H^+(aq) + C_3H_5O_3^-(aq)$; $K_a = \dfrac{[H^+][C_3H_5O_3^-]}{[HC_3H_5O_3]}$

$[H^+] = [C_3H_5O_3^-] = 10^{-2.44} = 3.63 \times 10^{-3} = 3.6 \times 10^{-3}\ M$

$[HC_3H_5O_3] = 0.10 - 3.63 \times 10^{-3} = 0.0964 = 0.096\ M$

$$K_a = \frac{(3.63 \times 10^{-3})^2}{(0.0964)} = 1.4 \times 10^{-4}$$

16.41 $HC_7H_5O_2(aq) \rightleftharpoons H^+(aq) + C_7H_5O_2^-(aq)$

initial   0.050 M 0   0

equil.   (0.050 - x) M x M x M

$$K_a = \frac{[H^+][C_7H_5O_2^-]}{[HC_7H_5O_2]} = \frac{x^2}{(0.050 - x)} \approx \frac{x^2}{0.050} = 6.5 \times 10^{-5}$$

$x^2 = 0.050\ (6.5 \times 10^{-5})$; $x = 1.8 \times 10^{-3}\ M = [H^+] = [H_3O^+] = [C_7H_5O_2^-]$

$[HC_7H_5O_2] = 0.050 - 0.0018 = 0.048\ M$

$$\frac{1.8 \times 10^{-3}\ M\ H^+}{0.050\ M\ HC_7H_5O_2} \times 100 = 3.6\%\ \text{ionization; the assumption is valid}$$

16.43   (a) $HN_3(aq) \rightleftharpoons H^+(aq) + N_3^-(aq)$

        initial   0.175 M 0   0

        equil.   (0.175 - x) M x M x M

$$K_a = \frac{[H^+][N_3^-]}{[HN_3]} = \frac{(x)(x)}{(0.175 - x)} \approx \frac{x^2}{0.175} = 1.9 \times 10^{-5}$$

$x^2 = 0.175 \,(1.9 \times 10^{-5}); \; x = [H^+] = 1.8 \times 10^{-3} \, M, \; pH = 2.74$

$$\frac{1.8 \times 10^{-3} \, M \, H^+}{0.175 \, M \, HN_3} \times 100 = 1.0\% \text{ ionization}; \quad \text{the assumption is valid}$$

      (b) $K_a = \dfrac{[H^+][C_3H_5O_2^-]}{[HC_3H_5O_2]} = \dfrac{(x)(x)}{(0.040 - x)} \approx \dfrac{x^2}{0.040} = 1.3 \times 10^{-5}$

$x^2 = 0.040 \,(1.3 \times 10^{-5}); \; x = [H^+] = 7.2 \times 10^{-4} \, M; \; pH = 3.14$

$$\frac{7.2 \times 10^{-4} \, M \, H^+}{0.040 \, M \, HC_3H_5O_2} \times 100 = 1.8\% \text{ ionization}; \quad \text{the assumption is valid}$$

16.45   (a) $HN_3(aq) \rightleftharpoons H^+(aq) + N_3^-(aq)$

        initial   0.400 M 0   0

        equil   (0.400 - x) M x M x M

$$K_a = \frac{[H^+][N_3^-]}{[HN_3]} = 1.9 \times 10^{-5}; \; \frac{x^2}{(0.400 - x)} \approx \frac{x^2}{0.400} = 1.9 \times 10^{-5}$$

$x = 0.00276 = 2.8 \times 10^{-3} \, M = [H^+]; \; \%\text{ ionization} = \dfrac{2.76 \times 10^{-3}}{0.400} \times 100 = 0.69\%$

      (b) $1.9 \times 10^{-5} \approx \dfrac{x^2}{0.100}; \; x = 0.00138 = 1.4 \times 10^{-3} \, M \, H^+$

$$\%\text{ ionization} = \frac{1.38 \times 10^{-3} \, M \, H^+}{0.100 \, M \, HN_3} \times 100 = 1.4\%$$

      (c) $1.9 \times 10^{-5} \approx \dfrac{x^2}{0.0400}; \; x = 8.72 \times 10^{-4} = 8.7 \times 10^{-4} \, M \, H^+$

$$\%\text{ ionization} = \frac{8.72 \times 10^{-4} \, M \, H^+}{0.0400 \, M \, HN_3} \times 100 = 2.2\%$$

Notice that a tenfold dilution [part (a) versus part (c)] leads to a slightly more than threefold increase in percent ionization.

16.47 $[H^+] = 0.094 \times [HX]_{initial} = 0.0188 = 0.019\ M$

(a) $HX(aq) \rightleftharpoons H^+(aq) + X^-(aq)$

     initial $0.200\ M$ 0   0

     equil. $(0.200 - 0.019)\ M$ $0.019\ M$ $0.019\ M$

$$K_a = \frac{[H^+][X^-]}{[HX]} = \frac{(0.0188)^2}{0.181} = 2.0 \times 10^{-3}$$

16.49   Let the weak acid be HX. $HX(aq) \rightleftharpoons H^+(aq) + X^-(aq)$

$$K_a = \frac{[H^+][X^-]}{[HX]}; \quad [H^+] = [X^-] = y; \quad K_a = \frac{y^2}{[HX] - y}; \quad \text{assume that \% ionization is small}$$

$$K_a = \frac{y^2}{[HX]}; \quad y = K_a^{1/2}[HX]^{1/2}$$

$$\% \text{ ionization} = \frac{y}{[HX]} \times 100 = \frac{K_a^{1/2}[HX]^{1/2}}{[HX]} \times 100 = \frac{K_a^{1/2}}{[HX]^{1/2}} \times 100$$

That is, percent ionization varies inversely as the square root of concentration HX.

16.51 $H_3C_6H_5O_7(aq) \rightleftharpoons H^+(aq) + H_2C_6H_5O_7^-(aq)$ $K_{a1} = 7.4 \times 10^{-4}$

       $H_2C_6H_5O_7^-(aq) \rightleftharpoons H^+(aq) + HC_6H_5O_7^{2-}(aq)$ $K_{a2} = 1.7 \times 10^{-5}$

       $HC_6H_5O_7^{2-}(aq) \rightleftharpoons H^+(aq) + C_6H_5O_7^{3-}(aq)$ $K_{a3} = 4.0 \times 10^{-7}$

To calculate the pH of a 0.050 *M* solution, assume initially that only the first ionization is important:

         $H_3C_6H_5O_7(aq) \rightleftharpoons H^+(aq) + H_2C_6H_5O_7^-(aq)$

initial $0.050\ M$ 0   0

equil. $(0.050 - x)\ M$ $x\ M$ $x\ M$

$$K_{a1} = \frac{[H^+][H_2C_6H_5O_7^-]}{[H_3C_6H_5O_7]} = \frac{x^2}{(0.050 - x)} = 7.4 \times 10^{-4}$$

$x^2 = (0.050 - x)(7.4 \times 10^{-4}); \quad x^2 \approx (0.050)(7.4 \times 10^{-4}); \quad x = 0.00608 = 6.1 \times 10^{-3}\ M$

Since this value for x is rather large in relation to 0.050, a better approximation for x can be obtained by substituting this first estimate into the expression for x^2, then solving again for x:

$$x^2 = (0.050 - x)(7.4 \times 10^{-4}) = (0.050 - 6.08 \times 10^{-3})(7.4 \times 10^{-4})$$
$$x^2 = 3.2 \times 10^{-5}; \quad x = 5.7 \times 10^{-3}\ M$$

The correction to the value of x, though not large, is significant. (This is the same result obtained from the quadratic formula.) Does the second ionization produce a significant additional concentration of H^+?

$$H_2C_6H_5O_7^-(aq) \rightleftharpoons H^+(aq) + HC_6H_5O_7^{2-}(aq)$$

initial 5.7×10^{-5} M 5.7×10^{-3} M   0

equil. $(5.7 \times 10^{-3} - y)$ $(5.7 \times 10^{-3} + y)$ y

$$K_{a2} = \frac{[H^+][HC_6H_5O_7^{2-}]}{[H_2C_6H_5O_7^-]} = 1.7 \times 10^{-5}; \quad \frac{(5.7 \times 10^{-3} + y)(y)}{(5.7 \times 10^{-3} - y)} = 1.7 \times 10^{-5}$$

Assume that y is small relative to 5.7×10^{-3} ; that is, that additional ionization of $H_2C_6H_5O_7^-$ is small, then

$$\frac{(5.7 \times 10^{-3})y}{(5.7 \times 10^{-3})} = 1.7 \times 10^{-5} \text{ M}; \quad y = 1.7 \times 10^{-5} \text{ M}$$

This value is indeed small compared to 5.7×10^{-3} M. This indicates that the second ionization can be neglected. pH is therefore $-\log [5.7 \times 10^{-3}] = 2.24$.

Weak Bases; the K_a-K_b Relationship; Acid-Base Properties of Salts

16.53   All Brønsted-Lowry bases contain at least one lone (nonbonded) pair of electrons to attract H^+.

16.55   (a) $(CH_3)_2NH(aq) + H_2O(l) \rightleftharpoons (CH_3)_2NH_2^+(aq) + OH^-(aq); \quad K_b = \dfrac{[(CH_3)_2NH_2^+][OH^-]}{[(CH_3)_2NH]}$

(b) $CO_3^{2-}(aq) + H_2O(l) \rightleftharpoons HCO_3^-(aq) + OH^-(aq); \quad K_b = \dfrac{[HCO_3^-][OH^-]}{[CO_3^{2-}]}$

(c) $CHO_2^-(aq) + H_2O(l) \rightleftharpoons HCHO_2(aq) + OH^-(aq); \quad K_b = \dfrac{[HCHO_2][OH^-]}{[CHO_2^-]}$

16.57 $C_2H_5NH_2(aq) + H_2O(l) \rightleftharpoons C_2H_5NH_3^+(aq) + OH^-(aq)$

initial   0.050 M   0   0

equil.   (0.050 - x) M   x M   x M

$$K_b = \frac{[C_2H_5NH_3^+][OH^-]}{[C_2H_5NH_2]} = \frac{(x)(x)}{(0.050 - x)} \approx \frac{x^2}{0.050} = 6.4 \times 10^{-4}$$

$x^2 = 0.050 \, (6.4 \times 10^{-4}); \quad x = [OH^-] = 5.7 \times 10^{-3}$ M; pH = 11.76

$$\frac{5.7 \times 10^{-3} \text{ M OH}^-}{0.050 \text{ M C}_2\text{H}_5\text{NH}_2} \times 100 = 11.3\% \text{ ionization; the assumption is \textbf{not} valid}$$

To obtain a more precise result, the K_b expression is rewritten in standard quadratic form and solved via the quadratic formula.

$$\frac{x^2}{0.050 - x} = 6.4 \times 10^{-4}; \quad x^2 + 6.4 \times 10^{-4} x - 3.2 \times 10^{-5} = 0$$

$$x = \frac{b \pm \sqrt{b^2 - 4ac}}{2a} = \frac{-6.4 \times 10^{-4} \pm \sqrt{(6.4 \times 10^{-4})^2 - 4(1)(-3.2 \times 10^{-5})}}{2}$$

$x = 5.346 \times 10^{-3} = 5.3 \times 10^{-3}\ M\ OH^-$; $pOH = 2.27$, $pH = 14.00 - pOH = 11.73$

Note that the pH values obtained using the two algebraic techniques are very similar.

16.59 (a) For a conjugate acid/conjugate base pair such as $C_6H_5OH/C_6H_5O^-$, K_b for the conjugate base is always K_w/K_a for the conjugate acid. K_b for the conjugate base can always be calculated from K_a for the conjugate acid, so a separate list of K_b values is not necessary.

(b) $K_b = K_w/K_a = 1.0 \times 10^{-14}/1.3 \times 10^{-10} = 7.7 \times 10^{-5}$

(c) K_b for phenolate $(7.7 \times 10^{-5}) > K_b$ for ammonia (1.8×10^{-5}).
Phenolate is a stronger base than NH_3.

16.61 When the solute in an aqueous solution is a salt, evaluate the acid/base properties of the component ions.

(a) NaCN is a soluble salt and thus a strong electrolyte. When it is dissolved in H_2O, it dissociates completely into Na^+ and CN^-. $[NaCN] = [Na^+] = [CN^-] = 0.10$ M. Na^+ is the conjugate acid of the strong base NaOH and thus does not influence the pH of the solution. CN^-, on the other hand, is the conjugate base of the weak acid HCN and **does** influence the pH of the solution. Like any other weak base, it hydrolyzes water to produce $OH^-(aq)$. Solve the equilibrium problem to determine $[OH^-]$.

$$CN^-(aq) + H_2O(l) \rightleftharpoons HCN(aq) + OH^-(aq)$$

initial	0.10 M	0	0
equil.	(0.10 - x) M	x M	x M

$$K_b \text{ for } CN^- = \frac{[HCN][OH^-]}{[CN^-]} = \frac{K_w}{K_a \text{ for HCN}} = \frac{1 \times 10^{-14}}{4.9 \times 10^{-10}} = 2.04 \times 10^{-5} = 2.0 \times 10^{-5}$$

$2.04 \times 10^{-5} = \dfrac{(x)(x)}{(0.10-x)}$; assume the percent of CN^- that hydrolyzes is small

$x^2 = 0.10\ (2.04 \times 10^{-5})$; $x = [OH^-] = 0.00143 = 1.4 \times 10^{-3}\ M$

$pOH = 2.85$; $pH = 14 - 2.85 = 11.15$

(b) $Na_2CO_3(aq) \rightarrow 2Na^+(aq) + CO_3^{2-}(aq)$

CO_3^{2-} is the conjugate base of HCO_3^- and its hydrolysis reaction will determine the $[OH^-]$ and pH of the solution (see similar explanation for NaCN in part (a)). We will assume the process $HCO_3^-(aq) + H_2O(l) \rightleftharpoons H_2CO_3(aq) + OH^-$ will not add significantly to the $[OH^-]$ in solution because $[HCO_3^-(aq)]$ is so small. Solve the equilibrium problem for $[OH^-]$.

$$CO_3^{2-}(aq) + H_2O(l) \rightleftharpoons HCO_3^-(aq) + OH^-(aq)$$

initial	0.080 M	0	0
equil.	(0.080 - x) M	x	x

$$K_b = \frac{[HCO_3^-][OH^-]}{[CO_3^{2-}]} = \frac{K_w}{K_a \text{ for } HCO_3^-} = \frac{1.0 \times 10^{-14}}{5.6 \times 10^{-11}} = 1.79 \times 10^{-4} = 1.8 \times 10^{-4}$$

$$1.8 \times 10^{-4} = \frac{x^2}{(0.080 - x)}; \quad x^2 = 0.080 \,(1.79 \times 10^{-4}); \quad x = 0.00378 = 3.8 \times 10^{-3} \, M \text{ OH}^-$$

(Assume x is small compared to 0.080); pOH = 2.42; pH = 14 - 2.42 = 11.58

$$\frac{3.8 \times 10^{-3} \, M \text{ OH}^-}{0.080 \, M \, CO_3^{2-}} \times 100 = 4.75\% \text{ hydrolysis; the assumption is valid.}$$

(c) For the two salts present, Na^+ and Ca^{2+} are negligible acids. NO_2^- is the conjugate base of HNO_2 and will determine the pH of the solution.

Calculate total $[NO_2^-]$ present initially.

$[NO_2^-]_{total}$ = $[NO_2^-]$ from $NaNO_2$ + $[NO_2^-]$ from $Ca(NO_2)_2$

$[NO_2^-]_{total}$ = 0.10 M + 2(0.20 M) = 0.50 M

The hydrolysis equilibrium is:

	$NO_2^-(aq) + H_2O(l)$	$\rightleftharpoons$	HNO_2 +	$OH^-(aq)$
initial	0.50 M		0	0
equil.	(0.50 - x) M		x M	x M

$$K_b = \frac{[HNO_2][OH^-]}{[NO_2^-]} = \frac{K_w}{K_a \text{ for } HNO_2} = \frac{1.0 \times 10^{-14}}{4.5 \times 10^{-4}} = 2.22 \times 10^{-11} = 2.2 \times 10^{-11}$$

$$2.2 \times 10^{-11} = \frac{x^2}{(0.50 - x)} \approx \frac{x^2}{0.50}; \quad x^2 = 0.50 \,(2.22 \times 10^{-11})$$

$x = 3.33 \times 10^{-6} = 3.3 \times 10^{-6} \, M \text{ OH}^-$; pOH = 5.48; pH = 14 - 5.48 = 8.52

16.63 (a) acidic; NH_4^+ is a weak acid, Br^- is negligible.

(b) acidic; Fe^{3+} is a highly charged metal cation and a Lewis acid; Cl^- is negligible.

(c) basic; CO_3^{2-} is the conjugate base of HCO_3^-; Na^+ is negligible.

(d) neutral; both K^+ and ClO_4^- are negligible.

(e) acidic; $HC_2O_4^-$ is amphoteric, but K_a for the acid dissociation (6.4×10^{-5}) is much greater than K_b for the base hydrolysis ($1.0 \times 10^{-14} / 5.9 \times 10^{-2}$ = 1.7×10^{-13}).

16.65 The solution will be basic because of the hydrolysis of the sorbate anion, $C_6H_7O_2^-$. Calculate the initial molarity of $C_6H_7O_2^-$.

$$\frac{4.93 \text{ g } KC_6H_7O_2}{0.500 \text{ L}} \times \frac{1 \text{ mol } KC_6H_7O_2}{150.2 \text{ g } KC_6H_7O_2} = 0.065646 = 0.0656 \, M \, KC_6H_7O_2$$

$[C_6H_7O_2^-] = [KC_6H_7O_2] = 0.0656 \, M$

$$C_6H_7O_2^-(aq) \ + \ H_2O(l) \ \rightleftharpoons \ HC_6H_7O_2(aq) \ + \ OH^-(aq)$$

initial	0.0656 M	0	0
equil.	(0.0656 - x) M	x M	x M

$$K_b = \frac{[HC_6H_7O_2][OH^-]}{[C_6H_7O_2^-]} = \frac{K_w}{K_a \text{ for } HC_6H_7O_2} = \frac{1.0 \times 10^{-14}}{1.7 \times 10^{-5}} = 5.88 \times 10^{-10} = 5.9 \times 10^{-10}$$

$$5.88 \times 10^{-10} = \frac{x^2}{0.0656 - x} \approx \frac{x^2}{0.0656}; \ x^2 = 0.0656 \,(5.88 \times 10^{-10})$$

$$x = [OH^-] = 6.21 \times 10^{-6} = 6.2 \times 10^{-6} \ M; \ pOH = 5.21; \ pH = 14 - pOH = 8.79$$

Acid-Base Character and Chemical Structure

16.67   (a)   As the electronegativity of the central atom (X) increases, more electron density is withdrawn from the X-O and O-H bonds, respectively. In water, the O-H bond is ionized to a greater extent and the strength of the oxyacid increases.

(b)   As the number of nonprotonated oxygen atoms in the molecule increases, they withdraw electron density from the other bonds in the molecule and the strength of the oxyacid increases.

16.69   (a) HNO_3 is a stronger acid than HNO_2 because it has one more nonprotonated oxygen atom, and thus a higher oxidation number on N.

(b)   For binary hydrides, acid strength increases going down a family, so H_2S is a stronger acid than H_2O.

(c) H_2SO_4 is a stronger acid because H^+ is much more tightly held by the anion HSO_4^-.

(d)   For oxyacids, the greater the electronegativity of the central atom, the stronger the acid, so H_2SO_4 is a stronger acid than H_2SeO_4.

(e) CCl_3COOH is stronger because the electronegative Cl atoms withdraw electron density from other parts of the molecule, which weakens the O-H bond and makes H^+ easier to remove.

16.71   (a) BrO^- (HClO is the stronger acid due to a more electronegative central atom, so BrO^- is the stronger base.)

(b) BrO^- ($HBrO_2$ has more nonprotonated O atoms and is the stronger acid, so BrO^- is the stronger base.)

(c) HPO_4^{2-} (larger negative charge, greater attraction for H^+)

16.73 (a)   True

        (b)   False. In a series of acids that have the same central atom, acid strength increases with the number of nonprotonated oxygen atoms bonded to the central atom.

        (c)   False. H_2Te is a stronger acid than H_2S because the H-Te bond is longer, weaker and more easily dissociated than the H-S bond.

Lewis Acids and Bases

16.75

Theory	**Acid**	**Base**
Arrhenius	forms H^+ ions in water	produces OH^- in water
Brønsted-Lowry	proton (H^+) donor	proton acceptor
Lewis	electron pair acceptor	electron pair donor

The Brønsted-Lowry theory is more general than Arrhenius's definition, because it is based on a unified model for the processes responsible for acidic or basic character, and it shows the relationships between these processes. The Lewis theory is more general still because it does not restrict the acidic species to compounds having ionizable hydrogen. Any substance that can be viewed as an electron-pair acceptor is defined as a Lewis acid.

16.77

	Lewis Acid	**Lewis Base**
(a)	$Fe(ClO_4)_3$ or Fe^{3+}	H_2O
(b)	H_2O	CN^-
(c)	BF_3	$(CH_3)_3N$
(d)	HIO	NH_2^-

16.79 (a) Cu^{2+}, higher cation charge

        (b) Fe^{3+}, higher cation charge

        (c) Al^{3+}, smaller cation radius, same charge

Additional Exercises

16.81 (a) $H_2O(l) \rightarrow H^+(aq) + OH^-(aq)$; $H^+(aq) + OH^-(aq) \rightarrow H_2O(l)$

        (b)   As OH^- is removed from solution, more H_2O must dissociate to replace it, increasing $[H^+]$ as well.

16.84 The equilibrium shifts in the forward direction with reaction between OH^- and H^+. The solution will be blue, corresponding to an excess of the conjugate base form, Bb^-, in solution.

16.87 A pH of 3.25 corresponds to $[H^+] = 5.6 \times 10^{-4}\ M$.

        (a)   Because HCl is a strong acid, $[HCl] = 5.62 \times 10^{-4}\ M = 5.6 \times 10^{-4}\ M$
            $(0.200\ L)(5.62 \times 10^{-4}\ mol/L) = 1.1 \times 10^{-4}\ mol\ HCl$

(b) K_a for $HC_7H_5O_2$ is 6.5×10^{-5} (benzoic acid, Appendix D)

$$\frac{(5.62 \times 10^{-4})^2}{x} = 6.5 \times 10^{-5}; \quad x = 4.86 \times 10^{-3} = 4.9 \times 10^{-3}\ M$$

$$0.200\ L \times \frac{4.86 \times 10^{-3}\ mol\ HC_7H_5O_2}{1\ L} = 9.7 \times 10^{-4}\ mol\ HC_7H_5O_2$$

(c)   HF has $K_a = 6.8 \times 10^{-4}$

$$\frac{(5.62 \times 10^{-4})^2}{x} = 6.8 \times 10^{-4}; \quad x = 4.64 \times 10^{-4} = 4.6 \times 10^{-4}\ M$$

This is the amount that remains in solution **after** ionization to form H^+ and F^-. The total HF is the sum of the ionized and nonionized portions:

$$4.64 \times 10^{-4}\ M + 5.62 \times 10^{-4}\ M = 1.026 \times 10^{-3} = 1.03 \times 10^{-3}\ M$$

$$0.200\ L \times \frac{1.03 \times 10^{-3}\ mol\ HF}{1\ L} = 2.06 \times 10^{-4}\ mol\ HF$$

16.90   Calculate the initial concentration of $HC_9H_7O_4$.

$$\frac{325\ mg}{1\ tablet} \times 2\ tablets \times \frac{1\ g}{1000\ mg} \times \frac{1\ mol\ HC_9H_7O_4}{180.2\ g\ HC_9H_7O_4} = 0.003607 = 0.00361\ mol\ HC_9H_7O_4$$

$$\frac{0.003607\ mol\ HC_9H_7O_4}{0.250\ L} = 0.01443 = 0.0144\ M\ HC_9H_7O_4$$

	$HC_9H_7O_4(aq)$	$\rightleftharpoons$	$C_9H_7O_4^-(aq)$	+	$H^+(aq)$
initial	0.0144 M		0 M		0 M
equil	(0.0144 - x)		x M		x M

$$K_a = 3.3 \times 10^{-4} = \frac{[H^+][C_7H_9O_4^-]}{[HC_7H_9O_4]} = \frac{x^2}{(0.0144 - x)}$$

Assuming x is small compared to 0.0144,

$$x^2 = 0.01443\ (3.3 \times 10^{-4}); \quad x = [H^+] = 2.2 \times 10^{-3}\ M$$

$$\frac{2.2 \times 10^{-3}\ M\ H^+}{0.01443\ M\ HC_9H_7O_4} \times 100 = 15\%\ ionization; \quad \text{the assumption is not valid}$$

Using the quadratic formula, $x^2 + 3.3 \times 10^{-4}\ x - 4.76 \times 10^{-6} = 0$

$$x = \frac{-3.3 \times 10^{-4} \pm \sqrt{(3.3 \times 10^{-4})^2 - 4(1)(-4.76 \times 10^{-6})}}{2(1)} = \frac{-3.3 \times 10^{-4} \pm \sqrt{1.915 \times 10^{-5}}}{2}$$

$x = 2.02 \times 10^{-3} = 2.0 \times 10^{-3}\ M\ H^+; \quad pH = -\log(2.02 \times 10^{-3}) = 2.69$

16.93   Considering the stepwise dissociation of H_3PO_4:

$$H_3PO_4(aq) \rightleftharpoons H^+(aq) + H_2PO_4^-(aq)$$

initial   0.100   0   0

equil.   (0.100 - x) M x   x

$$K_{a1} = \frac{[H^+][H_2PO_4^-]}{[H_3PO_4]} = \frac{x^2}{(0.100 - x)} = 7.5 \times 10^{-3}; \quad x^2 + 7.5 \times 10^{-3}\,x - 7.5 \times 10^{-4} = 0$$

$$x = \frac{-7.5 \times 10^{-3} \pm \sqrt{(7.5 \times 10^{-3})^2 - 4(1)(-7.5 \times 10^{-4})}}{2(1)} = \frac{-7.5 \times 10^{-3} \pm \sqrt{0.00306}}{2}$$

$x = 0.024\ M\ H^+$, $0.024\ M\ H_2PO_4^-$ available for further ionization

$$H_2PO_4^-(aq) \rightleftharpoons H^+(aq) + HPO_4^{2-}(aq)$$

initial   0.024 M 0.024 M 0 M

equil.   (0.024 - y) M (0.024 + y) M y M

$$K_{a2} = \frac{[H^+][HPO_4^{2-}]}{[H_2PO_4^-]} = \frac{(y)(0.024 + y)}{(0.024 - y)} = 6.2 \times 10^{-8}$$

Since K_{a2} is very small, assume x is small compared to 0.024 M.

$$6.2 \times 10^{-8} = \frac{0.024\,y}{0.024}; \quad y = [HPO_4^{2-}] = 6.2 \times 10^{-8}\ M$$

$[H_2PO_4^-] = (2.4 \times 10^{-2}\ M + 6.2 \times 10^{-8}\ M) \approx 2.4 \times 10^{-2}\ M$

$[H^+] = (2.4 \times 10^{-2}\ M + 6.2 \times 10^{-8}\ M) \approx 2.4 \times 10^{-2}\ M$

$[HPO_4^{2-}]$ available for further ionization $= 6.2 \times 10^{-8}\ M$

$$HPO_4^{2-} \rightleftharpoons H^+(aq) + PO_4^{3-}(aq)$$

initial $6.2 \times 10^{-8}\ M$ $2.4 \times 10^{-2}\ M$ 0 M

equil. $(6.2 \times 10^{-8} - z)\ M$ $(2.4 \times 10^{-2} + z)\ M$ z M

$$K_{a3} = \frac{[H^+][PO_4^{3-}]}{[HPO_4^{2-}]} = \frac{2.4 \times 10^{-2} + z)(z)}{(6.2 \times 10^{-8} - z)} = 4.2 \times 10^{-13}$$

Assuming z is small compared to 6.2×10^{-8} (and 2.4×10^{-2}),

$$\frac{(2.4 \times 10^{-2})(z)}{(6.2 \times 10^{-8})} = 4.2 \times 10^{-13}; \quad z = 1.1 \times 10^{-18}\ M\ PO_4^{3-}$$

The contribution of z to $[HPO_4^{2-}]$ and $[H^+]$ is negligible. In summary, after all dissociation steps have reached equilibrium:

$[H^+] = 2.4 \times 10^{-2}\ M$, $[H_2PO_4^-] = 2.4 \times 10^{-2}\ M$, $[HPO_4^{2-}] = 6.2 \times 10^{-8}\ M$,

$$[PO_4^{3-}] = 1.1 \times 10^{-18}\ M$$

Note that the first ionization step is the major source of H^+ and the others are important as sources of HPO_4^{2-} and PO_4^{3-}. The $[PO_4^{3-}]$ is **very** small at equilibrium.

16.96   Call each compound in the neutral form Q.

Then, $Q(aq) + H_2O(l) \rightleftharpoons QH^+(aq) + OH^-$. $K_b = [QH^+][OH^-]/[Q]$

The ratio in question is $[QH^+]/[Q]$, which equals $K_b/[OH^-]$ for each compound. At pH = 2.5, pOH = 11.5, $[OH^-]$ = antilog (-11.5) = 3.16×10^{-12} = 3×10^{-12} M. Now calculate $K_b/[OH^-]$ for each compound:

Nicotine $\dfrac{[QH^+]}{[Q]} = 7 \times 10^{-7}/3.16 \times 10^{-12} = 2 \times 10^5$

Caffeine $\dfrac{[QH^+]}{[Q]} = 4 \times 10^{-14}/3.16 \times 10^{-12} = 1 \times 10^{-2}$

Strychnine $\dfrac{[QH^+]}{[Q]} = 1 \times 10^{-6}/3.16 \times 10^{-12} = 3 \times 10^5$

Quinine $\dfrac{[QH^+]}{[Q]} = 1 \times 10^{-6}/3.16 \times 10^{-12} = 3.5 \times 10^5$

For all the compounds except caffeine the protonated form is much higher concentration than the neutral form. However, for caffeine, a very weak base, the neutral form dominates.

Integrative Exercises

16.99   At 25°C, $[H^+] = [OH^-] = 1.0 \times 10^{-7}$ M

$$\dfrac{1.0 \times 10^{-7}\ \text{mol}\ H^+}{1\ L\ H_2O} \times 0.0010\ L \times \dfrac{6.022 \times 10^{23}\ H^+\ \text{ions}}{\text{mol}\ H^+} = 6.0 \times 10^{13}\ H^+\ \text{ions}$$

16.101 Strategy: Use PV = nRT to calculate mol SO_2, and thus mol H_2SO_3 and M H_2SO_3. Solve the equilibrium problem to find $[H^+]$ and pH.

$$n = \dfrac{PV}{RT} = \dfrac{1.0\ \text{atm} \times 3.9\ L\ SO_2}{293\ K} \times \dfrac{K \bullet \text{mol}}{0.08206\ L \bullet \text{atm}} = 0.162 = 0.16\ \text{mol}\ SO_2$$

From the given reaction, mol SO_2 = mol H_2SO_3. 0.16 mol H_2SO_3/1.0 L = 0.16 M H_2SO_3

$H_2SO_3(aq) \rightleftharpoons H^+(aq) + HSO_3^-(aq)$ $K_{a1} = 1.7 \times 10^{-2}$

$HSO_3^-(aq) \rightleftharpoons H^+(aq) + SO_3^{2-}(aq)$ $K_{a2} = 6.4 \times 10^{-8}$

$$K_{a1} = 1.7 \times 10^{-2} = \dfrac{[H^+][HSO_3^-]}{[H_2SO_3]} = \dfrac{x^2}{0.162 - x};\ \text{since}\ K_{a1}\ \text{is relatively large, use the quadratic.}$$

$$x^2 + 1.7 \times 10^{-2}\,x - 2.75 \times 10^{-3} = 0;\ \ x = \dfrac{-1.7 \times 10^{-2} \pm \sqrt{(1.7 \times 10^{-2})^2 - 4(1)(-2.75 \times 10^{-3})}}{2}$$

$x = 0.0447 = 0.045\ M\ H^+$; pH = 1.35

Assumptions: 1) SO_2 is an ideal gas

2)   The volume of $SO_2(aq)$ is 1.0 L. That is, there is no change in volume when 3.9 L of $SO_2(g)$ dissolve in 1.0 L of water.

3)   Because K_{a2} is small, the second dissociation of H_2SO_3 does not contribute significantly to [H^+] and pH.

By using the quadratic formula, we avoided assuming that the % ionization of $H_2SO_3(aq)$ was small relative to the initial concentration of $H_2SO_3(aq)$.

16.104   The **apparent** molality of the solution is given by

$$1.90°C \times \frac{1\,m}{1.86°C} = 1.02\,m$$

The ionization equilibrium, and molalities of the various species are:

$$HF(aq) \rightleftharpoons H^+(aq) + F^-(aq)$$

equil.   1.00 - x m x m x m

The total molality is thus (1.00 - x + x + x) = 1.00 + x = 1.02. Thus, x = 0.02. For such a dilute aqueous solution, molality and molarity are essentially identical (Section 13.2). Thus, we can write

$$\text{% ionization} = \frac{[H^+]}{[HF]} \times 100 = \frac{0.02}{1.00} \times 100 = 2\%$$

(b)   Assuming $M \approx m$, $K_a = \dfrac{[H^+][F^-]}{[HF]} = \dfrac{(0.02)^2}{0.98} = 4 \times 10^{-4}$.

The precision of the K_a is not very good; to improve upon it, we would need to know the magnitude of the freezing point depression more precisely.

16.107 (a)   (i) $HCO_3^-(aq) \rightleftharpoons H^+(aq) + CO_3^{2-}(aq)$ $K_1 = K_{a2}$ for $H_2CO_3 = 5.6 \times 10^{-11}$

$H^+(aq) + OH^-(aq) \rightleftharpoons H_2O(l)$ $K_2 = 1/K_w = 1 \times 10^{14}$

$HCO_3^-(aq) + OH^-(aq) \rightleftharpoons CO_3^{2-}(aq) + H_2O(l)$ $K = K_1 \times K_2 = 5.6 \times 10^3$

(ii) $NH_4^+(aq) \rightleftharpoons H^+(aq) + NH_3(aq)$ $K_1 = K_a$ for $NH_4^+ = 5.6 \times 10^{-10}$

$CO_3^{2-}(aq) + H^+(aq) \rightleftharpoons HCO_3^-(aq)$ $K_2 = 1/K_{a2}$ for $H_2CO_3 = 1.8 \times 10^{10}$

$NH_4^+(aq) + CO_3^{2-}(aq) \rightleftharpoons HCO_3^-(aq) + NH_3(aq)$ $K = K_1 \times K_2 = 10$

(b)   Both (i) and (ii) have K > 1, although K = 10 is not **much** greater than 1. Both could be written with a single arrow. (This is true in general when a strong acid or strong base, $H^+(aq)$ or $OH^-(aq)$, is a reactant.)

17 Additional Aspects of Equilibria

Common-Ion Effect

17.1　(a)　The extent of dissociation of a weak electrolyte is decreased when a strong electrolyte containing an ion in common with the weak electrolyte is added to it.

　　　(b)　NaOCl

17.3　In general, when an acid is added to a solution, pH decreases; when a base is added to a solution, pH increases.

　　　(a)　pH increases; NO_2^- decreases the ionization of HNO_2 and decreases $[H^+]$.

　　　(b)　pH decreases; $CH_3NH_3^+$ decreases the ionization (hydrolysis) of CH_3NH_2 and decreases $[OH^-]$.

　　　(c)　pH increases; CHO_2^- decreases the ionization of $HCHO_2$ and decreases $[H^+]$.

　　　(d)　no change; Br^- is a negligible base and does not affect the 100% ionization of the strong acid HBr.

　　　(e)　pH decreases; the pertinent equilibrium is
$$C_2H_3O_2^-(aq) + H_2O(l) \rightleftharpoons HC_2H_3O_2 + OH^-(aq).$$
HCl reacts with $OH^-(aq)$, decreasing $[OH^-]$ and pH.

17.5　(a)

$$HC_3H_5O_2(aq) \rightleftharpoons H^+(aq) + C_3H_5O_2^-(aq)$$

i	0.16 M		0.080 M
c	$-x$	$+x$	$+x$
e	(0.16 - x) M	$+x$ M	(0.080 + x) M

$$K_a = 1.3 \times 10^{-5} = \frac{[H^+][C_3H_5O_2^-]}{[HC_3H_5O_2]} = \frac{(x)(0.080 + x)}{(0.16 - x)}$$

Assume x is small compared to 0.080 and 0.16.

$$1.3 \times 10^{-5} = \frac{0.080\,x}{0.16}; \quad x = 2.6 \times 10^{-5} = [H^+], \text{ pH} = 4.59$$

(b) $(CH_3)_3N(aq) + H_2O(l) \rightleftharpoons (CH_3)_3NH^+(aq) + OH^-(aq)$

i   0.15 M 0.12 M

c   -x   +x   +x

e   (0.15 - x) M (0.12 + x) M +x M

$$K_b = 6.4 \times 10^{-5} = \frac{[OH^-][(CH_3)_3NH^+]}{[(CH_3)_3N]} = \frac{(x)(0.12+x)}{(0.15-x)} \approx \frac{0.12\,x}{0.15}$$

$x = 8.0 \times 10^{-5} = [OH^-]$, pOH = 4.10, pH = 14.00 - 4.10 = 9.90

17.7 $HBu(aq) \rightleftharpoons H^+(aq) + Bu^-(aq)$ $K_a = \dfrac{[H^+][Bu^-]}{[HBu]} = 1.5 \times 10^{-5}$

equil (a)   0.10 - x M x M x M

equil (b)   0.10 - x M x M 0.050 + x M

(a) $K_a = 1.5 \times 10^{-5} = \dfrac{x^2}{0.10-x} \approx \dfrac{x^2}{0.10}$; $x = [H^+] = 1.225 \times 10^{-3} = 1.2 \times 10^{-3}\, M$

% ionization = $\dfrac{1.2 \times 10^{-3}\, M\, H^+}{0.10\, M\, HBu} \times 100 = 1.2\%$

(b) $K_a = 1.5 \times 10^{-5} = \dfrac{(x)(0.050+x)}{0.10-x} \approx \dfrac{0.050\,x}{0.10}$; $x = 3.0 \times 10^{-5}\, M\, H^+$

% ionization = $\dfrac{3.0 \times 10^{-5}\, M\, H^+}{0.10\, M\, HBu} \times 100 = 0.030\%$ ionization

Buffers

17.9 $HC_2H_3O_2$ and $NaC_2H_3O_2$ are a weak conjugate acid/conjugate base pair which act as a buffer because unionized $HC_2H_3O_2$ reacts with added base, while $C_2H_3O_2^-$ combines with added acid, leaving $[H^+]$ relatively unchanged. Although HCl and KCl are a conjugate acid/conjugate base pair, Cl^- is a negligible base. That is, it has no tendency to combine with added acid to form unionized HCl. Any added acid simply increases $[H^+]$ in an HCl/KCl mixture. In general, the conjugate bases of strong acids are negligible and mixtures of strong acids and their conjugate salts do not act as buffers.

17.11   Assume that % ionization is small in these buffers (Exercises 17.7 and 17.8).

(a) $K_a = \dfrac{[H^+][CHO_2^-]}{[HCHO_2]}$; $[H^+] = \dfrac{K_a[HCHO_2]}{[CHO_2^-]} \approx \dfrac{1.8 \times 10^{-4}\,(0.20)}{(0.15)}$

$[H^+] = 2.40 \times 10^{-4} = 2.4 \times 10^{-4}\, M$;   pH = -log (2.40×10^{-4}) = 3.62

(b)   mol = $M \times$ L; total volume = 85 mL + 95 mL = 180 mL = 0.180 L

$$[H^+] = \frac{K_a\,(0.16\, M \times 0.085\, L)/0.180\, L}{(0.15\, M \times 0.095\, L)/0.180\, L} = \frac{1.8 \times 10^{-4}\,(0.0136)}{0.01425}$$

(The total volume of the solution cancels in this calculation.)

$[H^+] = 1.718 \times 10^{-4} = 1.7 \times 10^{-4}\, M$; pH = -log (1.718×10^{-4}) = 3.77

17.13   (a) $HC_2H_3O_2(aq) \rightleftharpoons H^+(aq) + C_2H_3O_2^-(aq); \quad K_a = 1.8 \times 10^{-5} = \dfrac{[H^+][C_2H_3O_2^-]}{[HC_2H_3O_2]}$

$$[HC_2H_3O_2] = \frac{20.0 \text{ g } HC_2H_3O_2}{2.00 \text{ L soln}} \times \frac{1 \text{ mol } HC_2H_3O_2}{60.05 \text{ g } HC_2H_3O_2} = 0.167 \text{ } M$$

$$[C_2H_3O_2^-] = \frac{20.0 \text{ g } NaC_2H_3O_2}{2.00 \text{ L soln}} \times \frac{1 \text{ mol } NaC_2H_3O_2}{82.04 \text{ g } NaC_2H_3O_2} = 0.122 \text{ } M$$

$$[H^+] = \frac{K_a[HC_2H_3O_2]}{[C_2H_3O_2^-]} = \frac{1.8 \times 10^{-5}(0.167 - x)}{(0.122 + x)} \approx \frac{1.8 \times 10^{-5}(0.167)}{(0.122)}$$

$[H^+] = 2.4843 \times 10^{-5} = 2.5 \times 10^{-5} \text{ } M, \quad pH = 4.60$

    (b) $Na^+(aq) + C_2H_3O_2^-(aq) + H^+(aq) + Cl^-(aq) \rightarrow HC_2H_3O_2(aq) + Na^+(aq) + Cl^-(aq)$

    (c) $HC_2H_3O_2(aq) + Na^+(aq) + OH^-(aq) \rightarrow C_2H_3O_2^-(aq) + H_2O(l) + Na^+(aq)$

17.15   In this problem, $[BrO^-]$ is the unknown.

pH = 8.80, $[H^+] = 10^{-8.80} = 1.585 \times 10^{-9} = 1.6 \times 10^{-9} \text{ } M$

$[HBrO] = 0.200 - 1.6 \times 10^{-9} \approx 0.200 \text{ } M$

$K_a = 2.5 \times 10^{-9} = \dfrac{1.585 \times 10^{-9} [BrO^-]}{0.200}; \quad [BrO^-] = 0.3155 = 0.32 \text{ } M$

For 1.00 L, 0.32 mol NaBrO are needed.

17.17   (a) $K_a = \dfrac{[H^+][C_2H_3O_2^-]}{[HC_2H_3O_2]}; \quad [H^+] = \dfrac{K_a[HC_2H_3O_2]}{[C_2H_3O_2^-]}$

$$[H^+] \approx \frac{1.8 \times 10^{-5}(0.11)}{(0.15)} = 1.320 \times 10^{-5} = 1.3 \times 10^{-5} \text{ } M; \quad pH = 4.88$$

    (b) $HC_2H_3O_2(aq) + KOH(aq) \rightarrow C_2H_3O_2^-(aq) + H_2O(l) + K^+(aq)$

0.11 mol	0.02 mol	0.15 mol
-0.02 mol	-0.02 mol	+0.02 mol
0.09 mol	0 mol	0.17 mol

$$[H^+] = \frac{1.8 \times 10^{-5}(0.09 \text{ mol}/0.100 \text{ L})}{(0.17 \text{ mol}/0.100 \text{ L})} = 9.53 \times 10^{-6} = 1 \times 10^{-5} \text{ } M; \quad pH = 5.02 = 5.0$$

    (c) $C_2H_3O_2^-(aq) + HCl(aq) \rightarrow HC_2H_3O_2(aq) + Cl^-(aq)$

0.15 mol	0.02 mol	0.11 mol
-0.02 mol	-0.02 mol	+0.02 mol
0.13 mol	0 mol	0.13 mol

$$[H^+] = \frac{1.8 \times 10^{-5}(0.13 \text{ mol}/0.100 \text{ L})}{(0.13 \text{ mol}/0.100 \text{ L})} = 1.8 \times 10^{-5} \text{ } M; \quad pH = 4.74$$

17.19 $H_2CO_3(aq) \rightleftharpoons H^+(aq) + HCO_3^-(aq)$ $K_a = \dfrac{[H^+][HCO_3^-]}{[H_2CO_3]}$; $\dfrac{[HCO_3^-]}{[H_2CO_3]} = \dfrac{K_a}{[H^+]}$

(a) at pH = 7.4, $[H^+] = 10^{-7.4} = 4.0 \times 10^{-8}$ M ; $\dfrac{[HCO_3^-]}{[H_2CO_3]} = \dfrac{4.3 \times 10^{-7}}{4.0 \times 10^{-8}} = 11$

(b) at pH = 7.1, $[H^+] = 7.9 \times 10^{-8}$ M; $\dfrac{[HCO_3^-]}{[H_2CO_3]} = 5.4$

Acid-Base Titrations

17.21 (a) 40.0 mL $HNO_3 \times \dfrac{0.0350 \text{ mol } HNO_3}{1000 \text{ mL soln}} \times \dfrac{1 \text{ mol NaOH}}{1 \text{ mol } HNO_3} \times \dfrac{1000 \text{ mL soln}}{0.0350 \text{ mol NaOH}}$

$= 40.0$ mL NaOH soln

(b) 65.0 mL HBr $\times \dfrac{0.0620 \text{ } M \text{ HBr}}{1000 \text{ mL soln}} \times \dfrac{1 \text{ mol NaOH}}{1 \text{ mol HBr}} \times \dfrac{1000 \text{ mL soln}}{0.0350 \text{ mol NaOH}}$

$= 115$ mL NaOH soln

(c) $\dfrac{1.65 \text{ g HCl}}{1 \text{ L soln}} \times \dfrac{1 \text{ mol HCl}}{36.46 \text{ g HCl}} = 0.04526 = 0.0453 \text{ } M \text{ HCl}$

80.0 mL HCl $\times \dfrac{0.04526 \text{ mol HCl}}{1000 \text{ mL}} \times \dfrac{1 \text{ mol NaOH}}{1 \text{ mol HCL}} \times \dfrac{1000 \text{ mL soln}}{0.0350 \text{ mol NaOH}}$

$= 103.45 = 103$ mL NaOH soln

17.23 Construct a table similar to Table 17.1.

moles $H^+ = M_{HBr} \times L_{HBr} = 0.200 \text{ } M \times 0.0200 \text{ L} = 4.00 \times 10^{-3}$ mol

moles $OH^- = M_{NaOH} \times L_{NaOH} = 0.200 \text{ } M \times L_{NaOH}$

	mL_{HBr}	mL_{NaOH}	Total Volume	Moles H^+	Moles OH^-	Molarity Excess Ion	pH
(a)	20.0	15.0	35.0	4.00×10^{-3}	3.00×10^{-3}	$0.0286(H^+)$	1.544
(b)	20.0	19.9	39.9	4.00×10^{-3}	3.98×10^{-3}	$5 \times 10^{-4}(H^+)$	3.3
(c)	20.0	20.0	40.0	4.00×10^{-3}	4.00×10^{-3}	$1 \times 10^{-7}(H^+)$	7.0
(d)	20.0	20.1	40.1	4.00×10^{-3}	4.02×10^{-3}	$5 \times 10^{-4}(H^+)$	10.7
(e)	20.0	35.0	55.0	4.00×10^{-3}	7.00×10^{-3}	$0.0545(OH^-)$	12.737

molarity of excess ion = moles ion / total vol in L

(a) $\dfrac{4.00 \times 10^{-3} \text{ mol } H^+ - 3.00 \times 10^{-3} \text{ mol } OH^-}{0.0350 \text{ L}} = 0.0286 \text{ } M \text{ } H^+$

(b) $\dfrac{4.00 \times 10^{-3} \text{ mol } H^+ - 3.98 \times 10^{-3} \text{ mol } OH^-}{0.0339 \text{ L}} = 5.01 \times 10^{-4} = 5 \times 10^{-4} \text{ } M \text{ } H^+$

(c)   equivalence point, mol H^+ = mol OH^-

NaBr does not hydrolyze, so $[H^+]$ = $[OH^-]$ = 1×10^{-7} M

(d) $$\frac{4.02 \times 10^{-3}\, mol\, H^+ - 4.00 \times 10^{-3}\, mol\, OH^-}{0.041\, L} = 4.88 \times 10^{-4} = 5 \times 10^{-4}\, M\, OH^-$$

(e) $$\frac{7.00 \times 10^{-3}\, mol\, H^+ - 4.00 \times 10^{-3}\, mol\, OH^-}{0.0550\, L} = 0.054545 = 0.0545\, M\, OH^-$$

17.25 (a) The quantity of base required to reach the equivalence point is the same in the two titrations.

(b) The pH is higher initially in the titration of a weak acid.

(c) The pH is higher at the equivalence point in the titration of a weak acid.

(d) The pH in excess base is essentially the same for the two cases.

(e) In titrating a weak acid, one needs an indicator that changes at a higher pH than for the strong acid titration. The choice is more critical because the **change** in pH close to the equivalence point is smaller for the weak acid titration.

17.27 (a) At 0 mL, only weak acid, $HC_2H_3O_2$, is present in solution. Using the acid dissociation equilibrium

$$HC_2H_3O_2(aq) \rightleftharpoons H^+(aq) + C_2H_3O_2^-(aq)$$

initial   0.150 M 0   0

equil   0.150 - x M x M x M

$$K_a = \frac{[H^+][C_2H_3O_2^-]}{[HC_2H_3O_2]} = 1.8 \times 10^{-5}\ (Appendix\ D)$$

$$1.8 \times 10^{-5} = \frac{x^2}{(0.150 - x)} \approx \frac{x^2}{0.150};\ x^2 = 2.7 \times 10^{-6};\ x = [H^+] = 0.001643$$
$$= 1.6 \times 10^{-3};\ pH = 2.78$$

(b)-(f) Calculate the moles of each component after the acid-base reaction takes place. Moles $HC_2H_3O_2$ originally present = $M \times L$ = 0.150 $M \times$ 0.0500 L = 7.50×10^{-3} mol. Moles NaOH added = $M \times L$ = 0.0150 $M \times$ y mL.

$$NaOH(aq) + HC_2H_3O_2(aq) \rightarrow Na^+C_2H_3O_2^-(aq) + H_2O(l)$$

(0.150 $M \times$ 0.0250 L) =

(b) before rx 3.75×10^{-3} mol 7.50×10^{-3} mol

    after rx   0 3.75×10^{-3} mol 3.75×10^{-3} mol

(0.150 $M \times$ 0.0490 L) =

(c) before rx 7.35×10^{-3} mol 7.50×10^{-3} mol

    after rx   0 0.15×10^{-3} mol 7.35×10^{-3} mol

(0.150 $M \times$ 0.0500 L) =

(d) before rx 7.50×10^{-3} mol 7.50×10^{-3} mol

    after rx   0   0 7.50×10^{-3} mol

$$(0.150\ M \times 0.0510\ L) =$$

(e)   before rx 7.65×10^{-3} mol 7.50×10^{-3} mol

   after rx 0.15×10^{-3} mol   0 7.50×10^{-3} mol

$$(0.150\ M \times 0.0750\ L) =$$

(f)   before rx 11.25×10^{-3} mol 7.50×10^{-3} mol

   after rx 3.75×10^{-3} mol   0 7.50×10^{-3} mol

Calculate the molarity of each species (M = mol/L) and solve the appropriate equilibrium problem in each part.

(b) V_T = 50.0 mL $HC_2H_3O_2$ + 25.0 mL NaOH = 75.0 mL = 0.0750 L

$$[HC_2H_3O_2] = \frac{3.75 \times 10^{-3}\ \text{mol}}{0.0750} = 0.0500\ M$$

$$[C_2H_3O_2^-] = \frac{3.75 \times 10^{-3}\ \text{mol}}{0.0750} = 0.0500\ M$$

$$HC_2H_3O_2(aq) \rightleftharpoons H^+(aq)\ +\ C_2H_3O_2^-(aq)$$

equil $0.0500 - x\ M$ $x\ M$ $0.0500 + x\ M$

$$K_a = \frac{[H^+][C_2H_3O_2^-]}{[HC_2H_3O_2]};\quad [H^+] = \frac{K_a\,[HC_2H_3O_2]}{[C_2H_3O_2^-]}$$

$$[H^+] = \frac{1.8 \times 10^{-5}\,(0.0500 - x)}{(0.0500 + x)} = 1.8 \times 10^{-5}\ M\,H^+;\ \ pH = 4.74$$

(c) $$[HC_2H_3O_2] = \frac{0.15 \times 10^{-3}\ \text{mol}}{0.0990\ L} = 0.001515 = 1.5 \times 10^{-3}\ M$$

$$[C_2H_3O_2^-] = \frac{7.35 \times 10^{-3}\ \text{mol}}{0.0990\ L} = 0.07424 = 0.074\ M$$

$$[H^+] = \frac{1.8 \times 10^{-5}\,(1.515 \times 10^{-3} - x)}{(0.07424 + x)} \approx 3.7 \times 10^{-7}\ M\,H^+;\ \ pH = 6.43$$

(d)   At the equivalence point, only $C_3H_5O_2^-$ is present.

$$[C_2H_3O_2^-] = \frac{7.50 \times 10^{-3}\ \text{mol}}{0.100\ L} = 0.0750\ M$$

The pertinent equilibrium is the base hydrolysis of $C_2H_3O_2^-$.

$$C_2H_3O_2^-\,(aq) + H_2O(l) \rightleftharpoons HC_2H_3O_2\,(aq)\ + OH^-(aq)$$

initial   0.0750 M 0   0

equil $0.0750 - x\ M$ x   x

$$K_b = \frac{K_w}{K_a \text{ for } HC_2H_3O_2} = \frac{1.0 \times 10^{-14}}{1.8 \times 10^{-5}} = 5.56 \times 10^{-10} = 5.6 \times 10^{-10} = \frac{[HC_2H_3O_2][OH^-]}{[C_2H_3O_2^-]}$$

$$5.56 \times 10^{-10} = \frac{x^2}{0.0750 - x}; \quad x^2 \approx 5.56 \times 10^{-10} (0.0750); \quad x = 6.458 \times 10^{-6}$$

$$= 6.5 \times 10^{-10} \; M \; OH^-$$

$$pOH = -\log(6.458 \times 10^{-6}) = 5.19; \quad pH = 14.00 - pOH = 8.81$$

(e)   After the equivalence point, the excess strong base determines the pOH and pH. The $[OH^-]$ from the hydrolysis of $C_2H_3O_2^-$ is small and can be ignored.

$$[OH^-] = \frac{0.15 \times 10^{-3} \text{ mol}}{0.101 \text{ L}} = 1.485 \times 10^{-3} = 1.5 \times 10^{-3} \; M; \; pOH = 2.83$$

$$pH = 14.00 - 2.83 = 11.17$$

(f) $[OH^-] = \dfrac{3.75 \times 10^{-3} \text{ mol}}{0.125 \text{ L}} = 0.0300 \; M \; OH^-; \; pOH = 1.52; \; pH = 14.00 - 1.52 = 12.48$

17.29   The volume of 0.200 M HBr required in all cases equals the volume of base and the final volume = $2V_{base}$. The concentration of the salt produced at the equivalence point =

$$\frac{0.200 \, M \times V_{base}}{2V_{base}} = 0.100 \; M.$$

(a)   0.100 M NaBr, pH = 7.00

(b)   0.100 M HONH$_3^+$Br$^-$;   HONH$_3^+$(aq) $\rightleftharpoons$ H$^+$(aq) + HONH$_2$

                                         [equil]   0.100 - x   x   x

$$K_a = \frac{[H^+][HONH_2]}{[HONH_3^+]} = \frac{K_w}{K_b} = \frac{1.0 \times 10^{-14}}{1.1 \times 10^{-8}} = 9.09 \times 10^{-7} = 9.1 \times 10^{-7}$$

Assume x is small with respect to [salt].

$$K_a = x^2 / 0.100; \; x = [H^+] = 3.02 \times 10^{-4} = 3.0 \times 10^{-4} \; M, \; pH = 3.52$$

(c)   0.100 M C$_6$H$_5$NH$_3^+$Br$^-$.   Proceeding as in (b):

$$K_a = \frac{[H^+][C_6H_5NH_2]}{[C_6H_5NH_3^+]} = \frac{K_w}{K_b} = 2.33 \times 10^{-5} = 2.3 \times 10^{-5}$$

$$[H^+]^2 = 0.100(2.33 \times 10^{-5}); \; [H^+] = 1.52 \times 10^{-3} = 1.5 \times 10^{-3} \; M, \; pH = 2.82$$

Solubility Equilibria

17.31   (a)   The concentration of undissolved solid does not appear in the solubility produce expression because it is constant as long as there is solid present. Concentration is a ratio of moles solid to volume of the solid; solids occupy a specific volume not dependent on the solution volume. As the amount (moles) of solid changes, the volume changes proportionally, so that the ratio of moles solid to volume solid is constant.

(b) $K_{sp} = [Ag^+][I^-]$; $K_{sp} = [Ba^{2+}][CO_3^{2-}]$; $K_{sp} = [Cu^+]^2[S^{2-}]$

$K_{sp} = [Ce^{3+}][F^-]^3$; $K_{sp} = [Ca^{2+}]^3[PO_4^{3-}]^2$

17.33 (a) $CaF_2(s) \rightleftharpoons Ca^{2+}(aq) + 2F^-(aq)$; $K_{sp} = [Ca^{2+}][F^-]^2$

The molar solubility is the moles of CaF_2 that dissolve per liter of solution. Each mole of CaF_2 produces **1** mol $Ca^{2+}(aq)$ and **2** mol $F^-(aq)$.
$[Ca^{2+}] = 1.24 \times 10^{-3}$ M; $[F^-] = 2 \times 1.24 \times 10^{-3}$ $M = 2.48 \times 10^{-3}$ M
$K_{sp} = (1.24 \times 10^{-3})(2.48 \times 10^{-3})^2 = 7.63 \times 10^{-9}$

(b) $SrF_2(s) \rightleftharpoons Sr^{2+}(aq) + 2F^-(aq)$; $K_{sp} = [Sr^{2+}][F^-]^2$

Transform the gram solubility to molar solubility.

$$\frac{1.1 \times 10^{-2}\,g\,SrF_2}{0.100\,L} \times \frac{1\,mol\,SrF_2}{125.6\,g\,SrF_2} = 8.76 \times 10^{-4} = 8.8 \times 10^{-4}\,mol\,SrF_2/L$$

$[Sr^{2+}] = 8.76 \times 10^{-4}$ M; $[F^-] = 2(8.76 \times 10^{-4}$ $M)$

$K_{sp} = (8.76 \times 10^{-4})(2(8.76 \times 10^{-4}))^2 = 2.7 \times 10^{-9}$

(c) $Ba(IO_3)_2(s) \rightleftharpoons Ba^{2+}(aq) + 2IO_3^-(aq)$; $K_{sp} = [Ba^{2+}][IO_3^-]^2$

Since 1 mole of dissolved $Ba(IO_3)_2$ produces 1 mole of Ba^{2+}, the molar solubility of

$Ba(IO_3)_2 = [Ba^{2+}]$. Let $x = [Ba^{2+}]$; $[IO_3^-] = 2x$

$K_{sp} = 6.0 \times 10^{-10} = (x)(2x)^2$; $4x^3 = 6.0 \times 10^{-10}$; $x^3 = 1.5 \times 10^{-10}$; $x = 5.3 \times 10^{-4}$ M

The molar solubility of $Ba(IO_3)_2$ is 5.3×10^{-4} mol/L.

17.35 $CaC_2O_4(s) \rightleftharpoons Ca^{2+}(aq) + C_2O_4^{2-}(aq)$; $K_{sp} = [Ca^{2+}][C_2O_4^{2-}]$

$$[Ca^{2+}] = [C_2O_4^{2-}] = \frac{0.0061\,g\,CaC_2O_4}{1.00\,L\,soln} \times \frac{1\,mol\,CaC_2O_4}{128.1\,g\,CaC_2O_4} = 4.76 \times 10^{-5} = 4.8 \times 10^{-5}\,M$$

$K_{sp} = (4.76 \times 10^{-5}$ $M)(4.76 \times 10^{-5}$ $M) = 2.3 \times 10^{-9}$

17.37 (a) $AgBr(s) \rightleftharpoons Ag^+(aq) + Br^-(aq)$; $K_{sp} = [Ag^+][Br^-] = 5.0 \times 10^{-13}$
molar solubility $= x = [Ag^+] = [Br^-]$; $K_{sp} = x^2$
$x = (5.0 \times 10^{-13})^{1/2}$; $x = 7.1 \times 10^{-7}$ mol AgBr/L

(b) Molar solubility $= x = [Br^-]$; $[Ag^+] = 0.030$ $M + x$
$K_{sp} = (0.030 + x)(x) \approx 0.030(x)$
$5.0 \times 10^{-13} = 0.030(x)$; $x = 1.7 \times 10^{-11}$ mol AgBr/L

(c)   Molar solubility = x = $[Ag^+]$

There are two sources of Br^-: $NaBr(0.50\ M)$ and $AgBr(x\ M)$

$K_{sp} = (x)(0.50 + x)$; Assuming x is small compared to 0.50 M

$5.0 \times 10^{-13} = 0.50\ (x)$; $x \approx 1.0 \times 10^{-12}$ mol AgBr/L

17.39 $Cu(OH)_2(s) \rightleftharpoons Cu^{2+}(aq) + 2OH^-(aq)$; $K_{sp} = 2.2 \times 10^{-20}$

Since the $[OH^-]$ is set by the pH of the solution, the solubility of $Cu(OH)_2$ is just $[Cu^{2+}]$.

(a)   pH = 7.0, pOH = 14 - pH = 7.0, $[OH^-] = 10^{-pOH} = 1.0 \times 10^{-7}\ M$

$K_{sp} = 2.2 \times 10^{-20} = [Cu^{2+}](1.0 \times 10^{-7})^2$; $[Cu^{2+}] = \dfrac{2.2 \times 10^{-20}}{1.0 \times 10^{-14}} = 2.2 \times 10^{-6}\ M$

(In pure water, $[OH^-]$ from $Cu(OH)_2$ is similar to (OH^-) from the autoionization of water, resulting in a cubic equation for $[Cu^{2+}]$. The solubility of $Cu(OH)_2$ at pH = 7.0 is actually greater than the solubility in pure water.)

(b)   pH = 9.0, pOH = 5.0, $[OH^-] = 1.0 \times 10^{-5}$

$K_{sp} = 2.2 \times 10^{-20} = [Cu^{2+}][1.0 \times 10^{-5}]^2$; $[Cu^{2+}] = \dfrac{2.2 \times 10^{-20}}{1.0 \times 10^{-10}} = 2.2 \times 10^{-10}\ M$

(c)   pH = 11.0, pOH = 3.0, $[OH^-] = 1.0 \times 10^{-3}$

$K_{sp} = 2.2 \times 10^{-20} = [Cu^{2+}][1.0 \times 10^{-3}]^2$; $[Cu^{2+}] = \dfrac{2.2 \times 10^{-20}}{1.0 \times 10^{-6}} = 2.2 \times 10^{-14}\ M$

17.41 Let the molar solubility of CaF_2 = x. $K_{sp} = 3.9 \times 10^{-11}$

$$BaF_2 = y \quad K_{sp} = 1.0 \times 10^{-6}$$

$[F^-] = 0.010\ M + x + y$; assume x and y are small compared to 0.010.

CaF_2; $K_{sp} = [Ca^{2+}][F^-]^2$; $3.9 \times 10^{-11} \approx (x)\ (0.010)^2$; $x = 3.9 \times 10^{-7}$ mol CaF_2/L

BaF_2; $K_{sp} = [Ca^{2+}][F^-]^2$; $1.0 \times 10^{-6} \approx (y)\ (0.010)^2$; $y = 1.0 \times 10^{-2}$ mol BaF_2/L

$$\dfrac{[Ca^{2+}]}{[Ba^{2+}]} = \dfrac{3.9 \times 10^{-7}\ M}{1.0 \times 10^{-2}\ M} = 3.9 \times 10^{-5}\ \text{mol } Ca^{2+}\ /\ 1.0\ \text{mol } Ba^{2+}$$

17.43 $K_{sp} = [Ba^{2+}][MnO_4^-]^2 = 2.5 \times 10^{-10}$

$[MnO_4^-]^2 = 2.5 \times 10^{-10} / 2.0 \times 10^{-8} = 0.0125$; $[MnO_4^-] = \sqrt{0.0125} = 0.11\ M$

17.45 If the anion in the slightly soluble salt is the conjugate base of a strong acid, there will be no reaction.

(a) $MnS(s) + 2H^+(aq) \rightarrow H_2S(aq) + Mn^{2+}(aq)$

(b) $PbF_2(s) + 2H^+(aq) \rightarrow 2HF(aq) + Pb^{2+}(aq)$

(c) $AuCl_3(s) + H^+(aq) \rightarrow$ no reaction

(d) $Hg_2C_2O_4(s) + 2H^+(aq) \rightarrow H_2C_2O_4(aq) + Hg_2^{2+}(aq)$

(e) $CuBr(s) + H^+(aq) \rightarrow$ no reaction

17.47   The formation equilibrium is

$$Cu^{2+}(aq) + 4NH_3(aq) \rightleftharpoons Cu(NH_3)_4^{2+}(aq) \quad K_f = \frac{[Cu(NH_3)_4^{2+}]}{[Cu^{2+}][NH_3]^4} = 5 \times 10^{12}$$

Assuming that nearly all the Cu^{2+} is in the form $Cu(NH_3)_4^{2+}$

$[Cu(NH_3)_4^{2+}] = 1 \times 10^{-3} \ M; \ [Cu^{2+}] = x; \ [NH_3] = 0.10 \ M$

$$5 \times 10^{12} = \frac{(1 \times 10^{-3})}{x(0.10)^4}; \ x = 2 \times 10^{-12} \ M = [Cu^{2+}]$$

17.49
$$Ag \ I(s) \rightleftharpoons Ag^+(aq) + I^-(aq)$$
$$Ag^+(aq) + 2CN^-(aq \rightleftharpoons Ag(CN)_2^-(aq)$$
$$\overline{Ag \ I(s) + 2CN^-(aq) \rightleftharpoons Ag(CN)_2^-(aq) + I^-(aq)}$$

$$K = K_{sp} \times K_f = [Ag^+][I^-] \times \frac{[Ag(CN)_2^-]}{[Ag^+][CN^-]^2} = (8.3 \times 10^{-17})(1 \times 10^{-21}) = 8 \times 10^4$$

Precipitation; Qualitative Analysis

17.51   Precipitation conditions: will Q (see Chapter 15) exceed K_{sp} for the compound?

(a)   In base, Mn^{2+} can form $Mn(OH)_2(s)$.

$Mn(OH)_2(s) \rightleftharpoons Mn^{2+}(aq) + 2OH^-(aq); \ K_{sp} = [Mn^{2+}][OH^-]^2$

$Q = [Mn^{2+}][OH^-]^2; \ [Mn^{2+}] = 0.050 \ M; \ pOH = 6; \ [OH^-] = 10^{-6} = 1 \times 10^{-6} \ M$

$Q = (0.050)(1 \times 10^{-6})^2 = 5 \times 10^{-14}; \ K_{sp} = 1.9 \times 10^{-13}$ (Appendix D)

$Q < K_{sp}$, no $Mn(OH)_2$ precipitates.

(b) $Ag_2SO_4(s) \rightleftharpoons 2Ag^+(aq) + SO_4^{2-}(aq); \ K_{sp} = [Ag^+]^2[SO_4^{2-}]$

$$[Ag^+] = \frac{0.010 \ M \times 100 \ mL}{120 \ mL} = 8.33 \times 10^{-3} = 8.3 \times 10^{-3} \ M$$

$$[SO_4^{2-}] = \frac{0.050 \ M \times 20 \ mL}{120 \ mL} = 8.33 \times 10^{-3} = 8.3 \times 10^{-3} \ M$$

$Q = (8.33 \times 10^{-3})^2(8.33 \times 10^{-3}) = 5.8 \times 10^{-7}; \ K_{sp} = 1.4 \times 10^{-5}$

$Q < K_{sp}$, no Ag_2SO_4 precipitates.

17.53 $Ni(OH)_2(s) \rightleftharpoons Ni^{2+}(aq) + 2OH^-(aq); \ K_{sp} = [Ni^{2+}][OH^-]^2 = 1.6 \times 10^{-14}$

At equilibrium, $[Ni^{2+}][OH^-]^2 = 1.6 \times 10^{-14}$. Change $[Ni^{2+}]$ to mol/L and solve for $[OH^-]$.

$$\frac{1 \ \mu g \ Ni^{2+}}{1 \ L} \times \frac{1 \times 10^{-6} g}{1 \ \mu g} \times \frac{1 \ mol \ Ni^{2+}}{58.7 \ g \ Ni^{2+}} = 1.70 \times 10^{-8} = 2 \times 10^{-8} \ M \ Ni^{2+}$$

$1.6 \times 10^{-14} = (1.70 \times 10^{-8})[OH^-]^2; \ [OH^-]^2 = 9.39 \times 10^{-7}; \ [OH^-] = 9.69 \times 10^{-4}$
$$= 1 \times 10^{-3}$$

$pOH = 3.01; \ pH = 14.0 - 3.01 = 10.99 = 11.0$

17.55 Calculate $[I^-]$ needed to initiate precipitation of each ion. The cation that requires lower $[I^-]$ will precipitate first.

Ag^+: $K_{sp} = [Ag^+][I^-]$; $8.3 \times 10^{-17} = (2.0 \times 10^{-4})[I^-]$; $[I^-] = \dfrac{8.3 \times 10^{17}}{2.0 \times 10^{-4}} = 4.2 \times 10^{-13}$ M I^-

Pb^{2+}: $K_{sp} = [Pb^{2+}][I^-]^2$; $1.4 \times 10^{-8} = (1.5 \times 10^{-3})[I^-]^2$; $[I^-] = \left(\dfrac{1.4 \times 10^{-8}}{1.5 \times 10^{-3}}\right)^{1/2}$

$$= 3.1 \times 10^{-3} \ M \ I^-$$

Ag I will precipitate first, at $[I^-] = 4.2 \times 10^{-13}$ M.

17.57 The first two experiments eliminate Group 1 and 2 ions (Figure 17.22). The fact that no insoluble carbonates form in the filtrate from the third experiment rules out Group 4 ions. The ions which might be in the sample are those of Group 3, that is, Al^{3+}, Fe^{2+}, Zn^{2+}, Cr^{3+}, Ni^{2+}, Co^{2+}, or Mn^{2+}, and those of Group 5, NH_4^+, Na^+ or K^+.

17.59 (a) Make the solution acidic using 0.5 M HCl; saturate with H_2S. CdS will precipitate, ZnS will not.

     (b) Add excess base; $Fe(OH)_3(s)$ precipitates, but Cr^{3+} forms the soluble complex $Cr(OH)_4^-$.

     (c) Add $(NH_4)_2HPO_4$; Mg^{2+} precipitates as $MgNH_4PO_4$, K^+ remains in solution.

     (d) Add 6 M HCl, precipitate Ag^+ as AgCl(s).

17.61 (a) Because phosphoric acid is a weak acid, the concentration of free PO_4^{3-}(aq) in an aqueous phosphate solution is low except in strongly basic media. In less basic media, the solubility product of the phosphates that one wishes to precipitate is not exceeded.

     (b) K_{sp} for those cations in Group 3 is much larger. Thus, to exceed K_{sp} a higher $[S^{2-}]$ is required. This is achieved by making the solution more basic.

     (c) They should all redissolve in strongly acidic solution, e.g., in 12 M HCl (all the chlorides of Group 3 metals are soluble).

Additional Exercises

17.63 The equilibrium of interest is

$HC_5H_3O_3$(aq) $\rightleftharpoons$ H^+(aq) + $C_5H_3O_3^-$ (aq); $K_a = 6.76 \times 10^{-4} = \dfrac{[H^+][C_5H_3O_3^-]}{[HC_5H_3O_3]}$

Begin by calculating $[HC_5H_3O_3]$ and $[C_5H_3O_3^-]$ for each case.

     (a) $\dfrac{35.0 \text{ g } HC_5H_3O_3}{0.250 \text{ L soln}} \times \dfrac{1 \text{ mol } HC_5H_3O_3}{112.1 \text{ g } HC_5H_3O_3} = 1.249 = 1.25 \ M \ HC_5H_3O_3$

        $\dfrac{30.0 \text{ g } NaC_5H_3O_3}{0.250 \text{ L soln}} \times \dfrac{1 \text{ mol } NaC_5H_3O_3}{134.1 \text{ g } NaC_5H_3O_3} = 0.8949 = 0.895 \ M \ C_5H_3O_3^-$

$$[H^+] = \frac{K_a[HC_5H_3O_3]}{[C_5H_3O_3^-]} = \frac{6.76 \times 10^{-4}(1.249 - x)}{(0.8949 + x)} \approx \frac{6.76 \times 10^{-4}(1.249)}{(0.8949)}$$

$[H^+] = 9.43 \times 10^{-4}$ M, pH = 3.025

(b)   For dilution, $M_1V_1 = M_2V_2$

$$[HC_5H_3O_3] = \frac{0.250 \ M \times 30.0 \ mL}{125 \ mL} = 0.0600 \ M$$

$$[C_5H_3O_3^-] = \frac{0.220 \ M \times 20.0 \ mL}{125 \ mL} = 0.0352 \ M$$

$$[H^+] \approx \frac{6.76 \times 10^{-4}(0.0600)}{0.0352} = 1.15 \times 10^{-3} \ M, \ pH = 2.938$$

(yes, $[H^+]$ is < 5% of 0.0352 M)

(c)   0.0850 M × 0.500 L = 0.0425 mol $HC_5H_3O_3$

1.65 M × 0.0500 L = 0.0825 mol NaOH

	$HC_5H_3O_3$(aq)	+ NaOH(aq)	→	$NaC_5H_3O_3$(aq)	+ H_2O(l)
initial	0.0425 mol	0.0825 mol			
reaction	-0.0425 mol	-0.0425mol		+0.0425 mol	
after	0 mol	0.0400 mol		0.0425 mol	

The strong base NaOH dominates the pH; the contribution of $C_5H_3O_3^-$ is negligible. This combination would be "after the equivalence point" of a titration. The total volume is 0.550 L.

$$[OH^-] = \frac{0.0400 \ mol}{0.550 \ L} = 0.0727 \ M; \ pOH = 1.138, \ pH = 12.862$$

17.65 $K_a = \dfrac{[H^+][In^-]}{[HIn]}$; at pH = 4.68, [HIn] = [In$^-$]; $[H^+] = K_a$; pH = pK_a = 4.68

17.68   The pH of a buffer is centered around pK_a for its conjugate acid. For the bases in Table D.2, pK_a for the conjugate acids = 14 - pK_b. 14 - pK_b = 10.6; pK_b = 3.4, K_b = $10^{-3.4}$ = 4 × 10^{-4}. Select two bases with K_b values near 4 × 10^{-4}.

Methylamine, dimethylamine and ethylamine have K_b values closest to 4 × 10^{-4}, and ammonia and trimethylamine would probably also work. We will select methylamine and dimethylamine. (We could also select very weak acids with pK_a = 10.6, K_a = $10^{-10.6}$ ≈ 2.5 × 10^{-11}. Either HIO or HCO_3^- would be appropriate.)

In general, $BH^+(aq) \rightleftharpoons B(aq) + H^+(aq)$

$K_a = \dfrac{[B][H^+]}{[BH^+]}$; $[H^+] = \dfrac{K_a[BH^+]}{[B]}$; $[H^+] = 10^{-10.6} = 2.51 \times 10^{-11}$ M

For methylamine, $K_a = \dfrac{1.0 \times 10^{-14}}{4.4 \times 10^{-4}} = 2.272 \times 10^{-11} = 2.3 \times 10^{-11}$;

$\dfrac{[BH^+]}{[B]} = \dfrac{[H^+]}{K_a} = \dfrac{2.51 \times 10^{-11}}{2.27 \times 10^{-11}} = 1.1$

The ratio of $[CH_3NH_3^+]$ to $[CH_3NH_2]$ is 1.1 to 1.

For dimethylamine, $K_a = \dfrac{1.0 \times 10^{-14}}{5.4 \times 10^{-4}} = 1.852 \times 10^{-11} = 1.9 \times 10^{-11}$;

$\dfrac{[BH^+]}{[B]} = \dfrac{[H^+]}{K_a} = \dfrac{2.51 \times 10^{-11}}{1.85 \times 10^{-11}} = 1.4$

The ratio of $[(CH_3)_2NH_2^+]$ to $[(CH_3)_2NH]$ is 1.4 to 1. (The stronger base requires more of its conjugate acid to achieve a buffer of the same pH.)

17.71   (a) $\dfrac{0.4885 \text{ g KHP}}{0.100 \text{ L}} \times \dfrac{1 \text{ mol KHP}}{204.2 \text{ g KHP}} = 0.02392 = 0.0239$ M P^{2-} at the equivalence point

The pH at the equivalence point is determined by the hydrolysis of P^{2-}.

$P^{2-}(aq) + H_2O(l) \rightleftharpoons HP^-(aq) + OH^-(aq)$

$K_b = \dfrac{[HP^-][OH^-]}{[P^{2-}]} = \dfrac{K_w}{K_a \text{ for } HP^-} = \dfrac{1.0 \times 10^{-14}}{3.1 \times 10^{-6}} = 3.23 \times 10^{-9} = 3.2 \times 10^{-9}$

$3.23 \times 10^{-9} = \dfrac{x^2}{(0.02392 - x)} \approx \dfrac{x^2}{0.2392}$; $X = [OH^-] = 8.8 \times 10^{-6}$ M

pH = 14 - 5.06 = 8.94. From Figure 16.4, either phenolphthalein (pH 8.2 - 10.0) or thymol blue (pH 8.0 - 9.6) could be used to detect the equivalence point.

Phenolphthalein is usually the indicator of choice because the colorless to pink change is easier to see.

(b) $0.4885 \text{ g KHP} \times \dfrac{1 \text{ mol KHP}}{204.2 \text{ g KHP}} \times \dfrac{1 \text{ mol NaOH}}{1 \text{ mol KHP}} \times \dfrac{1}{0.03855 \text{ L NaOH}}$

$= 0.06206$ M NaOH

17.74   Assume that H_3PO_4 will react with NaOH in a stepwise fashion. (This is not unreasonable, since the three K_a values for H_3PO_4 are significantly different.)

	$H_3PO_4(aq)$	+ NaOH(aq)	$\rightarrow$ $H_2PO_4^-(aq)$	+ $Na^+(aq)$ + $H_2O(l)$
before	0.20 mol	0.30 mol	0 mol	
after	0 mol	0.10 mol	0.20 mol	

$$H_2PO_4^-(aq) + NaOH(aq) \rightarrow HPO_4^-(aq) + Na^+(aq) + H_2O(l)$$

before	0.20 mol	0.10 mol	0.25 mol
after	0.10 mol	0	0.35 mol

Thus, after all NaOH has reacted, the resulting 1.00 L solution is a buffer containing 0.10 mol $H_2PO_4^-$ and 0.35 mol HPO_4^{2-}. $H_2PO_4^-(aq) \rightleftharpoons H^+(aq) + HPO_4^{2-}(aq)$

$$K_a = 6.2 \times 10^{-8} = \frac{[HPO_4^{2-}][H^+]}{[H_2PO_4^-]}; \quad [H^+] = \frac{6.2 \times 10^{-8}(0.10\ M)}{0.35\ M} = 1.77 \times 10^{-8} = 1.8 \times 10^{-8}\ M;$$

$$pH = 7.75$$

17.77 $C_3H_5O_3^-$ will be formed by reaction of $HC_3H_5O_3$ with NaOH.
$0.1000\ M \times 0.05000\ L = 5.000 \times 10^{-3}$ mol $HC_3H_5O_3$; b = mol NaOH needed

$$HC_3H_5O_3 + NaOH \rightarrow C_3H_5O_3^- + H_2O + Na^+$$

initial	5.000×10^{-3}	b mol	
rx	-b mol	-b mol	+b mol
after rx	5.000×10^{-3} - b mol	0	b mol

$$K_a = \frac{[H^+][C_3H_5O_3^-]}{[HC_3H_5O_3]}; \quad K_a = 1.4 \times 10^{-4}; \quad [H^+] = 10^{-pH} = 10^{-3.50} = 3.16 \times 10^{-4} = 3.2 \times 10^{-4}\ M$$

Since solution volume is the same for $HC_3H_5O_3$ and $C_3H_5O_3^-$, we can use moles in the equation for $[H^+]$.

$$K_a = 1.4 \times 10^{-4} = \frac{3.16 \times 10^{-4}\ (b)}{(5.000 \times 10^{-3} - b)}$$

$0.4427\ (5.000 \times 10^{-3} - b) = b$, $2.214 \times 10^{-3} = 1.4427\ b$, $b = 1.53 \times 10^{-3}$

$$= 1.5 \times 10^{-3}\ mol\ OH^-$$

(The precision of K_a dictates that the result has 2 sig figs.)

Substituting this result into the K_a expression gives $[H^+] = 3.27 \times 10^{-4}$.

(Using 1.53×10^{-3} mol OH^- (3 sig figs) gives, $[H^+] = 3.16 \times 10^{-4}$, a more reassuring cross check.)

Calculate volume NaOH required from M = mol/L.

$$1.53 \times 10^{-3}\ mol\ OH^- \times \frac{1\ L}{1.000\ mol} \times \frac{1\ \mu L}{1 \times 10^{-6}\ L} = 1.5 \times 10^3\ \mu L\ (1.5\ mL)$$

17.80 **(a)** $K_{sp} = [Fe^{3+}][OH^-]^3 = 4 \times 10^{-38}$, $[Fe^{3+}] = x$, pH = 4.0, $[OH^-] = 1 \times 10^{-10}$

$(x)(1 \times 10^{-10})^3 = 4 \times 10^{-38}$, $x = 4 \times 10^{-8}\ M$

$$\frac{4 \times 10^{-8}\ mol\ Fe(OH)_3}{1\ L} \times \frac{107\ g\ Fe(OH)_3}{1\ mol\ Fe(OH)_3} = 4 \times 10^{-6}\ g\ Fe(OH)_3\ /\ L$$

(b) $[Fe^{3+}] = x$, pH $= 10.0$, $[OH^-] = 1 \times 10^{-4}$

$(x)(1 \times 10^{-4})^3 = 4 \times 10^{-38}$; $x = 4 \times 10^{-26}$ M

$$\frac{4 \times 10^{-26} \text{ mol Fe(OH)}_3}{1 \text{ L}} \times \frac{107 \text{ g Fe(OH)}_3}{1 \text{ mol Fe(OH)}_3} = 4 \times 10^{-24} \text{ g Fe(OH)}_3 / L$$

$Fe(OH)_3$ is much more soluble at the lower pH value.

17.83 $MgC_2O_4(s) \rightleftharpoons Mg^{2+}(aq) + C_2O_4^{2-}(aq)$

$K_{sp} = [Mg^{2+}][C_2O_4^{2-}] = 8.6 \times 10^{-5}$

If $[Mg^{2+}]$ is to be 3.0×10^{-2} M, $[C_2O_4^{2-}] = 8.6 \times 10^{-5} / 3.0 \times 10^{-2} = 2.87 \times 10^{-3} = 2.9 \times 10^{-3}$ M

The oxalate ion undergoes hydrolysis:

$C_2O_4^{2-}(aq) + H_2O(l) \rightleftharpoons HC_2O_4^-(aq) + OH^-(aq)$

$$K_b = \frac{[HC_2O_4^-][OH^-]}{[C_2O_4^{2-}]} = 1.0 \times 10^{-14} / 6.4 \times 10^{-5} = 1.56 \times 10^{-10} = 1.6 \times 10^{-10}$$

$[Mg^{2+}] = 3.0 \times 10^{-2}$ M, $[C_2O_4^{2-}] = 2.87 \times 10^{-3} = 2.9 \times 10^{-3}$ M

$[HC_2O_4^-] = (3.0 \times 10^{-2} - 2.87 \times 10^{-3})$ $M = 2.71 \times 10^{-2} = 2.7 \times 10^{-2}$ M

$[OH^-] = 1.56 \times 10^{-10} \times \dfrac{[C_2O_4^{2-}]}{[HC_2O_4^-]} = 1.56 \times 10^{-10} \times \dfrac{(2.87 \times 10^{-3})}{(2.71 \times 10^{-2})} = 1.65 \times 10^{-11}$

$= 1.7 \times 10^{-11}$ M; pH $= 3.22$

17.85 $PbCl_2(s) \rightleftharpoons Pb^{2+}(aq) + 2Cl^-(aq)$ $K_{sp} = 1.6 \times 10^{-5}$
$K_{sp} = 1.6 \times 10^{-5} = [Pb^{2+}][Cl^-]^2 = [Pb^{2+}][0.1]^2$; $[Pb^{2+}] = 1.6 \times 10^{-3}$ $M = 2 \times 10^{-3}$ M

17.88 $Zn(OH)_2(s) \rightleftharpoons Zn^{2+}(aq) + 2OH^-(aq)$ $K_{sp} = 1.2 \times 10^{-17}$
$Zn^{2+}(aq) + 4OH^-(aq) \rightleftharpoons Zn(OH)_4^{2-}(aq)$ $K_f = 4.6 \times 10^{17}$
$Zn(OH)_2(s) + 2OH^-(aq) \rightleftharpoons Zn(OH)_4^{2-}(aq)$ $K = K_{sp} \times K_f = 5.5$

$K = 5.52 = 5.5 = \dfrac{[Zn(OH)_4^{2-}]}{[OH^-]^2}$

If 0.010 mol $Zn(OH)_2$ dissolves, 0.010 mol $Zn(OH)_4^{2-}$ should be present at equilibrium.

$[OH^-]^2 = \dfrac{(0.010)}{5.52}$; $[OH^-] = 0.043$ M $[OH^-] \geq 0.043$ M or pH ≥ 12.63

Integrative Exercises

17.90   (a)   Complete ionic:

$H^+(aq) + Cl^-(aq) + Na^+(aq) + NO_2^-(aq) \rightarrow HNO_2(aq) + Na^+(aq) + Cl^-(aq)$

Na^+ and Cl^- are spectator ions.

Net ionic: $H^+(aq) + NO_2^-(aq) \rightleftharpoons HNO_2(aq)$

        (b)   The net ionic equation in part (a) is the reverse of the dissociation of HNO_2.

$$K = \frac{1}{K_a} = \frac{1}{4.5 \times 10^{-4}} = 2.22 \times 10^3 = 2.2 \times 10^3$$

        (c)   For Na^+ and Cl^-, this is just a dilution problem.

$M_1 V_1 = M_2 V_2$; V_2 is 50.0 mL + 50.0 mL = 100.0 mL

Cl^-: $\dfrac{0.15\, M \times 50.0\ \text{mL}}{100.0\ \text{mL}} = 0.075\, M$; Na^+: $\dfrac{0.15\, M \times 50.0\ \text{mL}}{100.0\ \text{mL}} = 0.075\, M$

H^+ and NO_2^- react to form HNO_2.   Since K >> 1, the reaction essentially goes to completion.

$$0.15\, M \times 0.0500\ \text{mL} = 7.5 \times 10^{-3}\ \text{mol}\ H^+$$
$$\underline{0.15\, M \times 0.0500\ \text{mL} = 7.5 \times 10^{-3}\ \text{mol}\ NO_2^-}$$
$$= 7.5 \times 10^{-3}\ \text{mol}\ HNO_2$$

Solve the weak acid problem to determine $[H^+]$, $[NO_2^-]$ and $[HNO_2]$ at equilibrium.

$$K_a = \frac{[H^+][NO_2^-]}{[HNO_2]}; \quad [H^+] = [NO_2^-] = x\, M; \quad [HNO_2] = \frac{(7.5 \times 10^{-3} - x)\ \text{mol}}{0.100\ \text{L}} = (0.075 - x)\, M$$

$$4.5 \times 10^{-4} = \frac{x^2}{(0.075 - x)} \approx \frac{x^2}{0.075}; \quad x = 5.8 \times 10^{-3}\, M$$

$$\frac{[H^+]}{[HNO_2]} \times 100 = \frac{5.8 \times 10^{-3}}{0.075} \times 100 = 7.8\%\ \text{dissociation}$$

Using the quadratic formula to determine x: $x^2 + 4.5 \times 10^{-4}\, x - 3.38 \times 10^{-5} = 0$

$$x = \frac{-4.5 \times 10^{-4} + \sqrt{(4.5 \times 10^{-4})^2 - 4(1)(-3.38 \times 10^{-5})}}{2}$$

$x = 5.58 \times 10^{-3} = 5.6 \times 10^{-3}\, M\ H^+$ and NO^{-2} (Note that $[H^+]$ is not very different from the value obtained using the assumption.) $[HNO_2] = 0.075 - 0.0056 = 0.069\, M$

In summary:

$[Na^+] = [Cl^-] = 0.075\, M, \quad [HNO_2] = 0.069\, M, \quad [H^+] = [NO_2^-] = 0.0056\, M$

17.92 $n = \dfrac{PV}{RT} = 735$ torr $\times \dfrac{1 \text{ atm}}{760 \text{ torr}} \times \dfrac{7.5 \text{ L}}{295 \text{ K}} \times \dfrac{\text{K} \cdot \text{mol}}{0.08206 \text{ L} \cdot \text{atm}} = 0.300 = 0.30$ mol NH_3

0.40 $M \times 0.50$ L = 0.20 mol HCl

	HCl(aq)	+	NH_3(g)	$\rightarrow$	NH_4^+(aq)	+	Cl^-(aq)
before	0.20 mol		0.30 mol				
after	0		0.10 mol		0.20 mol		0.20 mol

The solution will be a buffer because of the substantial concentrations of NH_3 and NH_4^+ present. Use K_a for NH_4^+ to describe the equilibrium.

$$NH_4^+(aq) \rightleftharpoons NH_3(aq) + H^+(aq)$$

equil.   0.20 - x   0.10 + x   x

$K_a = \dfrac{1.0 \times 10^{-14}}{1.8 \times 10^{-5}} = 5.56 \times 10^{-10} = 5.6 \times 10^{-10}$; $K_a = \dfrac{[NH_3][H^+]}{[NH_4^+]}$; $[H^+] = \dfrac{K_a[NH_4^+]}{[NH_3]}$

Since this expression contains a ratio of concentrations, volume will cancel and we can substitute moles directly. Assume x is small compared to 0.10 and 0.20.

$[H^+] = \dfrac{5.56 \times 10^{-10}(0.20)}{(0.10)} = 1.111 \times 10^{-9} = 1.1 \times 10^{-9}$ M, pH = 8.95

17.94 $\pi = MRT$, $M = \dfrac{\pi}{RT} = \dfrac{21 \text{ torr}}{298 \text{ K}} \times \dfrac{1 \text{ atm}}{760 \text{ torr}} \times \dfrac{\text{K} \cdot \text{mol}}{0.08206 \text{ L} \cdot \text{atm}} = 1.13 \times 10^{-3} = 1.1$ M

$SrSO_4$(s) $\rightleftharpoons$ Sr^{2+}(aq) + SO_4^{2-}(aq); $K_{sp} = [Sr^{2+}][SO_4^{2-}]$

The total particle concentration is 1.13×10^{-3} M. Each mole of $SrSO_4$ that dissolves produces 2 mol of ions, so $[Sr^{2+}] = [SO_4^{2-}] = 1.13 \times 10^{-3}$ M / 2 = 5.65×10^{-4} = 5.7×10^{-4} M.

$K_{sp} = (5.65 \times 10^{-4})^2 = 3.2 \times 10^{-7}$

18 Chemistry of the Environment

Earth's Atmosphere

18.1 (a) The temperature profile of the atmosphere (Figure 18.1) is the basis of its division into regions. The center of each peak or trough in the temperature profile corresponds to a new region.

(b) Troposphere, 0-12 km; stratosphere, 12-50 km; mesosphere, 50-85 km; thermosphere, 85-110 km.

18.3 $P_{Ar} = \chi_{Ar} \cdot P_{atm}$; $P_{Ar} = 0.00934 \, (765 \text{ torr}) = 7.15 \text{ torr}$

$P_{Ne} = \chi_{Ne} \cdot P_{atm}$; $P_{Ne} = 0.00001818 \, (765 \text{ torr}) = 0.0139 \text{ torr}$

18.5 $P_{Xe} = \chi_{Xe} \cdot P_{atm}$; $P_{Xe} = 8.7 \times 10^{-8} \, (0.94 \text{ atm}) = 8.178 \times 10^{-8} = 8.2 \times 10^{-8} \text{ atm}$

$$n_{Xe} \frac{P_{Xe}V}{RT} = \frac{8.178 \times 10^{-8} \text{ atm} \times 1.0 \text{ L}}{300 \text{ K}} \times \frac{\text{K} \cdot \text{mol}}{0.08206 \text{ L} \cdot \text{atm}} = 3.322 \times 10^{-9} = 3.3 \times 10^{-9} \text{ mol Xe}$$

$$3.322 \times 10^{-9} \text{ mol Xe} \times \frac{6.022 \times 10^{23} \text{ atoms}}{1 \text{ mol}} = 2.0 \times 10^{15} \text{ Xe atoms}$$

The Upper Atmosphere; Ozone

18.7 $$\frac{210 \times 10^3 \text{ J}}{1 \text{ mol}} \times \frac{1 \text{ mol}}{6.022 \times 10^{23} \text{ molecules}} = 3.487 \times 10^{-19} = 3.49 \times 10^{-19} \text{ J/molecule}$$

$\lambda = c/\nu$ We also have that $E = h\nu$, so $\nu = E/h$. Thus,

$$\lambda = \frac{hc}{E} = \frac{(6.626 \times 10^{-34} \text{ J} \cdot \text{sec}) \, (3.00 \times 10^8 \text{ m/sec})}{3.487 \times 10^{-19} \text{ J}} = 5.70 \times 10^{-7} \text{ m} = 570 \text{ nm}$$

18.9 The bond dissociation energy of N_2, 947 kJ/mol, is much higher than that of O_2, 495 kJ/mol. Photons with a wavelength short enough to photodissociate N_2 are not as abundant as the ultraviolet photons which lead to photodissociation of O_2. Also, N_2 does not absorb these photons as readily as O_2 so even if a short-wavelength photon is available, it may not be absorbed by an N_2 molecule.

18.11   (a)   Oxygen atoms exist longer at 120 km because there are fewer particles (atoms and molecules) at this altitude and thus fewer collisions and subsequent reactions that consume O atoms.

        (b)   Ozone is the primary absorber of high energy ultraviolet radiation in the 200-310 nm range. If this radiation were not absorbed in the stratosphere, plants and animals at the earth's surface would be seriously and adversely affected.

18.13   CFC stands for chlorofluorocarbon, a class of compounds that contain chlorine, fluorine and carbon. A common CFC is Freon-12, CF_2Cl_2.

18.15   (a) $HCl(g)$, $ClONO_2(g)$

        (b)   Neither HCl nor $ClONO_2$ react directly with ozone. The chlorine that is present in the "chlorine reservoir" does not participate in the destruction of ozone. Thus, the larger the "chlorine reservoir," the slower the rate of ozone depletion.

Chemistry of the Troposphere

18.17   (a)   CO binds with hemoglobin in the blood to block O_2 transport to the cells; people with CO poisoning suffocate from lack of O_2.

        (b) SO_2 is corrosive to lung tissue and contributes to higher levels of respiratory disease and shorter life expectancy, especially for people with other respiratory problems such as asthma. It also is a major source of acid rain, which damages forests and wildlife in natural waters.

        (c) O_3 is extremely reactive and toxic because of its ability to form free radicals upon reaction with organic molecules in the body. It is particularly dangerous for asthma suffers, exercisers and the elderly. O_3 can also react with organic compounds in polluted air to form peroxyacylnitrates, which cause eye irritation and breathing difficulties.

18.19   CO in unpolluted air is typically 0.05 ppm, whereas in urban air CO is about 10 ppm. A major source is automobile exhaust. SO_2 is less than 0.01 ppm in unpolluted air and on the order of 0.08 ppm in urban air. A major source is coal and oil-burning power plants, but there is also some SO_2 in auto exhausts. NO is about 0.01 ppm in unpolluted air and about 0.05 ppm in urban air. It comes mainly from auto exhausts.

18.21   All oxides of nonmetals produce acid solutions when dissolved in water. Sulfur oxides are produced naturally during volcanic eruptions and carbon oxides are products of combustion and metabolism. These dissolved gases cause rainwater to be naturally acidic.

18.23   Among the components of coal are sulfur-containing organic compounds. Combustion (oxidation) of these molecules produces $SO_2(g)$. Formation of $SO_3(g)$ requires further oxidation of SO_2 according to the reaction

$$2SO_2(g) + O_2(g) \rightleftharpoons 2\,SO_3(g)$$

Oxidation of $SO_2(g)$ to $SO_3(g)$ is significant but not complete, perhaps because C consumes most of the available $O_2(g)$.

18.25   (a)   Visible (Figure 6.4)

        (b) $E_{photon} = hc/\lambda = \dfrac{6.626 \times 10^{-34}\,\text{J}\cdot\text{s} \times 3.00 \times 10^8\,\text{m/s}}{420 \times 10^{-9}\,\text{m}} = 4.733 \times 10^{-19}$

$$= 4.73 \times 10^{-19}\ \text{J/photon}$$

$$\frac{4.733 \times 10^{-19}\,\text{J}}{1\ \text{photon}} \times \frac{6.022 \times 10^{23}\ \text{photons}}{1\ \text{mol}} \times \frac{1\ \text{kJ}}{1000\ \text{J}} = 285\ \text{kJ/mol}$$

18.27   Most of the energy entering the atmosphere from the sun is in the form of visible radiation, while most of the energy leaving the earth is in the form of infrared radiation. CO_2 is transparent to the incoming visible radiation, but absorbs the outgoing infrared radiation.

The World Ocean

18.29   A salinity of 5 denotes that there are 5 g of dry salt per kg of water.

$$\frac{5\ \text{g NaCl}}{1\ \text{kg soln}} \times \frac{1\ \text{kg soln}}{1\ \text{L soln}} \times \frac{1\ \text{mol NaCl}}{58\ \text{g NaCl}} \times \frac{1\ \text{mol Na}^+}{1\ \text{mol NaCl}} = 0.09\ M\ \text{Na}^+$$

18.31 1×10^{11} g Br $\times\ \dfrac{1 \times 10^3\,\text{g H}_2\text{O}}{0.067\ \text{g Br}} \times \dfrac{1\ \text{L H}_2\text{O}}{1 \times 10^3\,\text{g H}_2\text{O}} = 1.5 \times 10^{12}\ \text{L H}_2\text{O}$

Because the process is only 10% efficient, ten times this much, or 1.5×10^{13} L H_2O, must be processed.

Fresh Water

18.33   (a)   Decomposition of organic matter by aerobic bacteria depletes dissolved O_2. A low dissolved oxygen concentration indicates the presence of organic pollutants.

        (b)   According to Section 13.4, the solubility of $O_2(g)$ (or any gas) in water decreases with increasing temperature.

18.35   1.0 g $C_{18}H_{29}O_3S^-$ $\times\ \dfrac{1\ \text{mol C}_{18}\text{H}_{29}\text{O}_3\text{S}^-}{325\ \text{g C}_{18}\text{H}_{29}\text{O}_3\text{S}^-} \times \dfrac{51\ \text{mol O}_2}{2\ \text{mol C}_{18}\text{H}_{29}\text{O}_3\text{S}^-} \times \dfrac{32.0\ \text{g O}_2}{1\ \text{mol O}_2} = 2.5\ \text{g O}_2$

Notice that the mass of O_2 required is 2.5 times greater than the mass of biodegradable material.

18.37 $Ca^{2+}(aq) + 2HCO_3^-(aq) \rightarrow CaCO_3(s) + CO_2(g) + H_2O(l)$

18.39 $Ca(OH)_2$ is added to remove Ca^{2+} as $CaCO_3(s)$, and Na_2CO_3 removes the remaining Ca^{2+}.
$Ca^{2+}(aq) + 2HCO_3^-(aq) + [Ca^{2+}(aq) + 2OH^-(aq)] \rightarrow 2CaCO_3(s) + 2H_2O(l)$.   One mole
$Ca(OH)_2$ is needed for each 2 moles of $HCO_3^-(aq)$ present.   If there are 7.0×10^{-4} mol
$HCO_3^-(aq)$ per liter, we must add 3.5×10^{-4} mol $Ca(OH)_2$ per liter, or a total of 0.35 mol
$Ca(OH)_2$ for 10^3 L. This reaction removes 3.5×10^{-4} mol of the original Ca^{2+} from each
liter of solution, leaving 1.5×10^{-4} M $Ca^{2+}(aq)$. To remove this $Ca^{2+}(aq)$, we add
1.5×10^{-4} mol Na_2CO_3 per liter, or a total of 0.15 mol Na_2CO_3, forming $CaCO_3(s)$.

Additional Exercises

18.42   Avg. mol wt. at the surface = 40.0(0.20) + 16.0(0.35) + 32.0(0.45) = 28.0 = 28 g/mol.

Next, calculate the percentage composition at 200 km. The fractions can be "normalized" by
saying that the 0.45 fraction of O_2 is converted into <u>two</u> 0.45 fractions of O atoms, then dividing
by the total fractions, 0.20 + 0.35 + 0.45 + 0.45 = 1.45:

Avg. mol. wt. = $\dfrac{40.0(0.20) + 16.0(0.35) + 16.0(0.90)}{1.45}$ = 19 g/mol

18.44   (a)   The production of Cl atoms in the stratosphere is the result of the photodissociation
of a C–Cl bond in the chlorofluorocarbon molecule.

$$CF_2Cl_2(g) \xrightarrow{\ h\nu\ } CF_2Cl(g) + Cl(g)$$

According to Table 8.4, the bond dissociation energy of a C–Br bond is 276 kJ/mol,
while the value for a C–Cl bond is 328 kJ/mol. Photodissociation of $CBrF_3$ to form Br
atoms requires less energy than the production of Cl atoms and should occur readily
in the stratosphere.

(b) $CBrF_3(g) \xrightarrow{\ h\nu\ } CF_3(g) + Br(g)$ Also, under certain conditions
$BrO(g) + BrO(g) \longrightarrow Br_2O_2(g)$

$Br(g) + O_3(g) \longrightarrow BrO(g) + O_2(g)$ $Br_2O_2(g) + h\nu \longrightarrow O_2(g) + 2Br(g)$

18.47   From section 18.4:

$N_2(g) + O_2(g) \rightleftharpoons 2NO(g)$ $\Delta H = +180.8$ kJ   (1)
$2NO(g) + O_2(g) \rightleftharpoons 2NO_2(g)$ $\Delta H = -113.1$ kJ   (2)

In an endothermic reaction, heat is a reactant. As the temperature of the reaction increases,
the addition of heat favors formation of products and the value of K increases. The reverse
is true for exothermic reactions; as temperature increases, the value of K decreases. Thus,
K for reaction (1), which is endothermic, increases with increasing temperature and K for
reaction (2), which is exothermic, decreases with increasing temperature.

18.50 (a) According to Section 13.4, the solubility of gases in water decreases with increasing temperature. Thus, the solubility of $CO_2(g)$ in the ocean would decrease if the temperature of the ocean increased.

(b) If the solubility of $CO_2(g)$ in the ocean decreased because of global warming, more $CO_2(g)$ would be released into the atmosphere, perpetuating a cycle of increasing temperature and concomitant release of $CO_2(g)$ from the ocean.

18.52 (a) CO_3^{2-} is a relatively strong Brønsted base and produces OH^- in aqueous solution according to the hydrolysis reaction:

$$CO_3^{2-}(aq) + H_2O(l) \rightleftharpoons HCO_3^-(aq) + OH^-(aq), \quad K_b = 1.8 \times 10^{-4}$$

If $[OH^-(aq)]$ is sufficient to exceed K_{sp} for $Mg(OH)_2$, the solid will precipitate.

(b) $$\frac{150 \text{ mg Mg}^{2+}}{1 \text{ kg soln}} \times \frac{1 \text{ g Mg}^{2+}}{1000 \text{ mg Mg}^{2+}} \times \frac{1 \text{ kg soln}}{1 \text{ L soln}} \times \frac{1 \text{ mol Mg}^{2+}}{24.305 \text{ g Mg}^{2+}} = 6.172 \times 10^{-3}$$

$$= 6.17 \times 10^{-3} \text{ } M \text{ Mg}^{2+}$$

$$\frac{5.0 \text{ g Na}_2CO_3}{1.0 \text{ L soln}} \times \frac{1 \text{ mol CO}_3^{2-}}{106.0 \text{ g Na}_2CO_3} = 0.0472 = 0.047 \text{ } M \text{ CO}_3^{2-}$$

$$K_b = 1.8 \times 10^{-4} = \frac{[HCO_3^-][OH^-]}{[CO_3^{2-}]} \approx \frac{x^2}{0.0472}; \quad x = [OH^-] = 2.915 \times 10^{-3}$$

$$= 2.9 \times 10^{-3} \text{ } M$$

(This represents 6.2% hydrolysis, but the result will not be significantly different using the quadratic formula.)

$$Q = [Mg^{2+}][OH^-]^2 = (6.172 \times 10^{-3})(2.915 \times 10^{-3})^2 = 5.2 \times 10^{-8}$$

K_{sp} for $Mg(OH)_2 = 1.8 \times 10^{-11}$; $Q > K_{sp}$, so $Mg(OH)_2$ will precipitate.

18.54 (a) $Cl_2(g) + H_2O(l) \rightarrow HOCl(aq) + H^+(aq) + Cl^-(aq)$

(b) $Ca^{2+}(aq) + 2HCO_3^-(aq) \xrightarrow{\Delta} CaCO_3(s) + CO_2(g) + H_2O(l)$

Integrative Exercises

18.56 (a) $$0.021 \text{ ppm NO}_2 = \frac{0.021 \text{ mol NO}_2}{1 \times 10^6 \text{ mol air}} = 2.1 \times 10^{-8} = \chi_{NO_2}$$

$$P_{NO_2} = \chi_{NO_2} \cdot P_{atm} = 2.1 \times 10^{-8} \text{ (745 torr)} = 1.565 \times 10^{-5} = 1.6 \times 10^{-5} \text{ torr}$$

(b) $n = \dfrac{PV}{RT}$; molecules $= n \times \dfrac{6.022 \times 10^{23} \text{ molecules}}{\text{mol}} = \dfrac{PV}{RT} \times \dfrac{6.022 \times 10^{23} \text{ molecules}}{\text{mol}}$

$$V = 15 \text{ ft} \times 14 \text{ ft} \times 8 \text{ ft} \times \dfrac{12^3 \text{ in}^3}{\text{ft}^3} \times \dfrac{2.54^3 \text{ cm}^3}{\text{in}^3} \times \dfrac{1 \text{ L}}{1000 \text{ cm}^3} = 4.757 \times 10^4 = 5 \times 10^4 \text{ L}$$

$$1.565 \times 10^{-5} \text{ torr} \times \dfrac{1 \text{ atm}}{760 \text{ torr}} \times \dfrac{4.757 \times 10^4 \text{ L}}{293 \text{ K}} \times \dfrac{\text{K} \cdot \text{mol}}{0.08206 \text{ L} \cdot \text{atm}}$$

$$\times \dfrac{6.022 \times 10^{23} \text{ molecules}}{\text{mol}} = 2.453 \times 10^{19} = 2 \times 10^{19} \text{ molecules}$$

18.58   (a) $8{,}376{,}726 \text{ tons coal} \times \dfrac{83 \text{ ton C}}{100 \text{ ton coal}} \times \dfrac{44.01 \text{ ton CO}_2}{12.01 \text{ ton C}} = 2.5 \times 10^7 \text{ ton CO}_2$

                 $8{,}376{,}726 \text{ tons coal} \times \dfrac{2.5 \text{ ton S}}{100 \text{ ton coal}} \times \dfrac{64.07 \text{ ton SO}_2}{32.07 \text{ ton S}} = 4.18 \times 10^5$

                                                   $= 4.2 \times 10^5 \text{ ton SO}_2$

   (b) $CaO(s) + SO_2(g) \rightarrow CaSO_3(s)$

            $4.18 \times 10^5 \text{ ton SO}_2 \times \dfrac{55 \text{ ton SO}_2 \text{ removed}}{100 \text{ ton SO}_2 \text{ produced}} \times \dfrac{120.15 \text{ ton CaSO}_3}{64.07 \text{ ton SO}_2}$

                                                   $= 4.3 \times 10^5 \text{ ton CaSO}_3$

18.61   Osmotic pressure is determined by total concentration of dissolved particles. From Table 18.6, $M_T = 1.13$. The minimum pressure needed to initiate reverse osmosis is $\pi = MRT$, Equation [13.13]. Assuming 298 K,

$$\pi = MRT = \dfrac{1.13 \text{ mol}}{1 \text{ L}} \times 298 \text{ K} \times \dfrac{0.08206 \text{ L} \cdot \text{atm}}{\text{K} \cdot \text{mol}} = 27.6 \text{ atm}$$

18.64   Initial pressures: $P_{N_2} = 0.78 (1.5 \text{ atm}) = 1.17 = 1.2 \text{ atm}$; $P_{O_2} = 0.21 (1.5 \text{ atm}) = 0.315$

                                                                    $= 0.32 \text{ atm}$

    $P_{NO} = \dfrac{3200 \text{ atm}}{1 \times 10^6 \text{ atm}} = 1.5 \text{ atm} = 4.8 \times 10^{-3} \text{ atm}$

Calculate Q to determine which direction the reaction will proceed.

$$Q = \dfrac{P_{NO}^2}{P_{N_2} \times P_{O_2}} = \dfrac{(4.8 \times 10^{-3})^2}{(1.17)(0.315)} = 6.3 \times 10^{-5}, \; K_p = 1.0 \times 10^{-5}$$

$Q > K_p$, the reaction proceeds to the left.

	$N_2(g)$	+	$O_2(g)$	$\rightleftharpoons$	$2NO(g)$
initial	1.17 atm		0.315 atm		4.8×10^{-3} atm
change	+x		+x		-2x
equil	(1.17 + x)atm		(0.315 + x)atm		$(4.8 \times 10^{-3} - 2x)$ atm

$$K_p = 1.0 \times 10^{-5} = \frac{(4.8 \times 10^{-3} - 2x)^2}{(1.17 + x)(0.315 + x)}$$

Assume x is small compared to 1.17 and 0.315 (but not 4.8×10^{-3}).

P_{NO} at equilibrium = $(4.8 \times 10^{-3} - 2x) = y$

$$K_p = 1.0 \times 10^{-5} = \frac{y^2}{(1.17)(0.315)}; \quad y^2 = 3.686 \times 10^{-6}; \quad y = 1.92 \times 10^{-3} = 1.9 \times 10^{-3} \text{ atm}$$

Concentration, M = mol/L = n/V = p/RT;

$$[N_2] = \frac{1.17 \text{ atm}}{1173 \text{ K}} \times \frac{K \cdot mol}{0.08206 \text{ L} \cdot atm} = 1.2 \times 10^{-2} M$$

$$[O_2] = \frac{0.315 \text{ atm}}{1173 \text{ K}} \times \frac{K \cdot mol}{0.08206 \text{ L} \cdot atm} = 3.3 \times 10^{-3} M$$

$$[NO] = \frac{1.92 \times 10^{-3} \text{ atm}}{1173 \text{ K}} \times \frac{K \cdot mol}{0.0821 \text{ L} \cdot atm} = 2.0 \times 10^{-5} M \text{ (or 1300 ppm)}$$

Note that the effect of attaining equilibrium is to reduce the concentration of NO, without significantly changing the concentrations of N_2 and O_2.

18.66 (a) According to Table 18.1, the mole fraction of CO_2 in air is 0.000355.

$P_{CO_2} = \chi_{CO_2} \cdot P_{atm} = 0.000355 (1.00 \text{ atm}) = 3.55 \times 10^{-4} \text{ atm}$

$C_{CO_2} = kP_{CO_2} = 3.1 \times 10^{-2} M \text{/atm} \times 3.55 \times 10^{-4} \text{ atm} = 1.10 \times 10^{-5} = 1.1 \times 10^{-5} M$

(b) H_2CO_3 is a weak acid, so the $[H^+]$ is regulated by the equilibria:

$H_2CO_3(aq) \rightleftharpoons H^+(aq) + HCO_3^-(aq) \qquad K_{a1} = 4.3 \times 10^{-7}$

$HCO_3^-(aq) \rightleftharpoons H^+(aq) + CO_3^{2-}(aq) \qquad K_{a2} = 5.6 \times 10^{-11}$

Since the value of K_{a2} is small compared to K_{a1}, we will assume that most of the $H^+(aq)$ is produced by the first dissociation.

$K_{a1} = 4.3 \times 10^{-7} = \dfrac{[H^+][HCO_3^-]}{[H_2CO_3]}; \quad [H^+] = [HCO_3^-] = x, \; [H_2CO_3] = 1.1 \times 10^{-5} - x$

Since K_{a1} and $[H_2CO_3]$ have similar values, we cannot assume x is small compared to 1.1×10^{-5}.

$4.3 \times 10^{-7} = \dfrac{x^2}{(1.1 \times 10^{-5} - x)}; \quad 4.7 \times 10^{-12} - 4.3 \times 10^{-7} x = x^2$

$0 = x^2 + 4.3 \times 10^{-7} - 4.7 \times 10^{-12}$

$$x = \frac{-4.3 \times 10^{-7} \pm \sqrt{(4.3 \times 10^{-7})^2 - 4(1)(-4.73 \times 10^{-12})}}{2(1)}$$

$$x = \frac{-4.3 \times 10^{-7} \pm \sqrt{1.85 \times 10^{-13} + 1.89 \times 10^{-11}}}{2} = \frac{-4.3 \times 10^{-7} \pm 4.37 \times 10^{-6}}{2}$$

The negative result is meaningless; $x = 1.97 \times 10^{-6} = 2.0 \times 10^{-6}$ M H^+; pH = 5.71

Since this $[H^+]$ is quite small, the $[H^+]$ from the autoionization of water might be significant. Calculation shows that for $[H^+] = 2.0 \times 10^{-6}$ M, from H_2CO_3, $[H^+]$ from $H_2O = 5.2 \times 10^{-9}$ M, which we can ignore.

19 Chemical Thermodynamics

Spontaneity and Entropy

19.1   Spontaneous: a, d, e; nonspontaneous: b, c

19.3   (a) $NH_4NO_3(s)$ dissolves in water, as in a chemical cold pack. Naphthalene (moth balls) sublimes at room temperature.

(b)   Melting of a solid is spontaneous above its melting point but nonspontaneous below its melting point.

19.5   (a)   At or below $0°C$ (b)   above $0°C$

(c)   An increase in the entropy of the universe always accompanies a spontaneous process. When water freezes, ΔS_{sys} decreases, but because the reaction is exothermic, ΔS_{surr} increases. Below $0°C$, the increase in ΔS_{surr} predominates, and the process is spontaneous. Above $0°C$, the increase in ΔS_{surr} does not compensate for the decrease in ΔS_{sys}, and the process is not spontaneous.

19.7   (a)   E is a state function.

(b)   The quantities q and w depend on path.

(c)   If the process is *reversible* the system can move back and forth from one state to the other by the same path.

Entropy and the Second Law

19.9   (a)   If the entropy of the system increases, the final state is more disordered than the initial state. However, we cannot predict whether the process is spontaneous. A positive ΔS_{system} does not guarantee spontaneity. For an irreversible spontaneous process, $\Delta S_{univ} > 0$; ΔS_{system} may be positive or negative.

(b) ΔS is positive for Exercise 19.1 (a).

19.11   S increases in (a), (b) and (c); S decreases in (d).

19.13   (a) $Hg(l) \rightarrow Hg(s)$, ΔS is negative

        (b)   Fusion is melting, the opposite of freezing.

$\Delta H = -\Delta H_{fus} = -2.33$ kJ/mol $= -2.33 \times 10^3$ J/mol

$\Delta G = \Delta H - T\Delta S = 0$; $\Delta S = \Delta H/T$

$\Delta S = -2.33 \times 10^3$ J/mol $/ (273.15 - 38.9)$ K $= -9.95$ J/mol•K

19.15   (a) $\Delta S = nR \ln (V_2 / V_1) = 1.50$ mol $\times 8.314$ J/mol ($\ln 90.0$ L $/ 20.0$ L) $= +18.8$ J

        (b)   Yes. Expansion provides more possible positions and greater motional freedom for the gas molecules, so disorder or entropy of the system increases.

19.17   (a)   For a spontaneous process, the entropy of the universe increases; for a reversible process, the entropy of the universe does not change.

        (b)   In a reversible process, $\Delta S_{system} + \Delta S_{surroundings} = 0$. If ΔS_{system} is positive, $\Delta S_{surroundings}$ must be negative.

        (c)   Since $\Delta S_{universe}$ must be positive for a spontaneous process, $\Delta S_{surroundings}$ must be positive and greater than 75 J/K.

19.19   (a)   The entropy of a pure crystalline substance at absolute zero is zero.

        (b)   In *translational* motion, the entire molecule moves in a single direction; in *rotational* motion, the molecule rotates or spins around a fixed axis. *Vibrational* motion is reciprocating motion. The bonds within a molecule stretch and bend, but the average position of the atoms does not change.

        (c)   H—Cl ⟷   H——Cl ⟷   H-Cl

19.21   (a) $O_2(g)$ at 0.5 atm (larger volume and more motional freedom)

        (b) $Br_2(g)$ (gases have higher entropy due primarily to much larger volume)

        (c)   1 mol of $N_2(g)$ in 22.4 L (larger volume provides more motional freedom)

        (d) $CO_2(g)$ (more motional freedom)

19.23   (a) ΔS positive (moles of gas increase)

        (b) ΔS negative (increased order)

        (c) ΔS positive (gas produced, increased disorder)

        (d) ΔS negative (fewer moles of gas, increased order)

19.25   (a) $I_2(s)$, 116.73 J/mol•K; $I_2(g)$, 260.57 J/mol•K. In general, the gas phase of a substance has a larger $S°$ than the solid phase because of the greater volume and motional freedom of the molecules.

        (b)   NaBr(s), 86.82 J/mol•K; NaBr(aq), 141 J/mol•K. Ions in solution have much greater motional freedom than ions in a solid lattice.

(c)   C (diamond), 2.43 J/mol•K; C (graphite) 5.69 J/mol•K. Diamond is a network covalent solid with each C atom tetrahedrally bound to four other C atoms. Graphite consists of sheets of fused planar 6-membered rings with each C atom bound in a trigonal planar arrangement to three other C atoms. The internal entropy in graphite is greater because there is translational freedom among the planar sheets of C atoms while there is very little vibrational freedom within the network covalent diamond lattice.

(d)   1 mol of H_2(g), 130.58 J/K; 2 mol of H(g), 2(114.60) = 229.20 J/K. More particles have a greater number of degrees of freedom.

19.27 (a) $\Delta S° = S° \ C_2H_6(g) - S° \ C_2H_4(g) - S° \ H_2(g)$
              $= 229.5 - 219.4 - 130.58 = -120.5$ J/K

$\Delta S°$ is negative because there are fewer moles of gas in the products.

(b) $\Delta S° = 2S° \ NO_2(g) - \Delta S° \ N_2O_4(g) = 2(240.45) - 304.3 = +176.6$ J/K

$\Delta S°$ is positive because there are more moles of gas in the products.

(c) $\Delta S° = \Delta S° \ BeO(s) + \Delta S° \ H_2O(g) - \Delta S° \ Be(OH)_2(s)$
              $= 13.77 + 188.83 - 50.21 = +152.39$ J/K

$\Delta S°$ is positive because the product contains more total particles and more moles of gas.

(d) $\Delta S° = 2S° \ CO_2(g) + 4S° \ H_2O(g) - 2S° \ CH_3OH(g) - 3S° \ O_2(g)$
              $= 2(213.6) + 4(188.83) - 2(237.6) - 3(205.0) = +92.3$ J/K

$\Delta S°$ is positive because the product contains more total particles and more moles of gas.

(d) $\Delta S° = S° \ C_2H_6(g) + S° \ H_2(g) - 2S° \ CH_4(g)$
              $= 229.5 + 130.58 - 2(186.3) = -12.5$ J/K

$\Delta S°$ is very small because there are the same number of moles of gas in the products and reactants. The slight decrease is related to the relatively small S° value for H_2(g), which has fewer degrees of freedom than molecules with more than two atoms.

Gibbs Free Energy

19.29 (a) $\Delta G = \Delta H - T\Delta S$

(b)   If ΔG is positive, the process is nonspontaneous, but the reverse process is spontaneous.

(c)   There is no relationship between ΔG and rate of reaction. A spontaneous reaction, one with a $-\Delta G$, may occur at a very slow rate. For example: $2H_2(g) + O_2(g) \rightarrow 2H_2O(g)$, $\Delta G = -457$ kJ is very slow if not initiated by a spark.

19.31 $\Delta G^\circ = \Delta H^\circ - T\Delta S^\circ = -1.06 \times 10^5$ J $- 298$ K$(58$ J/K$) = -1.23 \times 10^5$ J $= -123$ kJ
The process is highly spontaneous. We do not expect $H_2O_2(g)$ to be stable at 298 K.

19.33   (a) $\Delta G^\circ = 2\Delta G^\circ$ HCl(g) $- [\Delta G^\circ$ H$_2$(g) $+ \Delta G^\circ$ Cl$_2$(g)]
          $= 2(-95.27$ kJ$) - 0 - 0 = -190.5$ kJ, spontaneous

(b) $\Delta G^\circ = \Delta G^\circ$ MgO(s) $+ 2\Delta G^\circ$ HCl(g) $- [\Delta G^\circ$ MgCl$_2$(s) $+ \Delta G^\circ$ H$_2$O(l)]
          $= -569.6 + 2(-95.27) - [-592.1 + (-237.13)] = +69.1$ kJ, nonspontaneous

(c) $\Delta G^\circ = \Delta G^\circ$ N$_2$H$_4$(g) $+ \Delta G^\circ$ H$_2$(g) $- 2\Delta G^\circ$ NH$_3$(g)
          $= 159.4 + 0 - 2(-16.66) = +192.7$ kJ, nonspontaneous

(d) $\Delta G^\circ = 2\Delta G^\circ$ NO(g) $+ \Delta G^\circ$ Cl$_2$(g) $- 2\Delta G^\circ$ NOCl(g)
          $= 2(86.71) + 0 - 2(66.3) = +40.8$ kJ, nonspontaneous

19.35   (a) ΔG is negative at low temperatures, positive at high temperatures. That is, the reaction proceeds in the forward direction spontaneously at lower temperatures but spontaneously reverses at higher temperatures.

(b) ΔG is positive at all temperatures. The reaction is nonspontaneous in the forward direction at all temperatures.

(c) ΔG is positive at low temperatures, negative at high temperatures. That is, the reaction will proceed spontaneously in the forward direction at high temperature.

19.37   At 330 K, $\Delta G < 0$; $\Delta G = \Delta H - T\Delta S < 0$
23 kJ $- 330$ K $(\Delta S) < 0$; 23 kJ < 330 K (ΔS); $\Delta S > 23$ kJ/300 K
$\Delta S > 0.070$ kJ/K or $\Delta S > +70$ J/K

19.39   (a) $\Delta G = \Delta H - T\Delta S$; $0 = 45$ kJ $- T(125$ J/K$)$; 45×10^3 J $= T(125$ J/K$)$
          $T = 45 \times 10^3$ J/125 J/K $= 360$ K

(b)   spontaneous

19.41   (a) $\Delta H^\circ = \Delta H^\circ$ CH$_3$OH(g) $- \Delta H^\circ$ CO(g) $- 2\Delta H^\circ$ H$_2$(g)
          $= -201.2 - (-110.5) - 2(0) = -90.7$ kJ

$\Delta S^\circ = S^\circ$ CH$_3$OH(g) $- S^\circ$ CO(g) $- 2S^\circ$ H$_2$(g)
          $= 237.6 - 197.9 - 2(130.58) = -221.46 = -221.5$ J/K

(b)   Since ΔS° is negative, $-T\Delta S$ will become more positive and ΔG° will become more positive with increasing temperature.

(c)   The data in Appendix C are tabulated at 298 K.

$$\Delta G° = \Delta G° \ CH_3OH(g) - \Delta G° \ CO(g) - 2\Delta G° \ H_2(g)$$
$$= -161.9 - (-137.2) - 2(0) = -24.7 \text{ kJ}$$

For comparison, $\Delta G° = \Delta H° - T\Delta S° = -90.7$ kJ $- (298 \text{ K})(-0.22146 \text{ kJ/K})$

$$\Delta G° = -90.7 \text{ kJ} + 66.0 \text{ kJ} = -24.7 \text{ kJ}$$

The reaction is spontaneous under standard conditions at 298 K.

(d) $\Delta G° = \Delta H° - T\Delta S° = -90.7$ kJ $- 500$ K$(-0.22146 \text{ kJ/K}) = +20.0$ kJ

The reaction is nonspontaneous under standard conditions at 500 K.

19.43   (a) $C_2H_2(g) + 5/2 \ O_2(g) \rightarrow 2CO_2(g) + H_2O(l)$

$$\Delta H° = 2\Delta H° \ CO_2(g) + \Delta H° \ H_2O(l) - \Delta H° \ C_2H_2(g) - 5/2\Delta H° \ O_2(g)$$
$$= 2(-393.5) - 285.83 - 226.7 = -1299.5 \text{ kJ produced/mol } C_2H_2 \text{ burned}$$

(b) $w_{max} = \Delta G° = 2\Delta G° \ CO_2(g) + \Delta G° \ H_2O(l) - \Delta G° \ C_2H_2(g) - \Delta G° \ O_2(g)$

$$= 2(-394.4) - 237.13 - 209.2 = -1235.1 \text{ kJ}$$

The negative sign indicates that the system does work on the surroundings; the system can accomplish a maximum of 1235.1 kJ of work on its surroundings.

Free Energy and Equilibrium

19.45   Use $\Delta G = \Delta G° + RT \ln Q$, where Q is the reaction quotient. When the numerator in Q increases, ln Q becomes more positive, in turn ΔG is more positive. When the denominator in Q becomes larger, Q decreases, ln Q becomes smaller, or more negative, and ΔG becomes smaller or more negative. (a) ΔG becomes more positive; (b) ΔG becomes more negative; (c) ΔG becomes more positive

19.47   (a) $\Delta G° = \Delta G° \ N_2O_4(g) - 2\Delta G° \ NO_2(g) = 98.28 - 2(51.84) = -5.40 \text{ kJ}$

(b) $\Delta G = \Delta G° + RT \ln P_{N_2O_4} / P_{NO_2}^2$

$$= -5.40 \text{ kJ} + \frac{8.314 \text{ J}}{1 \text{ K} \cdot \text{mol}} \times 298 \ln 0.10/(2.00)^2 = -14.54 \text{ kJ}$$

19.49 $\Delta G° = -RT \ln K_p$, Equation 19.20; $\ln K_p = -\Delta G°/RT$

(a) $\Delta G° = 2\Delta G° \ HI(g) - \Delta G° \ H_2(g) - \Delta G° \ I_2(g)$
$$= 2(1.30) - 19.37 = -16.77 \text{ kJ}$$

$$\ln K_p = \frac{-(-16.77 \text{ kJ}) \times 10^3 \text{ J/kJ}}{8.314 \text{ J/K} \times 298 \text{ K}} = 6.76876 = 6.769; \ K_p = 870$$

(b) $\Delta G° = \Delta G° \ C_2H_4(g) + \Delta G° \ H_2O(g) - \Delta G° \ C_2H_5OH(g)$
$$= 68.11 - 228.57 - (-168.5) = 8.04 = 8.0 \text{ kJ}$$

$$\ln K_p = \frac{-8.04 \text{ kJ} \times 10^3 \text{ J/kJ}}{8.314 \text{ J/K} \times 298 \text{ K}} = -3.24511 = -3.2; \ K_p = 0.04$$

(c) $\Delta G° = \Delta G° \; C_6H_6(g) - 3\Delta G° \; C_2H_2(g) = 129.7 - 3(209.2) = -497.9$ kJ

$\ln K_p = \dfrac{-\Delta G°}{RT} = \dfrac{-(-497.9 \text{ kJ}) \times 10^3 \text{ J/kJ}}{8.314 \text{ J/K} \times 298 \text{ K}} = 200.963 = 201.0; \quad K_p = 2 \times 10^{87}$

19.51 $K_p = P_{CO_2}$. Calculate $\Delta G°$ at the two temperatures using $\Delta G° = \Delta H° - T\Delta S°$ and then calculate K_p and P_{CO_2}.

$\Delta H° = \Delta H°_f \; CaO(s) + \Delta H°_f \; CO_2(g) - \Delta H°_f \; CaCO_3(s)$
$\quad\quad = -635.5 + -393.5 - (-1207.1) = +178.1$ kJ

$\Delta S° = S° \; CaO(s) + \Delta S° \; CO_2(g) - S° \; CaCO_3(s)$
$\quad\quad = 39.75 + 213.6 - 92.88 = +160.47$ J/K $= 0.1605$ kJ/K

(a) ΔG at 298 K $= 178.1$ kJ $- 298$ K $(0.16047$ kJ/K$) = 130.28 = 130.3$ kJ

$\ln K_p = \dfrac{-\Delta G°}{RT} = \dfrac{-130.28 \times 10^3 \text{ J}}{8.314 \text{ J/K} \times 298 \text{ K}} = -52.5837 = -52.58$

$K_p = 1.5 \times 10^{-23}; \quad P_{CO_2} = 1.5 \times 10^{-23}$ atm

(b) ΔG at 800 K $= 178.1$ kJ $- 800$ K $(0.16047$ kJ$) = 49.724 = 49.7$ kJ

$\ln K_p = \dfrac{-\Delta G°}{RT} = \dfrac{-49.724 \times 10^3 \text{ J}}{8.314 \text{ J/K} \times 800 \text{ K}} = -7.4759 = -7.48$

$K_p = 5.7 \times 10^{-4}; \quad P_{CO_2} = 5.7 \times 10^{-4}$ atm

19.53   (a) $\Delta G° = -RT \ln K_a = -(8.314)(298) \ln (6.5 \times 10^{-5}) = 23.89 = +23.9$ kJ

(b)   By definition $\Delta G = 0$, when the system is at equilibrium.

(c) $\Delta G = \Delta G° + RT \ln Q$

$\quad\quad = +23.87 \text{ kJ} + (8.314 \times 10^{-3})(298) \times \ln \dfrac{(3.0 \times 10^{-3})(2.0 \times 10^{-5})}{(0.10)} = -11.6$ kJ

The fact that ΔG is negative tells us that this particular mixture will move spontaneously in the direction of further ionization.

Additional Exercises

19.55   (a)   False. The essential question is whether the reaction proceeds far to the right before arriving at equilibrium. The position of equilibrium, which is the essential aspect, is not only dependent on ΔH but the entropy change as well.

(b)   True.

(c)   False. Spontaneity relates to the position of equilibrium in a process, not to the rate at which that equilibrium is approached.

(d)   True.

(e)   False. Such a process **might** be spontaneous, but would not necessarily be so. Spontaneous processes are those that are exothermic and/or that lead to increased disorder in the system.

19.59 (a)   Formation reactions are the synthesis of 1 mole of compound from elements in their standard states.

$1/2\ N_2(g) + 3/2\ H_2(g) \rightarrow NH_3(g)$

$C(s) + 2Cl_2(g) \rightarrow CCl_4(l)$

$K(s) + 1/2\ N_2(g) + 3/2\ O_2(g) \rightarrow KNO_3(s)$

In each of these formation reactions, there are fewer moles of gas in the products than the reactants, so we expect $\Delta S°$ to be negative. If $\Delta G_f° = \Delta H_f° - T\Delta S_f°$ and $\Delta S_f°$ are negative, $-T\Delta S_f°$ is positive and $\Delta G_f°$ is more positive than $\Delta H_f°$.

(b) $C(s) + 1/2\ O_2(g) \rightarrow CO(g)$

In this reaction, there are more moles of gas in products, $\Delta S_f°$ is positive, $-T\Delta S_f°$ is negative and $\Delta G_f°$ is more negative than $\Delta H_f°$.

19.61 $\Delta G = \Delta G° + RT\ \ln Q$

(a) $Q = \dfrac{P_{NH_3}^2}{P_{N_2} \times P_{H_2}^3} = \dfrac{(4.7)^2}{(8.5)(2.4)^3} = 0.188 = 0.19$

$\Delta G° = 2\Delta G°\ NH_3(g) - \Delta G°\ N_2(g) - 3\Delta G°\ H_2(g)$

$= 2(-16.66) - 0 - 3(0) = -33.32\ kJ$

$\Delta G = -33.32\ kJ + \dfrac{8.314 \times 10^{-3}\ kJ}{K\bullet mol} \times 298\ K \times \ln(0.188)$

$\Delta G = -33.32 - 4.14 = -37.46\ kJ$

(b) $Q = \dfrac{P_{N_2}^3 \times P_{H_2O}^4}{P_{N_2H_4}^2 \times P_{NO_2}^2} = \dfrac{(2.5)^3(1.2)^4}{(1.0 \times 10^{-3})^2(1.0 \times 10^{-3})^2} = 3.24 \times 10^{13} = 3.2 \times 10^{13}$

$\Delta G° = 3\Delta G_f°\ N_2(g) + 4\Delta G_f°\ H_2O(g) - 2\Delta G_f°\ N_2H_4(g) - 2\Delta G_f°\ NO_2(g)$

$= 0 + 4(-228.61) - 2(159.4) - 2(51.84) = -1336.9\ kJ$

$\Delta G = -1336.9\ kJ + 2.478\ \ln 3.24 \times 10^{13} = -1259.8\ kJ$

(c) $Q = \dfrac{P_{N_2} \times P_{H_2}^2}{P_{N_2H_4}} = \dfrac{(1 \times 10^{-3})(2 \times 10^{-4})^2}{6.0} = 6.67 \times 10^{-12} = 7 \times 10^{-12}$

$\Delta G° = \Delta G_f°\ N_2(g) + 2\Delta G_f°\ H_2(g) - \Delta G_f°\ N_2H_4$

$= 0 + 2(0) - 159.4 = -159.4\ kJ$

$\Delta G = -159.4\ kJ + 2.478\ \ln 6.67 \times 10^{-12} = -223.2\ kJ$

19.64 (a) $Ag_2O(s) \rightarrow 2Ag(s) + 1/2 \, O_2(g)$

(b) $\Delta G° = 2\Delta G_f° \, Ag(s) + 1/2 \, \Delta G_f° \, O_2(g) - \Delta G_f° \, Ag_2O(s)$

        $= 2(0) + 1/2 \, (0) - (-11.20) = +11.20 \text{ kJ}$

(c) $K_p = \sqrt{P_{O_2}}; \quad \Delta G° = -RT \ln K_p; \quad \ln K_p = -\Delta G°/RT$

        $\ln K_p = \dfrac{-11.20 \times 10^3 \text{ J}}{8.314 \text{ J/K} \times 298 \text{ K}} = -4.52055 = -4.521; \quad K_p = 1.0883 \times 10^{-2} = 1.09 \times 10^{-2}$

        $P_{O_2} = K_p^2 = 1.18 \times 10^{-4} \text{ atm} = 9.00 \times 10^{-2} \text{ torr}$

19.66 (a)   First calculate $\Delta G°$ for each reaction:

        For $C_6H_{12}O_6(s) + 6O_2(g) \rightleftharpoons 6CO_2(g) + 6H_2O(l)$ (A)

          $\Delta G° = 6(-237.13) + 6(-394.4) - (-910.4) = -2879 \text{ kJ}$

        For $C_6H_{12}O_6(s) \rightleftharpoons 2C_2H_5OH(l) + 2CO_2(g)$ (B)

          $\Delta G° = 2(-394.4) + 2(-174.8) - (-910.4) = -228 \text{ kJ}$

        For (A), $\ln K = 2879 \times 10^3/(8.314)(298) = 1162; \quad K = 5 \times 10^{504}$

        For (B), $\ln K = 228 \times 10^3/(8.314)(298) = 92.026 = 92.0; \quad K = 9 \times 10^{39}$

(b)   Both these values for K are unimaginably large. However, K for reaction (A) is larger, because $\Delta G°$ is more negative. The magnitude of the work that can be accomplished by coupling a reaction to its surroundings is measured by ΔG. According to the calculations above, considerably more work can in principle be obtained from reaction (A), because $\Delta G°$ is more negative.

19.69 $\Delta G°$ for the metabolism of glucose is:

        $6\Delta G° \, CO_2(g) + 6\Delta G° \, H_2O(l) - \Delta G° C_6H_{12}O_6(s) - 6\Delta G° \, O_2(g)$

        $\Delta G° = 6(-394.4) + 6(-237.13) - (-910.4 \text{ kJ} + 0) = -2878.8 \text{ kJ}$

        moles ATP $= -2878.8 \text{ kJ} \times 1 \text{ mol ATP} / (-30.5 \text{ kJ}) = 94.4 \text{ mol ATP} / \text{mol glucose}$

Integrative Exercises

19.73 (a) $O_2(g) \xrightarrow{h\nu} 2O(g); \Delta S$ increases because there are more moles of gas in the products.

(b) $O_2(g) + O(g) \rightarrow O_3(g), \Delta S$ decreases because there are fewer moles of gas in the products.

(c) ΔS increases as the gas molecules diffuse into the larger volume of the stratosphere; there are more possible positions and therefore more motional freedom.

(d) $SO_3(g) + H_2O(l) \rightarrow H_2SO_4(aq); \Delta S$ decreases because the motion of SO_3 molecules in aqueous solution is much more restricted than in the gas phase.

(e)   NaCl(aq) → NaCl(s) + H₂O(l); ΔS decreases as the mixture (seawater, greater disorder) is separated into pure substances (fewer possible arrangements, more order).

19.77   (a)   Ag(s) + 1/2 N₂(g) + 3/2 O₂(g) → AgNO₃(s); ΔS decreases because there are fewer moles of gas in the product.

(b) $\Delta G_f^{\circ} = \Delta H_f^{\circ} - T\Delta S_f^{\circ}$; $\Delta S_f^{\circ} = (\Delta G_f^{\circ} - \Delta H_f^{\circ}) / (-T) = (\Delta H_f^{\circ} - \Delta G_f^{\circ}) / T$

ΔS_f° = -124.4 kJ - (-33.4 kJ) / 298 K = -0.305 kJ/K = -305 J/K

ΔS_f° is relatively large and negative, as anticipated from part (a).

(c)   Dissolving of AgNO₃ can be expressed as

AgNO₃(s) → AgNO₃ (aq, 1 m)

$\Delta H^{\circ} = \Delta H_f^{\circ}$ AgNO₃(aq) - ΔH_f° AgNO₃(s) = -101.7 - (-124.4) = +22.7 kJ

$\Delta H^{\circ} = \Delta H_f^{\circ}$ MgSO₄(aq) - ΔH_f° MgSO₄(s) = -1374.8 - (-1283.7) = -91.1 kJ

Dissolving AgNO₃(s) is endothermic (+ΔH°), but dissolving MgSO₄(s) is exothermic (-ΔH°).

(d)   AgNO₃: $\Delta G^{\circ} = \Delta G_f^{\circ}$ AgNO₃(aq) - ΔG_f° AgNO₃(s) = -34.2 - (-33.4) = -0.8 kJ

$\Delta S^{\circ} = (\Delta H^{\circ} - \Delta G^{\circ}) / T$ = [22.7 kJ - (-0.8 kJ)] / 298 K = 0.0789 kJ/K = 78.9 J/K

MgSO₄: $\Delta G^{\circ} = \Delta G_f^{\circ}$ MgSO₄(aq) - ΔG_f° MgSO₄(s) = -1198.4 - (-1169.6) = -28.8 kJ

$\Delta S^{\circ} = (\Delta H^{\circ} - \Delta G^{\circ}) / T$ = [-91.1 kJ - (-28.8 kJ)] / 298 K = -0.209 kJ/K = -209 J/K

(e)   In general, we expect dissolving a crystalline solid to be accompanied by an increase in positional disorder and an increase in entropy; this is the case for AgNO₃ (ΔS° = + 78.9 J/K). However, for dissolving MgSO₄(s), there is a substantial decrease in entropy (ΔS = -209 J/K). According to Section 13.5, ion-pairing is a significant phenomenon in electrolyte solutions, particularly in concentrated solutions where the charges of the ions are greater than 1. According to Table 13.6, a 0.1 m MgSO₄ solution has a van't Hoff factor of 1.21. That is, for each mole of MgSO₄ that dissolves, there are only 1.21 moles of "particles" in solution instead of 2 moles of particles. For a 1 m solution, the factor is even smaller. Also, the exothermic enthalpy of mixing indicates substantial interactions between solute and solvent. Substantial ion-pairing coupled with ion-dipole interactions with H₂O molecules lead to a decrease in entropy for MgSO₄(aq) relative to MgSO₄(s).

19.80   The activated complex in Figure 14.12 is a single "particle" or entity that contains 4 atoms. It is formed from two molecules, A_2 and B_2, that must collide with exactly the correct energy and orientation to form the single entity. There are many fewer degrees of freedom for the activated complex than the separate reactant molecules, so the *entropy of activation* is negative.

20 Electrochemistry

Oxidation-Reduction Reactions

20.1   (a)   *Oxidation* is the loss of electrons.

(b)   The electrons appear on the products side (right side) of an oxidation half-reaction.

(c)   The *oxidant* is the reactant that is reduced; it gains the electrons that are lost by the substance being oxidized.

20.3   (a)   I is reduced from +5 to 0; C is oxidized from +2 to +4.

(b)   Hg is reduced from +2 to 0; N is oxidized from -2 to 0.

(c)   N is reduced from +5 to +2; S is oxidized from -2 to 0.

(d)   Cl is reduced from +4 to +3; O is oxidized from -1 to 0.

20.5   (a) $2PbS(s) + 3O_2(g) \; 2PbO(s) + 2SO_2(g)$

(b) O_2 is the oxidant, it undergoes reduction from the 0 to the -2 state. S is the reductant; it is oxidized from the -2 to the +4 state. The oxidation state of Pb is unchanged in the reaction.

20.7   (a) $Co^{2+}(aq) \rightarrow Co^{3+}(aq) + 1e^-$ (oxidation)

(b) $H_2O_2(aq) \rightarrow O_2(g) + 2H^+(aq) + 2e^-$ (oxidation)

(c) $ClO_3^-(aq) + 6H^+(aq) + 6e^- \rightarrow Cl^-(aq) + 3H_2O(l)$ (reduction)

(d) $4OH^-(aq) \rightarrow O_2(g) + 2H_2O(l) + 4e^-$ (oxidation)

(e) $SO_3^{2-}(aq) + 2OH^-(aq) \rightarrow SO_4^{2-}(aq) + H_2O(l) + 2e^-$ (oxidation)

20.9   (a) $Pb(OH)_4^{2-}(aq) + ClO^-(aq) \rightarrow PbO_2(s) + Cl^-(aq) + 2OH^-(aq) + H_2O(l)$

(b)   The half reactions are:

$3H_2O(l) + Tl_2O_3(s) + 4e^- \rightarrow 2TlOH(s) + 4OH^-(aq)$

$2[2NH_2OH(aq) + 2OH^-(aq) \rightarrow N_2(g) + 4H_2O(l) + 2e^-]$

Net: $\overline{Tl_2O_3(s) + 4NH_2OH(aq) \rightarrow 2TlOH(s) + 2N_2(g) + 5H_2O(l)}$

(c) $2[Cr_2O_7^{2-}(aq) + 14H^+(aq) + 6e^- \rightarrow 2Cr^{3+}(aq) + 7H_2O(l)]$

$3[CH_3OH(aq) + H_2O(l) \rightarrow HCO_2H(aq) + 4H^+(aq) + 4e^-]$

Net: $\overline{2Cr_2O_7^{2-}(aq) + 3CH_3OH(aq) + 16H^+(aq) \rightarrow 4Cr^{3+}(aq) + 3HCO_2H(aq) + 11H_2O(l)}$

(d) $2[MnO_4^-(aq) + 8H^+(aq) + 5e^- \rightarrow Mn^{2+}(aq) + 4H_2O(l)]$

                       $5[2Cl^-(aq) \rightarrow Cl_2(g) + 2e^-]$

Net: $2MnO_4^-(aq) + 10Cl^-(aq) + 16H^+(aq) \rightarrow 2Mn^{2+}(aq) + 5Cl_2(g) + 8H_2O(l)$

(e) $H_2O_2(aq) + 2e^- \rightarrow O_2(g) + 2H^+(aq)$

Since the reaction is in base, the H^+ can be "neutralized" by adding $2OH^-$ to each side of the equation to give $H_2O_2(aq) + 2OH^-(aq) + 2e^- \rightarrow O_2(g) + 2H_2O(l)$. The other half reaction is $2[ClO_2(aq) + e^- \rightarrow ClO_2^-(aq)]$.

Net: $H_2O_2(aq) + 2ClO_2(aq) + 2OH^-(aq) \rightarrow O_2(g) + 2ClO_2^-(aq) + 2H_2O(l)$

(f) $3[NO_2^-(aq) + H_2O(l) \rightarrow NO_3^-(aq) + 2H^+(aq) + 2e^-]$

           $Cr_2O_7^{2-}(aq) + 14H^+(aq) + 6e^- \rightarrow 2Cr^{3+}(aq) + 7H_2O(l)$

Net: $3NO_2^-(aq) + Cr_2O_7^{2-}(aq) + 8H^+(aq) \rightarrow 3NO_3^-(aq) + 2Cr^{3+}(aq) + 4H_2O(l)$

Voltaic Cells; Cell Potential

20.11   (a)   The reaction $Cu^{2+}(aq) + Zn(s) \rightarrow Cu(s) + Zn^{2+}(aq)$ is occurring in both Figures. In Figure 20.3, the reactants are in contact, and the concentrations of the ions in solution aren't specified. In Figure 20.4, the oxidation half-reaction and reduction half-reaction are occurring in separate compartments, joined by a porous connector. The concentrations of the two solutions are initially 1.0 *M*. In Figure 20.4, electrical current is isolated and flows through the voltmeter. In Figure 20.3, the flow of electrons cannot be isolated or utilized.

       (b)   In the cathode compartment of the voltaic cell in Figure 20.4, Cu^{2+} cations are reduced to Cu atoms, decreasing the number of positively charged particles in the compartment. Na^+ cations are drawn into the compartment to maintain charge balance as Cu^{2+} ions are removed.

20.13   (a) $Zn(s) \rightarrow Zn^{2+}(aq) + 2e^-$; $Ni^{2+}(aq) + 2e^- \rightarrow Ni(s)$

       (b)   Zn(s) is the anode;   Ni(s) is the cathode.

       (c)   Zn(s) is negative (-);   Ni(s) is positive (+).

       (d)   Electrons flow from the Zn(-) electrode toward the Ni(+) electrode.

       (e)   Cations migrate toward the Ni(s) cathode;   anions migrate toward the Zn(s) anode.

20.15   (a)   *Electromotive force*, emf, is the driving force that causes electrons to flow through the external circuit of a voltaic cell. It is the potential energy difference between an electron at the anode and an electron at the cathode.

       (b)   One *volt* is the potential energy difference required to impart 1 J of energy to a charge of 1 coulomb. 1 V = 1 J/C.

       (c)   The *standard cell potential* is the cell potential (emf) measured at standard conditions, 1 *M* aqueous solutions and gases at 1 atm pressure.

20.17 (a) A *standard reduction potential* is the relative potential of a reduction half-reaction measured at standard conditions, 1 M aqueous solutions and 1 atm gas pressure.

(b) The reference half-reaction for determining standard reduction potentials is $2H^+ (aq, 1\ M) + 2e^- \rightarrow H_2 (g, 1\ atm)$. By definition, E°_{red} for this half-reaction is 0.

(c) The reduction of $Ag^+(aq)$ to $Ag(s)$ is much more energetically favorable, because it has a substantially more positive E°_{red} (0.799 V) than the reduction of $Sn^{2+}(aq)$ to $Sn(s)$ (-0.136 V).

20.19 (a) The two half-reactions are:

$Tl^{3+}(aq) + 2e^- \rightarrow Tl^+(aq)$ cathode E°_{red} = ?

$2(Cr^{2+}(aq) \rightarrow Cr^{3+}(aq) + e^-)$ anode E°_{red} = -0.41

(b) $E^{\circ}_{cell} = E^{\circ}_{red}$ (cathode) - E°_{red} (anode); 1.19 V = E°_{red} - (-0.41 V);

E°_{red} = 1.19 V - 0.41 V = 0.78 V

(c)

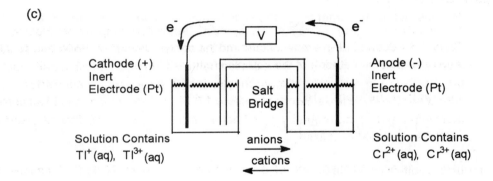

Note that because $Cr^{2+}(aq)$ is readily oxidized, it would be necessary to keep oxygen out of the right-hand cell compartment.

20.21 (a) $Cl_2(g) + 2e^- \rightarrow 2Cl^-(aq)$ E°_{red} = 1.359 V

       $2I^-(aq) \rightarrow I_2(s) + 2e^-$ E°_{red} = 0.536 V

E° = 1.359 V - 0.536 V = 0.823 V

(b) $2[Fe^{3+}(aq) + 1e^- \rightarrow Fe^{2+}(aq)]$ E°_{red} = 0.771 V

       $Hg(l) \rightarrow Hg^{2+}(aq) + 2e^-$ E°_{red} = 0.854 V

E° = 0.771V - 0.854 V = -0.083 V

(c) $Cu^+(aq) + 1e^- \rightarrow Cu^{\circ}(s)$ E°_{red} = 0.521 V

       $Cu^+(aq) \rightarrow Cu^{2+}(aq) + 1e^-$ E°_{red} = 0.153 V

E° = 0.521 V - 0.153 V = 0.368 V

20.23 (a)
$$Hg^{2+}(aq) + 2e^- \rightarrow Hg(l) \qquad E^{\circ}_{red} = 0.854 \text{ V}$$
$$Mn(s) \rightarrow Mn^{2+}(aq) + 2e^- \qquad E^{\circ}_{red} = -1.18 \text{ V}$$

$$Hg^{2+}(aq) + Mn(s) \rightarrow Hg(l) + Mn^{2+}(aq) \qquad E^{\circ} = 0.854 - (-1.18) = 2.03 \text{ V}$$

(b) The reverse of the two reactions above would produce the smallest (most negative) cell emf, -2.03 V. The combination with the smallest positive cell emf is

$$Hg^{2+}(aq) + 2e^- \rightarrow Hg(l) \qquad E^{\circ}_{red} = 0.854 \text{ V}$$
$$2Cu(s) \rightarrow 2Cu^+(aq) + 2e^- \qquad E^{\circ}_{red} = 0.521 \text{ V}$$

$$Hg^{2+}(aq) + 2Cu(s) \rightarrow Hg(l) + 2Cu^+(aq) \qquad E^{\circ} = 0.854 - 0.521 = 0.333 \text{ V}$$

20.25 The reduction half-reactions are:
$$Fe^{3+}(aq) + 1e^- \rightarrow Fe^{2+}(aq) \qquad E^{\circ}_{red} = 0.771 \text{ V}$$
$$H_2O_2(aq) + 2H^+(aq) + 2e^- \rightarrow 2H_2O(l) \qquad E^{\circ}_{red} = 1.776 \text{ V}$$

(a) Because it has the larger E°_{red}, the second half-reaction occurs more readily.

(b) Since the second half-reaction occurs more readily, it is the reduction half-reaction and occurs at the cathode of the complete cell.

(c) $Fe^{2+}(aq) \rightarrow Fe^{3+}(aq) + 1e^-$

(d) $E^{\circ}_{cell} = 1.776 \text{ V} - 0.771 \text{ V} = 1.005 \text{ V}$

20.27 The reduction half-reactions are:
$$Cu^{2+}(aq) + 2e^- \rightarrow Cu(s) \qquad E^{\circ} = 0.337 \text{ V}$$
$$Sn^{2+}(aq) + 2e^- \rightarrow Sn(s) \qquad E^{\circ} = -0.136 \text{ V}$$

(a) It is evident that Cu^{2+} is more readily reduced. Therefore, Cu serves as the cathode, Sn as the anode.

(b) The copper electrode gains mass as Cu is plated out, the Sn electrode loses mass as Sn is oxidized.

(c) The overall cell reaction is $Cu^{2+}(aq) + Sn(s) \rightarrow Cu(s) + Sn^{2+}(aq)$

(d) $E^{\circ} = 0.337 - (-0.136) = 0.473 \text{ V}$

Oxidizing and Reducing Agents; Spontaneity

20.29 (a) Top. The reduction half-reactions near the top of Table 20.1 are most likely to occur; a strong oxidant is most likely to be reduced.

(b) Left. An oxidant is reduced, so it is a reactant in a reduction half-reaction.

20.31 In each case, choose the half-reaction with the more positive reduction potential and with the given substance on the left.

(a) $Br_2(l)$ (1.065 V vs 0.536 V)   (b) $Ag^+(aq)$ (0.799 V vs 0.222 V)

(c) $Ce^{4+}(aq)$ (1.61 V vs 1.359 V)   (d) $MnO_4^-(aq,$ acidic) (1.51 V vs 1.33 V)

20.33 (a)   The strongest oxidizing agent is the species most readily reduced, as evidenced by a large, positive reduction potential. That species is H_2O_2. The weakest oxidizing agent is the species that least readily accepts an electron. We expect that it will be very difficult to reduce Zn(s); indeed, Zn(s) acts as a comparatively strong **reducing** agent. No potential is listed for reduction of Zn(s), but we can safely assume that it is less readily reduced than any of the other species present.

(b)   The strongest reducing agent is the species most easily oxidized (the largest negative reduction potential). Zn, $E_{red}^{\circ} = -0.76$ V, is the strongest reducing agent and F^-, $E_{red}^{\circ} = 2.87$ V, is the weakest.

20.35 Any of the **reduced** species in Table 20.1 from a half-reaction with a reduction potential more negative than -0.43 V will reduce Eu^{3+} to Eu^{2+}. These include Zn(s), $H_2(g)$, etc. Fe(s) is questionable.

20.37 (a)   The more positive the emf of a reaction the more spontaneous the reaction.

(b)   Reactions (a) and (c) in Exercise 20.21 are spontaneous.

20.39 $\Delta G^{\circ} = -nFE^{\circ}$

(a) $\Delta G^{\circ} = -2$ mol $e^- \times \dfrac{96,500 \text{ J}}{V \cdot \text{mol } e^-} \times \dfrac{1 \text{ kJ}}{1000 \text{ J}} \times 0.823$ V $= -159$ kJ

(b) $\Delta G^{\circ} = -2(96.5)(-0.083) = 16$ kJ

(c) $\Delta G^{\circ} = -1(96.5)(0.368) = -35.5$ kJ

20.41 (a) $2Fe^{2+}(aq) + S_2O_6^{2-}(aq) + 4H^+(aq) \rightarrow 2Fe^{3+}(aq) + 2H_2SO_3(aq)$

$E^{\circ} = 0.60$ V - 0.77 V = -0.17 V

$2Fe^{2+}(aq) + N_2O(aq) + 2H^+(aq) \rightarrow 2Fe^{3+}(aq) + N_2(g) + H_2O(l)$

$E^{\circ} = -1.77$ V - 0.77 V = -2.54 V

$Fe^{2+}(aq) + VO_2^+(aq) + 2H^+(aq) \rightarrow Fe^{3+}(aq) + VO^{2+}(aq) + H_2O(l)$

$E^{\circ} = 1.00$ V - 0.77 V = +0.23 V

(b) $\Delta G^{\circ} = -nFE^{\circ}$ For the first reaction,

$\Delta G^{\circ} = -2$ mol $\times \dfrac{96,500 \text{ J}}{1 \text{ V} \cdot \text{mol}} \times (-0.17 \text{ V}) = 3.3 \times 10^5$ J or 33 kJ

For the second reaction, $\Delta G^{\circ} = -2(96,500)(-2.54) = 4.90 \times 10^2$ kJ

For the third reaction, $\Delta G^{\circ} = -1(96,500)(0.23) = -22$ kJ

The Nernst Equation

20.43 $Zn(s) + 2H^+(aq) \rightarrow Zn^{2+}(aq) + H_2(g)$; $E = E^\circ - \dfrac{0.0592}{n} \log Q$; $Q = \dfrac{[Zn^{2+}]P_{H_2}}{[H^+]^2}$

   (a) P_{H_2} decreases, Q decreases, E increases

   (b)   No effect (Zn(s) does not appear in Q expression.)

   (c)   [H$^+$] increases, Q decreases, E increases

   (d)   No effect (NaNO$_3$ does not participate in the reaction.)

20.45   (a) $3[Mn^{2+}(aq) + 2e^- \rightarrow Mn(s)]$ $E^\circ_{red} = -1.18$ V

                 $2[Al(s) \rightarrow Al^{3+}(aq) + 3e^-]$ $E^\circ_{red} = -1.66$ V

   ───

                 $3Mn^{2+}(aq) + 2Al(s) \rightarrow 3Mn(s) + 2Al^{3+}(aq)$ $E^\circ = -1.18 - (-1.66) = 0.48$ V

   (b) $E = E^\circ - \dfrac{0.0592}{n} \log \dfrac{[Al^{3+}]^2}{[Mn^{2+}]^3}$; $n = 6$

         $E = 0.48 - \dfrac{0.0592}{6} \log \dfrac{(1.5)^2}{(0.10)^3} = 0.48 - \dfrac{0.0592}{6} \log (2.25 \times 10^3)$

         $E = 0.48 - \dfrac{0.0592}{6} (3.352) = 0.48 - 0.033 = 0.45$ V

20.47   (a) $4[Fe^{2+}(aq) \rightarrow Fe^{3+}(aq) + 1e^-]$ $E^\circ_{red} = 0.771$ V

             $O_2(g) + 4H^+(aq) + 4e^- \rightarrow 2H_2O(l)$ $E^\circ_{red} = 1.23$ V

   ───

             $4Fe^{2+}(aq) + O_2(g) + 4H^+(aq) \rightarrow 4Fe^{3+}(aq) + 2H_2O(l)$ $E^\circ = 1.23 - 0.771 = 0.46$ V

   (b) $E = E^\circ - \dfrac{0.0592}{n} \log \dfrac{[Fe^{3+}]^4}{[Fe^{2+}]^4[H^+]^4 P_{O_2}}$; $n = 4$, $[H^+] = 1.00 \times 10^{-3}$ M

         $E = 0.46 \text{ V} - \dfrac{0.0592}{4} \log \dfrac{(1.0 \times 10^{-3})^4}{(2.0)^4 (1.0 \times 10^{-3})^4 (0.50)} = 0.46 - \dfrac{0.0592}{4} \log (0.125)$

         $E = 0.46 - \dfrac{0.0592}{4} (-0.903) = 0.46 + 0.0134 = 0.47$ V

20.49 $E = E^\circ - \dfrac{0.0592}{2} \log \dfrac{[H_2][Zn^{2+}]}{[H^+]^2}$; $E^\circ = 0.0$ V $- (-0.763$ V$) = 0.763$ V

     $0.720 = 0.763 - \dfrac{0.0592}{2} \times (\log [H_2] [Zn^{2+}] - 2 \log [H^+]) = 0.763 - \dfrac{0.0592}{2} \times (1.00 - 2 \log [H^+])$

     $0.720 = 0.763 + 0.0296 + 0.0592 \log [H^+]$; $\log [H^+] = \dfrac{0.720 - 0.763 - 0.0296}{0.0592}$

     $\log [H^+] = -1.226 = -1.2$; $[H^+] = 0.0594 = 0.06$ M; pH = 1.2

20.51 $E° = \dfrac{0.0592\ V}{n} \log K$; $\log K = \dfrac{nE°}{0.0592\ V}$

   (a) $E° = 0.763 - 0.136 = 0.627\ V$; $n = 2$ ($Sn^{+2} + 2\ e^- \rightarrow Sn$)

         $\log K = \dfrac{2(0.627)}{0.0592} = 21.182 = 21.2$; $K = 1.52 \times 10^{21} = 2 \times 10^{21}$

   (b) $E° = 0.277 - (0) = 0.277\ V$; $n = 2$ ($2H^+ + 2\ e^- \rightarrow H_2$)

         $\log K = \dfrac{2(0.277)}{0.0592} = 9.358 = 9.36$; $K = 2.3 \times 10^9$

   (c) $E° = -1.065 - (-1.51) = 0.44\ V$; $n = 10$ ($2MnO_4^- + 10\ e^- \rightarrow 2Mn^{+2}$)

         $\log K = \dfrac{10(0.44)}{0.0592} = 74.324 \approx 74$; $K = 2.1 \times 10^{74} = 10^{74}$

20.53 $E° = \dfrac{0.0592}{n} \log K$; $\log K = \dfrac{nE°}{0.0592\ V}$

   (a) $\log K = \dfrac{1(0.35\ V)}{0.0592\ V} = 5.912 = 5.9$; $K = 8.2 \times 10^5 = 8 \times 10^5\ (= 10^6)$

   (b) $\log K = \dfrac{2(0.35\ V)}{0.0592\ V} = 11.824 = 12$; $K = 6.7 \times 10^{11} = 10^{12}$

   (c) $\log K = \dfrac{3(0.35\ V)}{0.0592\ V} = 17.736 = 18$; $K = 5.5 \times 10^{17} = 10^{18}$

Commercial Voltaic Cells

20.55   The overall cell reaction is:

   $Zn(s) + 2NH_4^+(aq) + 2MnO_2(s) \rightarrow Mn_2O_3(s) + 2NH_3(aq) + H_2O(l)$

   $120\ g\ Zn \times \dfrac{1\ mol\ Zn}{65.39\ g\ Zn} \times \dfrac{2\ mol\ MnO_2}{1\ mol\ Zn} \times \dfrac{86.94\ g\ MnO_2}{1\ mol\ MnO_2} = 319\ g\ MnO_2$

20.57   (a)   In discharge: $Cd(s) + NiO_2(s) + 2H_2O(l) \rightleftharpoons Cd(OH)_2(s) + Ni(OH)_2(s)$

         In charging, the reverse reaction occurs.

   (b) $E° = 0.49\ V - (-0.76\ V) = 1.25\ V$

20.59 $E_{red}°$ for Cd (-0.40 V) is less negative than $E_{red}°$ for Zn (-0.76 V), so E_{cell} will have a smaller (less positive) value.

Electrolysis; Electrical Work

20.61   (a)   The products are different because in aqueous electrolysis water is reduced in preference to Mg^{2+}.

(b) $MgCl_2(l) \rightarrow Mg(l) + Cl_2(g)$

$2Cl^-(aq) + 2H_2O(l) \rightarrow Cl_2(g) + H_2(g) + 2OH^-(aq)$

The aqueous solution electrolysis is entirely analogous to that for NaCl(aq), Section 20.8.

(c) $Mg^{2+}(aq) + 2e^- \rightarrow Mg(s)$ $E^\circ_{red} = -2.37$ V

       $2Cl^-(aq) \rightarrow Cl_2(g) + 2e^-$ $E^\circ_{red} = 1.359$ V

      $MgCl_2(aq) \rightarrow Mg(s) + Cl_2(g)$ $E^\circ = -2.37 - 1.359 = -3.73$ V

      $H_2O(l) + 2e^- \rightarrow H_2(g) + 2OH^-(aq)$ $E^\circ_{red} = -0.83$ V

       $2Cl^-(aq) \rightarrow Cl_2(g) + 2e^-$ $E^\circ_{red} = 1.359$ V

$2Cl^-(aq) + 2H_2O(l) \rightarrow Cl_2(g) + H_2(g) + 2OH^-(aq)$ $E^\circ = -0.83 - 1.359 = -2.19$ V

The minus signs mean that voltage must be applied in order for the reaction to occur.

20.63

Cathode Reaction:

$Ni^{2+} + 2e^- \longrightarrow Ni$

Anode Reaction:

$2Cl^- \longrightarrow Cl_2 + 2e^-$

anions $\longrightarrow$

$\longleftarrow$ cations

Chlorine is produced in preference to oxidation of water because of a large overvoltage for O_2 formation.

20.65   Coulombs = amps•s, since this is a 3e$^-$ reduction, each mole of Cr(s) requires 3 Faradays.

(a) $13.5 \text{ A} \times 3.00 \text{ d} \times \dfrac{24 \text{ hr}}{1 \text{ d}} \times \dfrac{60 \text{ min}}{1 \text{ hr}} \times \dfrac{60 \text{ s}}{1 \text{ min}} \times \dfrac{1 \text{ C}}{1 \text{ amp•s}} \times \dfrac{1 \text{ F}}{96,500 \text{ C}}$

$\times \dfrac{1 \text{ mol Cr}}{3 \text{ F}} \times \dfrac{52.00 \text{ g Cr}}{1 \text{ mol Cr}} = 629$ g Cr(s)

(b) $1.00 \text{ mol Cr} \times \dfrac{3 \text{ F}}{1 \text{ mol Cr}} \times \dfrac{96,500 \text{ C}}{\text{F}} \times \dfrac{1 \text{ amp•s}}{1 \text{ C}} \times \dfrac{1}{12.0 \text{ hr}} \times \dfrac{1 \text{ hr}}{60 \text{ min}} \times \dfrac{1 \text{ min}}{60 \text{ s}}$

$= 6.70$ A

20.67 (a) $7.50 \text{ amp} \times 100 \text{ min} \times \dfrac{60 \text{ s}}{1 \text{ min}} \times \dfrac{1 \text{ C}}{1 \text{ amp•s}} \times \dfrac{1 \text{ Faraday}}{96,500 \text{ C}} \times \dfrac{1 \text{ mol Cl}_2}{2 \text{ Faraday}}$

$\times \dfrac{22.400 \text{ L Cl}_2}{1 \text{ mol Cl}_2} = 5.22$ L Cl_2

(b)   From the balanced equation (section 20.8), we see that 2 mol NaOH are formed per mol Cl_2. Proceeding as in (a), but replacing the last factor by (2 mol NaOH/1 mol Cl_2), we obtain 0.466 mol NaOH.

20.69   For a voltaic cell at standard conditions, $w_{max} = \Delta G° = -nFE°$.

$$I_2(s) + 2e^- \rightarrow 2I^-(aq) \qquad\qquad E°_{red} = 0.536 \text{ V}$$

$$Sn(s) \rightarrow Sn^{2+}(aq) + 2e^- \qquad\qquad E°_{red} = -0.136 \text{ V}$$

$$I_2(s) + Sn(s) \rightarrow 2I^-(aq) + Sn^{2+}(aq) \qquad E° = 0.536 - (-0.136) = 0.672 \text{ V}$$

$$w_{max} = -2(96.5)(0.672) = -129.7 = -130 \text{ kJ/mol Sn}$$

$$\frac{129.7 \text{ kJ}}{\text{mol Sn(s)}} \times 0.460 \text{ mol Sn} = -59.7 \text{ kJ}$$

20.71   (a) $90{,}000 \text{ amp} \times 16 \text{ hr} \times \dfrac{3600 \text{ s}}{1 \text{ hr}} \times \dfrac{1 \text{ C}}{1 \text{ amp} \cdot \text{s}} \times \dfrac{1 \text{ Faraday}}{96{,}500 \text{ C}} \times \dfrac{1 \text{ mol Mg}}{2 \text{ Faraday}}$

$\times \dfrac{24.3 \text{ g Mg}}{1 \text{ mol Mg}} \times 0.50 = 3.3 \times 10^5 \text{ g Mg}$

(b) $\text{Coulombs} = 9 \times 10^4 \text{ amp} \times 16 \text{ hr} \times \dfrac{3600 \text{ s}}{1 \text{ hr}} \times \dfrac{1 \text{ C}}{1 \text{ amp} \cdot \text{s}} = 5.184 \times 10^9 = 5.2 \times 10^9 \text{ C}$

$\text{kilowatt-hours} = 5.184 \times 10^9 \text{ C} \times 4.20 \text{ V} \times \dfrac{1 \text{ J}}{1 \text{ C} \cdot \text{V}} \times \dfrac{1 \text{ kWh}}{3.6 \times 10^6 \text{ J}} = 6.0 \times 10^3 \text{ kWh}$

Corrosion

20.73   (a)   anode: $Fe(s) \rightarrow Fe^{2+}(aq) + 2e^-$

cathode: $O_2(g) + 4H^+(aq) + 4e^- \rightarrow 2H_2O(l)$

(b) $2Fe^{2+}(aq) + 3H_2O(l) + 3H_2O(l) \rightarrow Fe_2O_3 \cdot 3H_2O(s) + 6H^+(aq) + 2e^-$

$O_2(g) + 4H^+(aq) + 4e^- \rightarrow 2H_2O(l)$

(Multiply the oxidation half-reaction by two to balance electrons and obtain the overall balanced reaction.)

20.75   No. To act as a sacrificial anode, a metal must have a more negative reduction potential (be easier to oxidize) than Fe^{2+}. $E°_{red}$ Sn^{2+} = -0.14 V, $E°_{red}$ Fe^{2+} = -0.44 V. Tin gives protection by providing a complete cover for iron.

20.77   The pH of an aqueous medium is an important factor in corrosion. Note from Section 20.10 that an increased [H^+] shifts the equilibrium to the right. When [H^+] is depressed by the presence of a base, the reduction of O_2 is less favorable, and corrosion slows down. The added amines serve the function of keeping [H^+] low.

Additional Exercises

20.79 (a)

$$2HBrO_3(aq) + 10H^+(aq) + 10e^- \rightarrow Br_2(l) + 6H_2O(l)$$
$$10[Fe^{2+}(aq) \qquad\qquad \rightarrow Fe^{3+}(aq) + 1e^-]$$

$$10Fe^{2+}(aq) + 2HBrO_3(aq) + 10H^+(aq) \rightarrow 10Fe^{3+}(aq) + Br_2(l) + 6H_2O(l)$$

(b)

$$2[NO_3^-(aq) + 4H^+(aq) + 3e^- \rightarrow NO(g) + 2H_2O(l)]$$
$$3[Cu(s) \qquad\qquad \rightarrow Cu^{2+}(aq) + 2e^-]$$

$$3Cu(s) + 2NO_3^-(aq) + 8H^+(aq) \rightarrow 3Cu^{2+}(aq) + 2NO(g) + 4H_2O(l)$$

(c)

$$3IO_3^-(aq) + 18H^+(aq) + 16e^- \rightarrow I_3^-(aq) + 9H_2O(l)$$
$$\qquad\qquad + 18OH^-(aq) \qquad \rightarrow \qquad\qquad + 18OH^-(aq)$$

$$3IO_3^-(aq) + 9H_2O(l) + 16e^- \rightarrow I_3^-(aq) + 18OH^-(aq)$$
$$8[3I^-(aq) \qquad\qquad \rightarrow I_3^-(aq) + 2e^-]$$

$$1/3\ [3IO_3^-(aq) + 24I^-(aq) + 9H_2O(l) \rightarrow 9I_3^-(aq) + 18OH^-(aq)]$$
$$IO_3^-(aq) + 8I^-(aq) + 3H_2O(l) \rightarrow 3I_3^-(aq) + 6OH^-(aq)$$

20.82 (a)

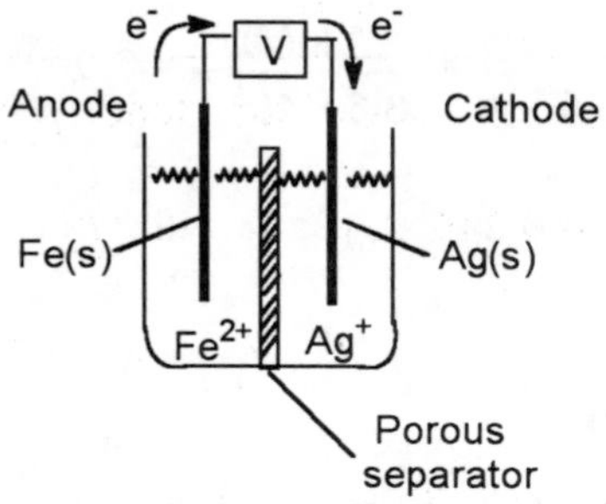

$$Fe(s) \rightarrow Fe^{2+}(aq) + 2e^-$$
$$2Ag^+(aq) + 2e^- \rightarrow 2Ag(s)$$

$$Fe(s) + 2Ag^+(aq) \rightarrow Fe^{2+}(aq) + 2Ag(s)$$

(b)

$$Zn(s) \rightarrow Zn^{2+}(s) + 2e^-$$
$$2H^+(aq) + 2e^- \rightarrow H_2(g)$$

$$Zn(s) + 2H^+(aq) \rightarrow Zn^{2+}(aq) + H_2(g)$$

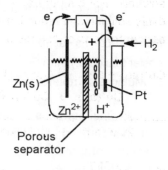

(c) $Cu|Cu^{2+}||ClO_3^-, Cl^-|Pt$ Here, both the oxidized and reduced forms of the cathode solution are in the same phase, so we separate them by a comma, and then indicate an inert electrode.

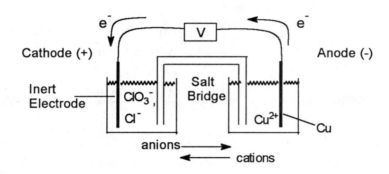

20.85   We need in each case to determine whether $E°$ is positive (spontaneous) or negative (nonspontaneous).

(a) $E° = 0.672$ V, spontaneous   (b) $E° = -0.82$ V, nonspontaneous

(c) $E° = 0.55$ V, spontaneous   (d) $E° = 0.12$ V, nonspontaneous

20.88   The cell half-reactions are: $Cr(s) \rightarrow Cr^{3+}(aq, Conc\ 1) + 3e^-$

$$Cr^{3+}(aq, Conc\ 2) + 3e^- \rightarrow Cr(s)$$

$$Cr^{3+}(aq, Conc\ 2) \rightarrow Cr^{3+}(aq, Conc\ 1)$$

$E = 0 - \dfrac{0.0592}{3} \log \dfrac{[Cr^{3+}]_1}{[Cr^{3+}]_2}$; $(E° = 0)$ Let $[Cr^{3+}]_1 = 0.040\ M$, $[Cr^{3+}]_2 = 2.0\ M$

$E = -\dfrac{0.0592}{3} \log \dfrac{0.040}{2.0} = +0.034$ V

The fact that E is positive means that the cell reaction is spontaneous as written: Cr^{3+} (0.040 *M*) $\rightarrow$ Cr^{3+} (2.0 *M*). This means that the cathode is in the compartment containing Cr^{3+}(2.0 *M*); the anode is in the compartment containing the dilute (0.040 *M*) solution. For concentration cells in general, oxidation occurs in the compartment with the dilute solution and reduction in the compartment with the concentrated solution, so that the concentrations of cations in the two compartments become more similar.

20.91 $Cu^+(aq) + 1e^- \rightarrow Cu^{2+}(aq)$ $E°_{red} = -0.153$ V

$\qquad Cu^+(aq) \rightarrow Cu°(s) + 1e^-$ $E°_{red} = -0.521$ V

$\qquad 2Cu^+(aq) \rightarrow Cu°(s) + Cu^{2+}(aq)$ $E° = -0.153 - (-0.521) = 0.368$ V

$E° = \dfrac{0.0592}{n} \log K$; $\log K = \dfrac{nE°}{0.0592} = \dfrac{1(0.368)}{0.0592} = 6.216 = 6.22$

$K = 10^{6.216} = 1.6 \times 10^6$

20.94 (a) $1.0 \text{ lb Al} \times \dfrac{453.6 \text{ g}}{1 \text{ lb}} \times \dfrac{1 \text{ mol Al}}{26.98 \text{ g Al}} \times \dfrac{3 \text{ F}}{1 \text{ mol Al}} \times \dfrac{96{,}500 \text{ C}}{\text{F}} \times \dfrac{1 \text{ amp} \cdot \text{s}}{\text{C}} \times \dfrac{1}{11.2 \text{ A}}$

$\times \dfrac{1 \text{ min}}{60 \text{ s}} \times \dfrac{1 \text{ hr}}{60 \text{ min}} = 1.2 \times 10^2 \text{ hr}$

(b) $\text{Coulombs} = 453.6 \text{ g Al} \times \dfrac{1 \text{ mol Al}}{26.98 \text{ g Al}} \times \dfrac{3 \text{ F}}{1 \text{ mol Al}} \times \dfrac{96{,}500 \text{ C}}{\text{F}} = 4.867 \times 10^6$

$= 4.9 \times 10^6 \text{ C}$

$\text{kWh} = 4.867 \times 10^6 \text{ C} \times 6.0 \text{ V} \times \dfrac{1 \text{ J}}{\text{C} \cdot \text{V}} \times \dfrac{1 \text{ kWh}}{3.6 \times 10^6 \text{ J}} = 8.1 \text{ kWh}$

This is the amount of electrical power expended if the cell is 100% efficient. 8.1 kWh/0.40 = **20 kWh** required to produce 1.0 lb of Al.

20.97 (a) $Ni^{2+}(aq) + 2e^- \rightarrow Ni(s) \qquad E^{\circ}_{red} = -0.28 \text{ V}$

$2Br^-(aq) \rightarrow Br_2(l) + 2e^- \qquad E^{\circ}_{red} = 1.065 \text{ V}$

$E^{\circ} = -0.28 - 1.065 = -1.35 \text{ V}; \quad E_{min} = 1.35 \text{ V}$

(b) A larger voltage than the calculated E_{min} is required to overcome cell resistance, if the reaction is to proceed at a reasonable rate. There may also be an overvoltage for reduction of Br^-(aq) at Pt, although this should not be large.

20.100 The ship's hull should be made negative. By keeping an excess of electrons in the metal of the ship, the tendency for iron to undergo oxidation, with release of electrons, is reduced. The ship, as a negatively charged "electrode," becomes the site of reduction, rather than oxidation, in an electrolytic process.

Integrative Exercises

20.105 (a) $NO_3^-(aq) + 4H^+(aq) + 3e^- \rightarrow NO(g) + 2H_2O(l) \qquad E^{\circ}_{red} = 0.96 \text{ V}$

$Au(s) \rightarrow Au^{3+}(aq) + 3e^- \qquad\qquad\qquad E^{\circ}_{red} = 1.498 \text{ V}$

$Au(s) + NO_3^-(aq) + 4H^+(aq) \rightarrow Au^{3+}(aq) + NO(g) + 2H_2O(l)$

$E^{\circ} = 0.96 - 1.498 = -0.54 \text{ V}; \quad E^{\circ}$ is negative, the reaction is not spontaneous.

(b) $3[2H^+(aq) + 2e^- \rightarrow H_2(g)] \qquad\qquad E^{\circ}_{red} = 0.000 \text{ V}$

$2[Au(s) + 4Cl^-(aq) \rightarrow AuCl_4^-(aq) + 3e^-] \qquad E^{\circ}_{red} = 1.002 \text{ V}$

$2Au(s) + 6H^+(aq) + 8Cl^-(aq) \rightarrow 2AuCl_4^-(aq) + 3H_2(g)$

$E^{\circ} = 0.000 - 1.002 = -1.002 \text{ V}; \quad E^{\circ}$ is negative, the reaction is not spontaneous.

(c) $NO_3^-(aq) + 4H^+(aq) + 3e^- \rightarrow NO(g) + 2H_2O(l)$ $E^{\circ}_{red} = 0.96 \text{ V}$

                   $Au(s) + 4Cl^-(aq) \rightarrow AuCl_4^-(aq) + 3e^-$ $E^{\circ}_{red} = 1.002 \text{ V}$

$$Au(s) + NO_3^-(aq) + 4Cl^-(aq) + 4H^+(aq) \rightarrow AuCl_4^-(aq) + NO(g) + 2H_2O(l)$$

$E^{\circ} = 0.96 - 1.002 = -0.04$; E° is small but negative, the process is not spontaneous.

(d) $E = E^{\circ} - \dfrac{0.0592}{3} \log \dfrac{[AuCl_4^-] P_{NO}}{[NO_3^-][Cl^-]^4[H^+]^4}$

If $[H^+]$, $[Cl^-]$ and $[NO_3^-]$ are much greater than 1.0 *M*, the log term is negative and the correction to E° is positive. If the correction term is greater than 0.042 V, the value of E is positive and the reaction at nonstandard conditions is spontaneous.

20.108 $AgSCN(s) + e^- \rightarrow Ag(s) + SCN^-(aq)$ $E^{\circ}_{red} = 0.0895 \text{ V}$

              $Ag(s) \rightarrow Ag^+(aq) + e^-$ $E^{\circ}_{red} = 0.799 \text{ V}$

$$AgSCN(s) \rightarrow Ag^+(aq) + SCN^-(aq) \quad E^{\circ} = 0.0895 - 0.799 = -0.710 \text{ V}$$

$E^{\circ} = \dfrac{0.0592}{n} \log K_{sp}$; $\log K_{sp} = \dfrac{(-0.710)(1)}{0.0592} = -11.993 = -12.0$

$K_{sp} = 10^{-11.993} = 1.02 \times 10^{-12} = 1 \times 10^{-12}$

21 Nuclear Chemistry

Radioactivity

21.1 p = protons, n = neutrons, e = electrons; number of protons = atomic number; number of neutrons = mass number - atomic number

(a) $^{59}_{28}Ni$: 28p, 31n   (b) ^{94}Zr: 40p, 54n   (c) ^{18}O: 8p, 10n

21.3 (a) ^1_1p or ^1_1H (b) ^0_1e (c) $^0_{-1}\beta$ or $^0_{-1}e$

21.5 (a) $^{214}_{83}Bi \rightarrow ^{214}_{84}Po + ^0_{-1}e$ (b) $^{195}_{79}Au + ^0_{-1}e$ (orbital electron) $\rightarrow ^{195}_{78}Pt$

(c) $^{38}_{19}K \rightarrow ^{38}_{18}Ar + ^0_1e$ (d) $^{242}_{94}Pu \rightarrow ^{238}_{92}U + ^4_2He$

21.7 The total mass number change is (235-207) = 28. Since each α particle accompanies a change of -4 in mass number, whereas emission of a β particle does not correspond to a mass change, there are 7 α particle emissions. The change in atomic number in the series is 10. Each α particle results in an atomic number lower by two. The 7 α particle emissions alone would cause a decrease of 14 in atomic number. Each β particle emission raises the atomic number by one. To obtain the observed lowering of 10 in the series, there must be 4 β emissions.

21.9 (a) $^{24}_{11}Na \rightarrow ^{24}_{12}Mg + ^0_{-1}e$;   a β particle is produced

(b) $^{188}_{80}Hg \rightarrow ^{188}_{79}Au + ^0_1e$;   a positron is produced

(c) $^{122}_{53}I \rightarrow ^{122}_{54}Xe + ^0_{-1}e$;   a β particle is produced

(d) $^{242}_{94}Pu \rightarrow ^{238}_{92}U + ^4_2He$;   an α particle is produced

Nuclear Stability

21.11 (a) ^8_5B - low neutron/proton ratio, positron emission

(b) $^{68}_{29}Cu$ - high neutron/proton ratio, beta emission

(c) $^{241}_{93}Np$ - high neutron/proton ratio, beta emission

(Even though ^{241}Np has an atomic number ≥ 84, the most common decay pathway for nuclides with neutron/proton ratios higher than the isotope listed on the periodic chart is beta decay.)

(d) $^{39}_{17}Cl$ - high neutron/proton ratio, beta emission

21.13   (a)   It is on the edge - slightly low neutron/proton ratio; could be a positron emitter or undergo orbital electron capture.

(b)   No - somewhat high neutron/proton ratio; beta emitter.

(c)   Yes.

(d)   No - high atomic number; alpha emitter.

21.15   Use the criteria listed in Table 21.3.

(a)   Stable: $^{39}_{19}K$ odd proton, even neutron more abundant than odd proton, odd neutron; 20 neutrons is a magic number.

(b)   Stable: $^{209}_{83}Bi$ odd proton, even neutron more abundant than odd proton, odd neutron; 126 neutrons is a magic number.

(c)   Stable: $^{25}_{12}Mg$ even though $^{24}_{10}Ne$ is an even proton, even neutron nuclide, it has a very high neutron/proton ratio and lies outside the band of stability.

21.17   (a) $^{4}_{2}He$ (c) $^{40}_{20}Ca$ (e) $^{208}_{82}Pb$

(d) $^{58}_{28}Ni$ has a magic number of protons, but not neutrons.

21.19   Radioactive:

$^{14}_{8}O$, $^{115}_{52}Te$ - low neutron/proton ratio; $^{208}_{84}Po$ - atomic number ≥ 84

$^{84}_{34}Se$ - even though this is an even proton, even neutron nuclide with a magic number of neutrons, the neutron/proton ratio is so high that the nuclide is radioactive.

Stable: $^{32}_{16}S$ - even proton, even neutron, stable neutron/proton ratio

Nuclear Transmutations

21.21   Protons and alpha particles are positively charged and must be moving very fast to overcome electrostatic forces which would repel them from the target nucleus. Neutrons are electrically neutral and not repelled by the nucleus.

21.23   (a) $^{252}_{98}Cf + ^{10}_{5}B \rightarrow 3\,^{1}_{0}n + ^{259}_{103}Lw$ (b) $^{2}_{1}H + ^{3}_{2}He \rightarrow ^{4}_{2}He + ^{1}_{1}H$

(c) $^{1}_{1}H + ^{11}_{5}B \rightarrow 3\,^{4}_{2}He$ (d) $^{122}_{53}I \rightarrow ^{122}_{54}Xe + ^{0}_{-1}e$ (e) $^{59}_{26}Fe \rightarrow ^{0}_{-1}e + ^{59}_{27}Co$

21.25   (a) $^{14}_{7}N + ^{1}_{1}H \rightarrow ^{11}_{6}C + ^{4}_{2}He$ (b) $^{14}_{7}N + ^{4}_{2}He \rightarrow ^{17}_{8}O + ^{1}_{1}H$

(c) $^{59}_{26}Fe + ^{4}_{2}He \rightarrow ^{63}_{29}Cu + ^{0}_{-1}e$

Rates of Radioactive Decay

21.27 Chemical reactions do not affect the character of atomic nuclei. The energy changes involved in chemical reactions are much too small to allow us to alter nuclear properties via chemical processes. Therefore, the nuclei that are formed in a nuclear reaction will continue to emit radioactivity regardless of any chemical changes we bring to bear. However, we can hope to use chemical means to separate radioactive substances, or remove them from foods or a portion of the environment.

21.29 After 12.3 yr, one half-life, there are $(1/2)48.0 = 24.0$ mg. 49.2 yr is exactly four half-lives. There are then $(48.0)(1/2)^4 = 3.0$ mg tritium remaining.

21.31 Using Equation [21.19],

$$k = \frac{-1}{t} \ln \frac{N_t}{N_o} = \frac{-1}{1.00 \text{ yr}} \times \ln \frac{2921}{3012} = 0.03068 = 0.0307 \text{ yr}^{-1}$$

Using Equation [21.20],

$$t_{1/2} = 0.693/k = 0.693/(0.03068 \text{ yr}^{-1}) = 22.6 \text{ yr}$$

21.33 $k = 0.693 / t_{1/2} = 0.693/27.8 \text{ d} = 0.02493 = 0.0249 \text{ d}^{-1}$

$$t = \frac{-1}{k} \ln \frac{N_t}{N_o} = \frac{-1}{0.02493 \text{ d}^{-1}} \ln \frac{0.75}{1.85} = 36.2 \text{ d}$$

21.35 $^{226}_{88}\text{Ra} \rightarrow {}^{222}_{86}\text{Rn} + {}^4_2\text{He}$

1 α particle is produced for each ^{226}Ra that decays. Calculate the mass of ^{226}Ra remaining after 1.0 min, calculate by subtraction the mass that has decayed, and use Avogadro's number to get the number of $^4_2$He particles.

Calculate k in min^{-1}. $1622 \text{ yr} \times \dfrac{365 \text{ d}}{1 \text{ yr}} \times \dfrac{24 \text{ hr}}{1 \text{ d}} \times \dfrac{60 \text{ min}}{1 \text{ hr}} = 8.525 \times 10^8 \text{ min}$

$$k = \frac{0.693}{t_{1/2}} = \frac{0.693}{8.525 \times 10^8 \text{ min}} = 8.129 \times 10^{-10} \text{ min}^{-1}$$

$\ln \dfrac{N_t}{N_o} = -kt = (-8.129 \times 10^{-10} \text{ min}^{-1})(1.0 \text{ min}) = -8.129 \times 10^{-10}$

$\dfrac{N_t}{N_o} = (1.000 - 8.129 \times 10^{-10})$; (don't round here!)

$N_t = 5.0 \times 10^{-3} \text{ g} (1.00 - 8.129 \times 10^{-10})$ The amount that decays is $N_o - N_t$:

$$5.0 \times 10^{-3} \text{ g} - [5.0 \times 10^{-3} (1.00 - 8.129 \times 10^{-10})] = 5.0 \times 10^{-3} \text{ g} (8.129 \times 10^{-10})$$

$$= 4.065 \times 10^{-12} = 4.1 \times 10^{-12} \text{ g Ra}$$

$$[N_o - N_t] = 4.065 \times 10^{-12} \text{ g Ra} \times \frac{1 \text{ mol Ra}}{226.0 \text{ g Ra}} \times \frac{6.022 \times 10^{23} \text{ Ra atoms}}{1 \text{ mol Ra}} \times \frac{1 \, {}^4_2\text{He}}{1 \text{ Ra atom}}$$

$$= 1.1 \times 10^{10} \, \alpha \text{ particles emitted in 1 min}$$

21.37 Calculate k in yr^{-1} and solve Equation 21.19 for t. $N_0 = 15.2/\text{min/g}$, $N_t = 8.9/\text{min/g}$

$k = 0.693/t_{1/2} = 0.693/5.73 \times 10^3 \text{ yr} = 1.209 \times 10^{-4} = 1.21 \times 10^{-4} \text{ yr}^{-1}$

$t = \dfrac{-1}{k} \ln \dfrac{N_t}{N_0} = \dfrac{-1}{1.209 \times 10^{-4}} \ln \dfrac{8.9}{15.2} = 4.43 \times 10^3 \text{ yr}$

21.39 Follow the procedure outlined in Sample Exercise 21.7. The original quantity of ^{238}U is 50.0 mg plus the amount that gave rise to 14.0 mg of ^{206}Pb. This amount is 14.0(238/206) = 16.2 mg.

$k = 0.693/4.5 \times 10^9 \text{ yr} = 1.54 \times 10^{-10} = 1.5 \times 10^{-10} \text{ yr}^{-1}$

$t = \dfrac{-1}{k} \ln \dfrac{N_t}{N_0} = \dfrac{-1}{1.54 \times 10^{-10} \text{ yr}^{-1}} \ln \dfrac{50.0}{66.2} = 1.8 \times 10^9 \text{ yr}$

Energy Changes

21.41 $\Delta E = c^2 \Delta m$; $\Delta m = \Delta E/c^2$; $1 \text{ J} = \text{kg} \cdot \text{m}^2/\text{s}^2$

$\Delta m = \dfrac{393.5 \times 10^3 \text{ kg} \cdot \text{m}^2/\text{s}^2}{(2.9979 \times 10^8 \text{ m/s})^2} \times \dfrac{1000 \text{ g}}{1 \text{ kg}} = 4.378 \times 10^{-9} \text{ g}$

21.43 $\Delta m = 8(1.0072765 \text{ amu}) + 8(1.0086655 \text{ amu}) - 15.99052 \text{ amu} = 0.137016$
$= 0.13702 \text{ amu}$

$\Delta E = (2.99795 \times 10^8 \text{ m/s})^2 \times 0.137016 \text{ amu} \times \dfrac{1 \text{ g}}{6.02214 \times 10^{23} \text{ amu}} \times \dfrac{1 \text{ kg}}{1 \times 10^3 \text{ g}}$

$= 2.04489 \times 10^{-11} = 2.0449 \times 10^{-11} \text{ J} / {}^{18}O \text{ nucleus required}$

$2.04489 \times 10^{-11} \dfrac{\text{J}}{\text{nucleus}} \times \dfrac{6.02214 \times 10^{23} \text{ atoms}}{\text{mol}} = 1.2315 \times 10^{13} \text{ J/mol } {}^{18}O$

21.45 In each case, calculate the mass defect, total nuclear binding energy and then binding energy per nucleon.

(a) $3(1.0072765) + 4(1.0086655) = 7.0564915 \text{ amu}$

mass defect $= 7.0564915 - 7.01600 = 0.04049 \text{ amu}$

$\Delta E = 0.04049 \text{ amu} \times \dfrac{1 \text{ g}}{6.02214 \times 10^{23} \text{ amu}} \times \dfrac{1 \text{ kg}}{1000 \text{ g}} \times \dfrac{8.98768 \times 10^{16} \text{ m}^2}{\text{s}^2}$

$= 6.0429 \times 10^{-12} = 6.043 \times 10^{-12} \text{ J}$

binding energy/nucleon $= 6.0429 \times 10^{-12} \text{ J} / 7 = 8.633 \times 10^{-13} \text{ J/nucleon}$

(b) $30(1.0072765) + 34(1.0086655) = 64.512922$ amu

mass defect $= 64.512922 - 63.92914 = 0.58378$ amu

$$\Delta E = 0.58378 \text{ amu} \times \frac{1 \text{ g}}{6.02214 \times 10^{23} \text{ amu}} \times \frac{1 \text{ kg}}{1000 \text{ g}} \times \frac{8.98768 \times 10^{16} \text{ m}^2}{\text{s}^2}$$

$$= 8.71256 \times 10^{-11} = 8.7126 \times 10^{-11} \text{ J}$$

binding energy/ nucleon $= 8.71256 \times 10^{-11}$ J $/ 64 = 1.3613 \times 10^{-12}$ J/nucleon

(c) $90(1.0072765) + 142(1.0086655) = 233.885386$ amu

mass defect $= 233.885386 - 232.0382 = 1.8472$ amu

$$\Delta E = 1.8472 \text{ amu} \times \frac{1 \text{ g}}{6.02214 \times 10^{23} \text{ amu}} \times \frac{1 \text{ kg}}{1000 \text{ g}} \times \frac{8.98768 \times 10^{16} \text{ m}^2}{\text{s}^2}$$

$$= 2.75683 \times 10^{-10} = 2.7568 \times 10^{-10} \text{ J}$$

binding energy/nucleon $= 2.75683 \times 10^{-10}$ J $/ 232 = 1.1883 \times 10^{-12}$ J/nucleon

21.47 $\dfrac{1.07 \times 10^{16} \text{ kJ}}{1 \text{ min}} \times \dfrac{60 \text{ min}}{1 \text{ hr}} \times \dfrac{24 \text{ hr}}{1.0 \text{ day}} = 1.541 \times 10^{19} \dfrac{\text{kJ}}{\text{day}} = 1.5 \times 10^{22}$ J/day

$$\Delta m = \frac{1.541 \times 10^{22} \text{ kg} \cdot \text{m}^2/\text{s}^2/\text{d}}{(2.998 \times 10^8 \text{ m/s})^2} = 1.714 \times 10^5 = 1.7 \times 10^5 \text{ kg/d}$$

Now calculate the mass change in the given nuclear reaction:

$\Delta m = 140.9140 + 91.9218 + 2(1.00867) - 235.0439 = -0.19076$ amu.

Converting from atoms to moles and amu to grams, it requires 235 g ^{235}U to produce energy equivalent to a change in mass of 0.191 g.

Now 0.10% of 1.714×10^5 kg is 1.714×10^2 kg $= 1.714 \times 10^5$ g. Then,

$$1.714 \times 10^5 \text{ g} \times \frac{235 \text{ g} \ ^{235}\text{U}}{0.191 \text{ g}} = 2.11 \times 10^8 \text{ g} \ ^{235}\text{U}$$

This is about 230 tons of ^{235}U **per day**.

21.49   We can use Figure 21.13 to see that the binding energy per nucleon (which gives rise to the mass defect) is greatest for nuclei of mass numbers around 50. Thus (a) $^{59}_{27}$Co should possess the greatest mass defect per nucleon.

Effects and Uses of Radioisotopes

21.51 The ^{59}Fe would be incorporated into the diet component, which in turn is fed to the rabbits. After a time blood samples could be removed from the animals, the red blood cells separated, and the radioactivity of the sample measured. If the iron in the dietary compound has been incorporated into blood hemoglobin, the blood cell sample should show beta emission. Samples could be taken at various times to determine the rate of iron uptake, rate of loss of the iron from the blood, and so forth.

21.53 (a)   *Control rods* control neutron flux so that there are enough neutrons to sustain the chain reaction but not so many that the core overheats.

     (b)   A *moderator* slows neutrons so that they are more easily captured by fissioning nuclei.

21.55 (a) $^{235}_{92}U + ^{1}_{0}n \rightarrow ^{160}_{62}Sm + ^{72}_{30}Zn + 4 \, ^{1}_{0}n$ (b) $^{239}_{94}Pu + ^{1}_{0}n \rightarrow ^{144}_{58}Ce + ^{94}_{36}Kr + 2 \, ^{1}_{0}n$

21.57 The extremely high temperature is required to overcome the electrostatic charge repulsions between the nuclei so that they come together to react.

21.59 •OH is a free radical; it contains an unpaired (free) electron, which makes it an extremely reactive species. (As an odd electron molecule, it violates the octet rule.) It can react with almost any particle (atom, molecule, ion) to acquire an electron and become OH^-. This often starts a disruptive chain of reactions, each producing a different free radical.

Hydroxide ion, OH^-, on the other hand, will be attracted to cations or the positive end of a polar molecule. Its most common reaction is ubiquitous and innocuous: $H^+ + OH^- \rightarrow H_2O$. The acid-base reactions of OH^- are usually much less disruptive to the organism than the chain of redox reactions initiated by •OH radical.

21.61 (a) $1 \, Ci = 3.7 \times 10^{10}$ disintegrations(dis)/s; $1 \, Bq = 1$ dis/s

$$8.7 \, mCi \times \frac{1 \, Ci}{1000 \, mCi} \times \frac{3.7 \times 10^{10} \, dis/s}{Ci} = 3.22 \times 10^8 = 3.2 \times 10^8 \, dis/s = 3.2 \times 10^8 \, Bq$$

     (b) $1 \, rad = 1 \times 10^{-2}$ J/kg; $1 \, Gy = 1$ J/kg $= 100$ rad. From part (a), the activity of the source is 3.2×10^8 dis/s.

$$3.22 \times 10^8 \, dis/s \times 2.0 \, s \times 0.65 \times \frac{9.12 \times 10^{-13} \, J}{dis} \times \frac{1}{0.250 \, kg} = 1.53 \times 10^{-3}$$

$$= 1.5 \times 10^{-3} \, J/kg$$

$$1.5 \times 10^{-3} \, J/kg \times \frac{1 \, rad}{1 \times 10^{-2} \, J/kg} \times \frac{1000 \, mrad}{rad} = 1.5 \times 10^2 \, mrad$$

$$1.5 \times 10^{-3} \, J/kg \times \frac{1 \, Gy}{1 \, J/kg} = 1.5 \times 10^{-3} \, Gy$$

     (c)   rem = rad (RBE); Sv = Gy (RBE) = 100 rem

mrem = 1.53×10^2 mrad (9.5) = 1.45×10^3 = 1.5×10^3 mrem (or 1.5 rem)

Sv = 1.53×10^{-3} Gy (9.5) = 1.45×10^{-2} = 1.5×10^{-2} Sv

Additional Exercises

21.63 $^{222}_{86}Rn \rightarrow X + 3\,^4_2He + 2\,^0_{-1}\beta$

This corresponds to a reduction in mass number of $(3 \times 4 =) 12$ and a reduction in atomic number of $(3 \times 2 - 2) = 4$. The stable nucleus is $^{210}_{82}Pb$. [This is part of the sequence in Figure 21.4.]

21.65 The most massive radionuclides will have the highest neutron/proton ratios. Thus, they are most likely to decay by a process that lowers this ratio, beta emission. The least massive nuclides, on the other hand, will decay by a process that increases the neutron/proton ratio, positron emission or orbital electron capture.

21.68 This is similar to Exercises 21.35 and 21.36.

$^{209}_{84}Po \rightarrow {}^{205}_{82}Pb + {}^4_2He$

Each ^{209}Po nucleus that decays is 1 disintegration. Calculate the mass of ^{209}Po remaining after 1.0 s, calculate by subtraction the mass that has decayed, and use Avogadro's number to get the number of nuclei that have decayed.

Calculate k in s^{-1}. $105\ yr \times \dfrac{365\ d}{1\ yr} \times \dfrac{24\ hr}{1\ d} \times \dfrac{60\ min}{1\ hr} \times \dfrac{60\ s}{1\ min} = 3.311 \times 10^9 = 3.31 \times 10^9\ s$

$k = 0.693\,/\,t_{1/2} = 0.693/3.311 \times 10^9\ s = 2.093 \times 10^{-10} = 2.09 \times 10^{-10}\ s^{-1}$

$\ln(N_t\,/\,N_o) - kt = -(2.093 \times 10^{-10}\ s^{-1})(1.0\ s) = -2.093 \times 10^{-10}$

$N_t\,/\,N_o = e^{-2.093 \times 10^{-10}} = (1.00 - 2.093 \times 10^{-10}); \quad N_t = 1.0 \times 10^{-12}\ g\ (1.000 - 2.093 \times 10^{-10})$

The amount that decays is $N_o - N_t$:

$1.0 \times 10^{-12}\ g - [1.0 \times 10^{-12}\ g\ (1.000 - 2.093 \times 10^{-10})] = 1.0 \times 10^{-12}\ g\ (2.093 \times 10^{-10})$

$= 2.093 \times 10^{-22} = 2.1 \times 10^{-22}\ g\ Po$

$N_o - N_t = 2.093 \times 10^{-22}\ g\ Po \times \dfrac{1\ mol\ Po}{209\ g\ Po} \times \dfrac{6.022 \times 10^{23}\ Po\ atoms}{1\ mol\ Po} = 0.603 = 0.60\ dis$

This is 0.60 disintegrations in 1.0 s, or approximately 1 dis/s.

$0.603\ dis/s \times \dfrac{1\ Ci}{3.7 \times 10^{10}\ dis/s} = 1.6 \times 10^{-10}\ Ci\ (2.7 \times 10^{-11}\ Ci\ based\ on\ 1\ dis/s.)$

21.70 1×10^{-6} curie $\times \dfrac{3.7 \times 10^{10} \text{ dis/s}}{\text{curie}} = 3.7 \times 10^{4}$ dis/s

rate = 3.7×10^{4} nuclei/s = kN

$k = \dfrac{0.693}{t_{1/2}} = \dfrac{0.693}{28.8 \text{ yr}} \times \dfrac{1 \text{ yr}}{365 \times 24 \times 3600 \text{ sec}} = 7.630 \times 10^{-10} = 7.63 \times 10^{-10} \text{ s}^{-1}$

3.7×10^{4} nuclei/s = $(7.63 \times 10^{-10}/\text{s})$ N;   N = $4.849 \times 10^{13} = 4.8 \times 10^{13}$ nuclei

mass ^{90}Sr = 4.849×10^{13} nuclei $\times \dfrac{90 \text{ g Sr}}{6.022 \times 10^{23} \text{ nuclei}} = 7.2 \times 10^{-9}$ g Sr

21.73   First, calculate k in s^{-1}

$k = \dfrac{0.693}{12.26 \text{ yr}} \times \dfrac{1 \text{ yr}}{365 \text{ d}} \times \dfrac{1 \text{ d}}{24 \text{ hr}} \times \dfrac{1 \text{ hr}}{3600 \text{ sec}} = 1.7924 \times 10^{-9} = 1.792 \times 10^{-9} \text{ s}^{-1}$

From Equation [21.18], $1.50 \times 10^{3} \text{ s}^{-1} = (1.7924 \times 10^{-9} \text{ s}^{-1})(\text{N})$;

N = $8.369 \times 10^{11} = 8.37 \times 10^{11}$ In 26.0 g of water, there are

26.0 g $H_2O \times \dfrac{1 \text{ mol } H_2O}{18.02 \text{ g } H_2O} \times \dfrac{6.022 \times 10^{23} \, H_2O}{1 \text{ mol } H_2O} \times \dfrac{2 \text{ H}}{1 \, H_2O} = 1.738 \times 10^{24}$

$= 1.74 \times 10^{24}$ H atoms

The mole fraction of $^{3}_{1}$H atoms in the sample is thus

$8.369 \times 10^{11}/1.738 \times 10^{24} = 4.816 \times 10^{-13} = 4.82 \times 10^{-13}$

21.75 $\Delta m = \Delta E/c^{2}$; $\Delta m = \dfrac{3.9 \times 10^{26} \text{ kg} \cdot \text{m/s}^{2}}{(3.00 \times 10^{8} \text{ m/s})^{2}} = 4.3 \times 10^{9}$ kg

This is the mass lost in one second.

Integrative Exercises

21.78   Calculate N, the number of ^{36}Cl nuclei, the value of k in s^{-1}, and the activity in dis/s.

49.5 mg NaClO$_4$ $\times \dfrac{1 \text{ g}}{1000 \text{ mg}} \times \dfrac{1 \text{ mol NaClO}_4}{122.44 \text{ g NaClO}_4} \times \dfrac{1 \text{ mol Cl}}{1 \text{ mol NaClO}_4} \times \dfrac{6.022 \times 10^{23} \text{ Cl atoms}}{\text{mol Cl}}$

$\times \dfrac{31 \, ^{36}\text{Cl atoms}}{100 \text{ Cl atoms}} = 7.547 \times 10^{19} = 7.55 \times 10^{19} \, ^{36}\text{Cl atoms}$

$k = 0.693/t_{1/2} = \dfrac{0.693}{3.1 \times 10^{5} \text{ yr}} \times \dfrac{1}{365 \times 24 \times 3600 \text{ s}} = 7.09 \times 10^{-14} = 7.1 \times 10^{-14} \text{ s}^{-1}$

rate = kN = $(7.09 \times 10^{-14} \text{ s}^{-1})(7.547 \times 10^{19}$ nuclei$) = 5.35 \times 10^{6} = 5.4 \times 10^{6}$ dis/s

21.80   (a) $0.18 \, \text{Ci} \times \dfrac{3.7 \times 10^{10} \, \text{dis/s}}{\text{Ci}} \times \dfrac{3600 \, \text{s}}{\text{Li}} \times \dfrac{24 \, \text{hr}}{\text{d}} \times 235 \, \text{d} = 1.35 \times 10^{17}$

$$= 1.4 \times 10^{17} \, \alpha \text{ particles}$$

(b) $P = nRT/V = 1.35 \times 10^{17} \, \text{He atoms} \times \dfrac{1 \, \text{mol He}}{6.022 \times 10^{23} \, \text{atoms}} \times \dfrac{295 \, \text{K}}{0.0150 \, \text{L}} \times \dfrac{0.08206 \, \text{L} \cdot \text{atm}}{\text{K} \cdot \text{mol}}$

$$= 3.62 \times 10^{-4} = 3.6 \times 10^{-4} \, \text{atm} = 0.28 \, \text{torr}$$

21.83   Determine the wavelengths of the photons by first calculating the energy equivalent of the mass of an electron or positron. (Since **two** photons are formed by annihilation of **two** particles of equal mass, we need to calculate the energy equivalent of just one particle.) The mass of an electron is 9.109×10^{-31} kg.

$\Delta E = (9.109 \times 10^{-31} \, \text{kg}) \times (2.998 \times 10^8 \, \text{m/s})^2 = 8.187 \times 10^{-14} \, \text{J}$

Also, $\Delta E = h\nu$; $\Delta E = hc/\lambda$; $\lambda = hc/\Delta E$

$\lambda = \dfrac{(6.626 \times 10^{-34} \, \text{J} \cdot \text{s})(2.998 \times 10^8 \, \text{m/s})}{8.187 \times 10^{-14} \, \text{J}} = 2.426 \times 10^{-12} \, \text{m} = 2.426 \times 10^{-3} \, \text{nm}$

This is a very short wavelength indeed; it lies at the short wavelength end of the range of observed gamma ray wavelengths (see Figure 6.4).

22 Chemistry of the Nonmetals

Periodic Trends and Chemical Reactions

22.1 Metals: Sr, Ce, Rh; nonmetals: Se, Kr; metalloid: Sb

22.3 Bismuth should be most metallic. It has a metallic luster and a relatively low ionization energy. Bi_2O_3 is soluble in acid but not in base, characteristic of the basic properties of the oxide of a metal rather than the oxide of a nonmetal.

22.5 (a)   Cl   (b) K

(c)   K in the gas phase (lowest ionization energy), Li in aqueous solution (most positive $E°$ value)

(d)   Ne; Ne and Ar are difficult to compare because they do not form compounds and their radii are not measured in the same way as other elements. However, Ne is several rows to the right of C and surely has a smaller atomic radius. The next smallest is C.

(e)   C

22.7 (a)   Nitrogen is too small to accommodate five fluorine atoms about it. The P and As atoms are larger. Furthermore, P and As have available 3d and 4d orbitals, respectively, to form hybrid orbitals that can accommodate more than an octet of electrons about the central atom.

(b)   Si does not readily form π bonds, which would be necessary to satisfy the octet rule for both atoms in SiO.

(c)   A reducing agent is a substance that readily loses electrons. As has a lower electronegativity than N; that is, it more readily gives up electrons to an acceptor and is more easily oxidized.

22.9 (a) $NaNH_2(s) + H_2O(l) \rightarrow NH_3(aq) + Na^+(aq) + OH^-(aq)$

(b) $2C_3H_7OH(l) + 9O_2(g) \rightarrow 6CO_2(g) + 8H_2O(l)$

(c) $NiO(s) + C(s) \rightarrow CO(g) + Ni(s)$ or $2NiO(s) + C(s) \rightarrow CO_2(g) + 2Ni(s)$

(d) $AlP(s) + 3H_2O(l) \rightarrow PH_3(g) + Al(OH)_3(s)$

(e) $Na_2S(s) + 2HCl(aq) \rightarrow H_2S(g) + 2Na^+(aq) + 2Cl^-(aq)$

Hydrogen, the Noble Gases, and the Halogens

22.11 ^1_1H - protium; ^2_1H - deuterium; ^3_1H - tritium

22.13 Like other elements in group 1A, hydrogen has only one valence electron. Like other elements in group 7A, hydrogen needs only one electron to complete its valence shell. The most common oxidation number of H is +1, like the group 1A elements; H can also exist in the -1 oxidation state, a state common to the group 7A elements.

22.15 (a) $Mg(s) + 2H^+(aq) \rightarrow Mg^{2+}(aq) + H_2(g)$

      (b) $C(s) + H_2O(g) \xrightarrow{1000\,^\circ C} CO(g) + H_2(g)$

      (c) $CH_4(g) + H_2O(g) \xrightarrow{1100\,^\circ C} CO(g) + 3H_2(g)$

22.17 (a) $NaH(s) + H_2O(l) \rightarrow NaOH(aq) + H_2(g)$

      (b) $Fe(s) + H_2SO_4(aq) \rightarrow Fe^{2+}(aq) + H_2(g) + SO_4^{2-}(aq)$

      (c) $H_2(g) + Br_2(g) \rightarrow 2HBr(g)$

      (d) $Na(l) + H_2(g) \rightarrow 2NaH(s)$

      (e) $PbO(s) + H_2(g) \xrightarrow{\Delta} Pb(s) + H_2O(g)$

22.19 (a)   Ionic (metal hydride)   (b)   molecular (nonmetal hydride)

      (c)   metallic (nonstoichiometric transition metal hydride)

22.21 Xenon is larger, and can more readily accommodate an expanded octet. More important is the lower ionization energy of xenon; because the valence electrons are a greater average distance from the nucleus, they are more readily promoted to a state in which the Xe atom can form bonds with fluorine.

22.23 (a) IO_3^-, +5   (b) H<u>Br</u>O$_3$, +5   (c) <u>BrF</u>$_3$; Br, +3; F, -1

      (d)   NaO<u>Cl</u>, +1   (e) H<u>I</u>O$_2$, +3   (f) <u>Xe</u>O$_3$, +6

22.25 (a)   potassium chlorate   (b) calcium iodate   (c) aluminum chloride

      (d)   bromic acid   (e) paraperiodic acid   (f) xenon tetrafluoride

22.27 (a)   Van der Waals intermolecular attractive forces increase with increasing numbers of electrons in the atoms.

      (b) F_2 reacts with water: $F_2(g) + H_2O(l) \rightarrow 2HF(aq) + O_2(g)$. That is, fluorine is too strong an oxidizing agent to exist in water.

      (c)   HF has extensive hydrogen bonding.

    (d)   Oxidizing power is related to electronegativity. Electronegativity decreases in the order given.

22.29 (a) $Br_2(l) + 2OH^-(aq) \rightarrow BrO^-(aq) + Br^-(aq) + H_2O(l)$

       (b) $Cl_2(g) + 2Br^-(aq) \rightarrow Br_2(l) + 2Cl^-(aq)$

       (c) $Br_2(l) + H_2O_2(aq) \rightarrow 2Br^-(aq) + O_2(g) + 2H^+(aq)$

22.31 (a) $PBr_5(l) + 4H_2O(l) \rightarrow H_3PO_4(aq) + 5H^+(aq) + 5Br^-(aq)$

       (b) $IF_5(l) + 3H_2O(l) \rightarrow H^+(aq) + IO_3^-(aq) + 5HF(aq)$

       (c) $SiBr_4(l) + 4H_2O(l) \rightarrow Si(OH)_4(s) + 4H^+(aq) + 4Br^-(aq)$

       (d) $2F_2(g) + 2H_2O(l) \rightarrow 4HF(aq) + O_2(g)$

       (e) $2ClO_2(g) + H_2O(l) \rightarrow H^+(aq) + ClO_3^-(aq) + HClO_2(aq)$

       (f) $HI(g) \rightarrow H^+(aq) + I^-(aq)$

22.33 (a)   linear   (b) square-planar   (c) trigonal pyramidal

       (d)   octahedral about the central iodine   (e) square-planar

Oxygen and the Group 6A Elements

22.35 (a)   As an oxidizing agent in steel-making; to bleach pulp and paper; in oxyacetylene torches; in medicine to assist in breathing

       (b)   Synthesis of pharmaceuticals, lubricants and other organic compounds where C=C bonds are cleaved; in water treatment

22.37 (a) $CaO(s) + H_2O(l) \rightarrow Ca^{2+}(aq) + 2OH^-(aq)$

       (b) $Al_2O_3(s) + 6H^+(aq) \rightarrow 2Al^{3+}(aq) + 3H_2O(l)$

       (c) $Na_2O_2(s) + 2H_2O(l) \rightarrow 2Na^+(aq) + 2OH^-(aq) + H_2O_2(aq)$

       (d) $N_2O_3(g) + H_2O(l) \rightarrow 2HNO_2(aq)$

       (e) $2KO_2(s) + 2H_2O(l) \rightarrow 2K^+(aq) + 2OH^-(aq) + O_2(g) + H_2O_2(aq)$

       (f) $NO(g) + O_3(g) \rightarrow NO_2(g) + O_2(g)$

22.39 (a)   Neutral   (b) acidic (oxide of a nonmetal)

       (c)   basic (oxide of a metal)   (d) amphoteric

22.41 (a) $\underline{Se}O_3$, +6   (b) $H_6\underline{Te}O_6$, +6   (c) $Zn\underline{Se}O_4$, +6   (d) $\underline{S}F_4$, +4

       (e) $H_2\underline{S}$, -2   (f) $H_2\underline{S}O_3$, +4

22.43 (a)   Potassium thiosulfate   (b) aluminum sulfide   (c) sodium hydrogen selenite

       (d)   selenium hexafluoride

22.45 The half reaction for oxidation in all these cases is:

$H_2S(aq) \rightarrow S(s) + 2H^+ + 2e^-$ (The product could be written as $S_8(s)$, but this is not necessary. In fact it is not necessarily the case that S_8 would be formed, rather than some other allotropic form of the element.)

(a) $2Fe^{3+}(aq) + H_2S(aq) \rightarrow 2Fe^{2+}(aq) + S(s) + 2H^+(aq)$

(b) $Br_2(l) + H_2S(aq) \rightarrow 2Br^-(aq) + S(s) + 2H^+(aq)$

(c) $2MnO_4^-(aq) + 6H^+(aq) + 5H_2S(aq) \rightarrow 2Mn^{2+}(aq) + 5S(s) + 8H_2O(l)$

(d) $2NO_3^-(aq) + H_2S(aq) + 2H^+(aq) \rightarrow 2NO_2(aq) + S(s) + 2H_2O(l)$

22.47   (a)

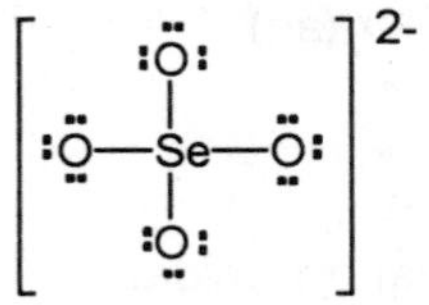

tetrahedral

(b)

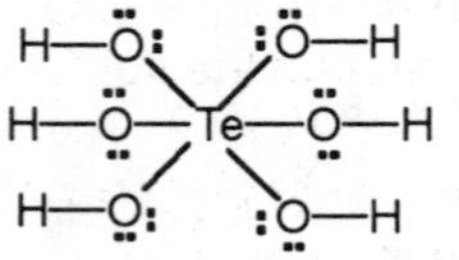

octahedral

(c)

bent

(d)

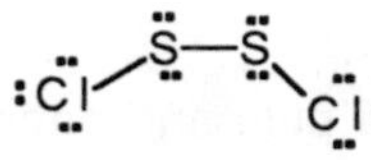

bent (free rotation around S-S bond)

(e)

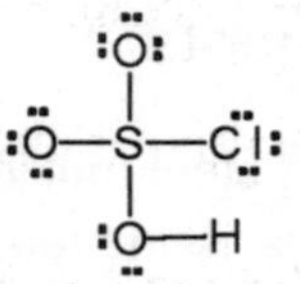

tetrahedral

22.49   (a) $SeO_2(s) + H_2O(l) \rightarrow H_2SeO_3(aq) \rightleftharpoons H^+(aq) + HSeO_3^-(aq)$

(b) $ZnS(s) + 2H^+(aq) \rightarrow Zn^{2+}(aq) + H_2S(g)$

(c) $8SO_3^{2-}(aq) + S_8(s) \rightarrow 8S_2O_3^{2-}(aq)$

(d) $Se(s) + 2H_2SO_4(l) \xrightarrow{\Delta} SeO_2(g) + 2SO_2(g) + 2H_2O(g)$

(e) $SO_3(aq) + H_2SO_4(l) \rightarrow H_2S_2O_7(l)$

(f) $H_2SeO_3(aq) + H_2O_2(aq) \rightarrow 2H^+(aq) + SeO_4^{2-}(aq) + H_2O(l)$

Nitrogen and the Group 5A Elements

22.51   (a) $H\underline{N}O_2$, +3   (b) $\underline{N}_2H_4$, -2   (c) KC$\underline{N}$, -3   (d) Na$\underline{N}O_3$, +5

         (e) $\underline{N}H_4Cl$, -3   (f) Li$_3\underline{N}$, -3

22.53   (a)

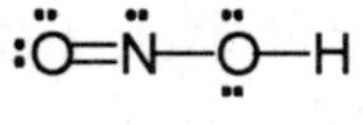

The molecule is bent around the central oxygen and nitrogen atoms; the four atoms need not lie in a plane.

(b) $\left[\ddot{N}=N=\ddot{N}:\right]^{-} \longleftrightarrow \left[:N\equiv N-\ddot{N}:\right]^{-} \longleftrightarrow \left[:\ddot{N}-N\equiv N:\right]^{-}$

The molecule is linear.

(c)
$$\left[\begin{array}{cc} H & H \\ | & | \\ H-N-N: \\ | & | \\ H & H \end{array}\right]^{+}$$

(d)
$$\left[\begin{array}{c} :\ddot{O}: \\ | \\ :\ddot{O}-N=\ddot{O} \end{array}\right]^{-}$$

The geometry is tetrahedral around the left nitrogen, trigonal pyramidal around the right.

(three equivalent resonance forms) The ion is trigonal planar.

22.55 (a) $Mg_3N_2(s) + 6H_2O(l) \rightarrow 3Mg(OH)_2(s) + 2NH_3(aq)$

(b) $2NO(g) + O_2(g) \rightarrow 2NO_2(g)$

(c) $4NH_3(g) + 3O_2(g) \xrightarrow{\Delta} 2N_2(g) + 6H_2O(g)$

(d) $NaNH_2(s) + H_2O(l) \rightarrow Na^+(aq) + OH^-(aq) + NH_3(aq)$

22.57 (a) $4Zn(s) + 2NO_3^-(aq) + 10H^+(aq) \rightarrow 4Zn^{2+}(aq) + N_2O(g) + 5H_2O(l)$

(b) $4NO_3^-(aq) + S(s) + 4H^+(aq) \rightarrow 4NO_2(g) + SO_2(g) + 2H_2O(l)$

(or $6NO_3^-(aq) + S(s) + 4H^+(aq) \rightarrow 6NO_2(g) + SO_4^{2-}(aq) + 2H_2O(l)$

(c) $2NO_3^-(aq) + 3SO_2(g) + 2H_2O(l) \rightarrow 2NO(g) + 3SO_4^{2-}(aq) + 4H^+(aq)$

22.59 (a) $2NO_3^-(aq) + 12H^+(aq) + 10e^- \rightarrow N_2(g) + 6H_2O(l)$ $E_{red}^{\circ} = +1.25$ V

(b) $2NH_4^+(aq) \rightarrow N_2(g) + 8H^+(aq) + 6e^-$ $E_{red}^{\circ} = 0.27$ V

22.61 (a) sodium phosphide   (b) arsenic acid   (c) tetraphosphorus decaoxide

(d) arsenic pentafluoride

22.63 (a) $H_3\underline{P}O_4$, +5 (b) $H_3\underline{As}O_3$, +3 (c) $\underline{Sb}_2\underline{S}_3$, +3 (d) $Ca(H_2\underline{P}O_4)_2$, +5 (e) $K_3\underline{P}$, -3

22.65 (a) Phosphorus is a larger atom and can more easily accommodate five surrounding atoms and an expanded octet of electrons than nitrogen can. Also, P has energetically "available" 3d orbitals which participate in the bonding, but nitrogen does not.

(b) Only one of the three hydrogens in H_3PO_2 is bonded to oxygen. The other two are bonded directly to phosphorus and are not easily ionized because the P-H bond is not very polar.

(c) PH_3 is a weaker base than H_2O (PH_4^+ is a stronger acid than H_3O^+). Any attempt to add H^+ to PH_3 in the presence of H_2O merely causes protonation of H_2O.

(d)   Antimony is more metallic in character than phosphorus. Because it is larger than P it has less attraction for additional electrons (it is a weaker Lewis acid). Thus, the reaction stops before complete hydrolysis has occurred.

(e)   White phosphorus consists of P_4 molecules, with P-P-P bond angles of $60°$. Each P atom has four VSEPR pairs of electrons, so the predicted electron pair geometry is tetrahedral and the preferred bond angle is $109°$. Because of the severely strained bond angles in P_4 molecules, white phosphorus is highly reactive.

22.67   (a) $2Ca_3(PO_4)_2(s) + 6SiO_2(s) + 10C(s) \rightarrow P_4(g) + 6CaSiO_3(l) + 10CO_2(g)$

        (b) $3H_2O(l) + PCl_3(l) \rightarrow H_3PO_3(aq) + 3H^+(aq) + 3Cl^-(aq)$

        (c) $6Cl_2(g) + P_4(s) \rightarrow 4PCl_3(l)$

        (d) $P_4O_{10}(s) + 6H_2O(l) \rightarrow 4H_3PO_4(aq)$

Carbon, the Other Group 4A Elements, and Boron

22.69   (a)   HCN   (b) SiC   (c) $CaCO_3$ (d) CaC_2

22.71   (a) $\left[:C \equiv N: \right]^-$ (b) $:C \equiv O:$ (c) $\left[:C \equiv C: \right]^{2-}$

        (d) $\ddot{S} = C = \ddot{S}$ (e) $\ddot{O} = C = \ddot{O}$ (f)

one of three equivalent
resonance structures

22.73   (a) $ZnCO_3(s) \xrightarrow{\Delta} ZnO(s) + CO_2(g)$

        (b) $BaC_2(s) + 2H_2O(l) \rightarrow Ba^{2+}(aq) + 2OH^-(aq) + C_2H_2(g)$

        (c) $C_2H_4(g) + 3O_2(g) \rightarrow 2CO_2(g) + 2H_2O(g)$

        (d) $2CH_3OH(l) + 3O_2(g) \rightarrow 2CO_2(g) + 4H_2O(g)$

        (e) $NaCN(s) + H^+(aq) \rightarrow Na^+(aq) + HCN(g)$

22.75   (a) $2CH_4(g) + 2NH_3(g) + 3O_2(g) \xrightarrow[\text{cat}]{800°C} 2HCN(g) + 6H_2O(g)$

        (b) $NaHCO_3(s) + H^+(aq) \rightarrow CO_2(g) + H_2O(l) + Na^+(aq)$

        (c) $2BaCO_3(s) + O_2(g) + 2SO_2(g) \rightarrow 2BaSO_4(s) + 2CO_2(g)$

22.77   (a) $\underline{Si}O_2$, +4   (b) $Ge\underline{Cl}_4$, +4   (c) $Na\underline{B}H_4$, +3   (d) $\underline{Sn}Cl_2$, +2   (e) $\underline{B}_2H_6$, -3

22.79   (a)   Carbon   (b) lead   (c) silicon

22.81 $GeCl_4(l) + 2H_2O(g) \rightarrow GeO_2(s) + 4HCl(g)$

        $SiCl_4(g) + 2H_2O(g) \rightarrow SiO_2(s) + 4HCl(g)$

22.83   (a) $SiO_4{}^{4-}$ (b) $SiO_3{}^{2-}$ (c) $SiO_3{}^{2-}$

22.85   (a)   Diborane (Figure 22.55 and below) has bridging H atoms linking the two B atoms. The structure of ethane shown below has the C atoms bound directly, with no bridging atoms.

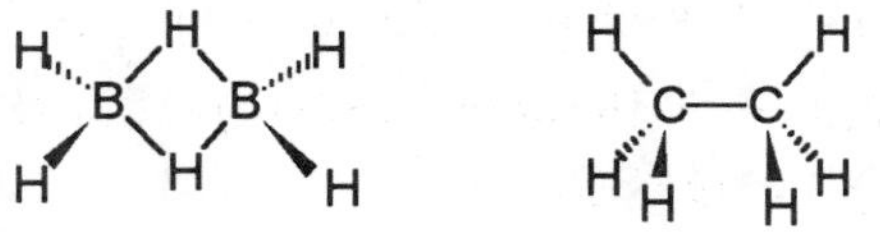

        (b) B_2H_6 is an electron deficient molecule. It has 12 valence electrons, while C_2H_6 has 14 valence electrons. The 6 valence electron pairs in B_2H_6 are all involved in B-H sigma bonding, so the only way to satisfy the octet rule at B is to have the bridging H atoms shown in Figure 22.55.

        (c)   A hydride ion, H^-, has two electrons while an H atom has one. The term *hydridic* indicates that the H atoms in B_2H_6 have more than the usual amount of electron density for a covalently bound H atom.

Additional Exercises

22.87   First find how many moles of gas occupy 1 ft³ at STP.

$$1\ ft^3 \times \left(\frac{12\ in}{1\ ft}\right)^3 \times \left(\frac{2.54\ cm}{1\ in}\right)^3 \times \frac{1\ L}{10^3\ cm^3} = 28.317 = 28.3\ L$$

$$\text{at STP}\ n = \frac{(1\ atm)(28.317\ L)}{(0.08206\ L\cdot atm/mol\cdot K)(273\ K)} = 1.264 = 1.26\ mol/ft^3$$

        (a) $2.2 \times 10^8\ kg\ H_2 \times \dfrac{1000\ g}{1\ kg} \times \dfrac{1\ mol\ H_2}{2.016\ g\ H_2} \times \dfrac{1\ ft^3}{1.264\ mol} = 8.6 \times 10^{10}\ ft^3\ H_2$

        (b) $2.4 \times 10^{10}\ kg\ N_2 \times \dfrac{1000\ g}{1\ kg} \times \dfrac{1\ mol\ N_2}{28.01\ g\ N_2} \times \dfrac{1\ ft^3}{1.264\ mol} = 6.8 \times 10^{11}\ ft^3\ N_2$

        (c) $1.7 \times 10^{10}\ kg\ O_2 \times \dfrac{1000\ g}{1\ kg} \times \dfrac{1\ mol\ O_2}{32.00\ g\ O_2} \times \dfrac{1\ ft^3}{1.264\ mol} = 4.2 \times 10^{11}\ ft^3\ O_2$

22.90 $2XeO_3(s) \rightarrow 2Xe(g) + 3O_2(g)$

$$0.654\ g\ XeO_3 \times \frac{1\ mol\ XeO_3}{179.1\ g\ XeO_3} \times \frac{5\ mol\ gas}{2\ mol\ XeO_3} = 9.129 \times 10^{-3} = 9.13 \times 10^{-3}\ mol\ gas$$

$$P = \frac{(9.129 \times 10^{-3}\ mol)(0.08206\ L\cdot atm/mol\cdot K)(321\ K)}{0.452\ L} = 0.532\ atm$$

22.93 $I_2 < F_2 < Br_2 < Cl_2$ The lower-than-expected value for F_2 can be ascribed to repulsions between the unshared electron pairs on the fluorine atoms at the short F-F distance necessary to give good overlap for bonding.

22.96   (a) $H_2SO_4 - H_2O \rightarrow SO_3$ (b) $2HClO_3 - H_2O \rightarrow Cl_2O_5$

        (c) $2HNO_2 - H_2O \rightarrow N_2O_3$ (d) $H_2CO_3 - H_2O \rightarrow CO_2$

        (e) $2H_3PO_4 - 3H_2O \rightarrow P_2O_5$

22.99

$$S(g) + O_2(g) \rightarrow SO_2(g) \qquad\qquad \Delta H = -296.9 \text{ kJ} \qquad (1)$$

$$SO_2(g) + 1/2\ O_2(g) \rightarrow SO_3(g) \qquad \Delta H = -98.3 \text{ kJ} \qquad (2)$$

$$SO_3(g) + H_2O(l) \rightarrow H_2SO_4(aq) \qquad \Delta H = -130 \text{ kJ} \qquad (3)$$

_____

$$S(g) + 3/2\ O_2(g) + H_2O(l) \rightarrow H_2SO_4(aq) \qquad \Delta H = -525 \text{ kJ}$$

$$1 \text{ ton } H_2SO_4 \times \frac{2000 \text{ lb}}{\text{ton}} \times \frac{453.6 \text{ g}}{1 \text{ lb}} \times \frac{1 \text{ mol } H_2SO_4}{98.09 \text{ g}} \times \frac{-525 \text{ kJ}}{\text{mol } H_2SO_4}$$

$$= -4.86 \times 10^6 \text{ kJ of heat/ton } H_2SO_4$$

22.102 $(CH_3)_2N_2H_2(g) + 2N_2O_4(g) \rightarrow 2CO_2(g) + 3N_2(g) + 4H_2O(g)$

$$4.0 \text{ tons } (CH_3)_2N_2H_2 \times \frac{2000 \text{ lb}}{1 \text{ ton}} \times \frac{453.6 \text{ g}}{1 \text{ lb}} \times \frac{1 \text{ mol } (CH_3)_2N_2H_2}{60.10 \text{ g } (CH_3)_2N_2H_2}$$

$$\times \frac{2 \text{ mol } N_2O_4}{1 \text{ mol } (CH_3)_2N_2H_2} \times \frac{92.02 \text{ g } N_2O_4}{1 \text{ mol } N_2O_4} \times \frac{1 \text{ lb}}{453.6 \text{ g}} \times \frac{1 \text{ ton}}{2000 \text{ lb}} = 12 \text{ tons } N_2O_4$$

22.105   The equation for this reaction is:

$$CaC_2(s) + 2H_2O(l) \rightarrow Ca^{2+}(aq) + 2OH^-(aq) + C_2H_2(g)$$

$$10.0 \text{ g } CaC_2 \times \frac{1 \text{ mol } CaC_2}{64.10 \text{ g } CaC_2} \times \frac{1 \text{ mol } C_2H_2}{1 \text{ mol } CaC_2} = 0.1562 = 0.156 \text{ mol } C_2H_2(g)$$

$$V = 0.1562 \text{ mol } C_2H_2 \times \frac{0.08206 \text{ L} \cdot \text{atm}}{\text{mol} \cdot \text{K}} \times \frac{300 \text{ K}}{(720/760) \text{ atm}} = 4.06 \text{ L}$$

22.108   (a) $2[5e^- + MnO_4^-(aq) + 8H^+(aq) \rightarrow Mn^{2+}(aq) + 4H_2O(l)]$

                       $5[H_2O_2(aq) \rightarrow O_2(g) + 2H^+(aq) + 2e^-]$

_____

$$2MnO_4^-(aq) + 5H_2O_2(aq) + 6H^+(aq) \rightarrow 2Mn^{2+}(aq) + 5O_2(g) + 8H_2O(l)$$

        (b) $2[Fe^{2+}(aq) \rightarrow Fe^{3+}(aq) + e^-]$

             $H_2O_2(aq) + 2H^+(aq) + 2e^- \rightarrow 2H_2O(l)$

_____

$$2Fe^{2+}(aq) + H_2O_2(aq) + 2H^+(aq) \rightarrow 2Fe^{3+}(aq) + 2H_2O(l)$$

(c)
$$2 I^-(aq) \rightarrow I_2(s) + 2e^-$$
$$H_2O_2(aq) + 2H^+(aq) + 2e^- \rightarrow 2H_2O(l)$$

$$2 I^-(aq) + H_2O_2(aq) + 2H^+(aq) \rightarrow I_2(s) + 2H_2O(l)$$

(d)
$$MnO_2(s) + 4H^+(aq) + 2e^- \rightarrow Mn^{2+}(aq) + 2H_2O(l)$$
$$H_2O_2(aq) \rightarrow O_2(g) + 2H^+(aq) + 2e^-$$

$$MnO_2(s) + 2H^+(aq) + H_2O_2(aq) \rightarrow Mn^{2+}(aq) + 2H_2O(l) + O_2(g)$$

(e)
$$2 I^-(aq) \rightarrow I_2(s) + 2e^-$$
$$O_3(g) + H_2O(l) + 2e^- \rightarrow O_2(g) + 2OH^-(aq)$$

$$2 I^-(aq) + O_3(g) + H_2O(l) \rightarrow O_2(g) + I_2(s) + 2OH^-(aq)$$

22.111

In both structures there are unshared pairs on all oxygens to give octets and the geometry around each P is approximately tetrahedral.

Integrative Exercises

22.114 First calculate the molar solubility of Cl_2 in water.

$$n = \frac{1(0.310 \text{ L})}{\frac{0.08206 \text{ L} \cdot \text{atm}}{1 \text{ mol} \cdot \text{K}} \times 273 \text{ K}} = 0.01384 = 0.0138 \text{ mol } Cl_2; \quad M = \frac{0.01384 \text{ mol}}{0.100 \text{ L}} = 0.1384 = 0.138 \text{ } M$$

$$K = \frac{[Cl^-][HOCl][H^+]}{[Cl_2]} = 4.7 \times 10^{-4}$$

$[Cl^-] = [HOCl] = [H^+]$ Let this quantity = x. Then, $\dfrac{x^3}{(0.1384 - x)} = 4.7 \times 10^{-4}$

Assuming that x is small compared with 0.1384:

$x^3 = (0.1384)(4.7 \times 10^{-4}) = 6.504 \times 10^{-5}$; x = 0.0402 = 0.040 M

We can correct the denominator using this value, to get a better estimate of x:

$\dfrac{x^3}{0.1384 - 0.0402} = 4.7 \times 10^{-4}$; x = 0.0359 = 0.036 M

One more round of approximation gives x = 0.0364 = 0.036 M. This is the equilibrium concentration of HClO.

22.118 (a) $SO_2(g) + 2H_2S(s) \rightarrow 3S(s) + 2H_2O(g)$ or, if we assume S_8 is the product,

$8SO_2(g) + 16H_2S(g) \rightarrow 3S_8(s) + 16H_2O(g)$.

(b) $2000 \text{ lb coal} \times \dfrac{0.035 \text{ lb S}}{1 \text{ lb coal}} \times \dfrac{453.6 \text{ g S}}{1 \text{ lb S}} \times \dfrac{1 \text{ mol S}}{32.07 \text{ g S}} \times \dfrac{1 \text{ mol SO}_2}{1 \text{ mol S}} \times \dfrac{2 \text{ mol H}_2\text{S}}{1 \text{ mol SO}_2}$

$= 1.98 \times 10^3 = 2.0 \times 10^3 \text{ mol H}_2\text{S}$

$V = \dfrac{1.98 \times 10^3 \text{ mol } (0.08206 \text{ L} \cdot \text{atm/mol} \cdot \text{K})(300 \text{ K})}{(740/760) \text{ atm}} = 5.01 \times 10^4 = 5.0 \times 10^4 \text{ L}$

(c) $1.98 \times 10^3 \text{ mol H}_2\text{S} \times \dfrac{3 \text{ mol S}}{2 \text{ mol H}_2\text{S}} \times \dfrac{32.07 \text{ g S}}{1 \text{ mol S}} = 9.5 \times 10^4 \text{ g S}$

This is about 210 lb S per ton of coal combusted. (However, two-thirds of this comes from the H_2S, which was presumably also obtained from coal.)

23 Metals and Metallurgy

Metallurgy

23.1 The important sources of iron are **hematite** (Fe_2O_3) and **magnetite** (Fe_3O_4). The major source of aluminum is **bauxite** ($Al_2O_3 \cdot xH_2O$). In ores, iron is present as the +3 ion, or in both the +2 and +3 states, as in magnetite. Aluminum is always present in the +3 oxidation state.

23.3 An ore consists of a little bit of the stuff we want, (chalcopyrite, $CuFeS_2$) and lots of other junk (gangue).

23.5 (a) $2PbS(s) + 3O_2(s) \rightarrow 2PbO(s) + 2SO_2(g)$

     (b) $PbCO_3(s) \xrightarrow{\Delta} PbO(s) + CO_2(g)$

     (c) $WO_3(s) + 3H_2(g) \rightarrow W(s) + 3H_2O(g)$

     (d) $ZnO(s) + CO(g) \rightarrow Zn(l) + CO_2(g)$

23.7 (a) $SO_3(g)$

     (b) $CO(g)$ provides a reducing environment for the transformation of Pb^{2+} to Pb^0.

     (c) $PbSO_4(s) \rightarrow PbO(s) + SO_3(g)$

         $PbO(s) + CO(g) \rightarrow Pb(s) + CO_2(g)$

23.9 $FeO(s) + H_2(g) \rightarrow Fe(s) + H_2O(g)$
$FeO(s) + CO(g) \rightarrow Fe(s) + CO_2(g)$

$Fe_2O_3(s) + 3H_2(g) \rightarrow 2Fe(s) + 3H_2O(g)$
$Fe_2O_3(s) + 3CO(g) \rightarrow 2Fe(s) + 3CO_2(g)$

23.11 (a) Air serves primarily to oxidize coke (C) to CO, the main reducing agent in the blast furnace. This exothermic reaction also provides heat for the furnace.

         $2C(s) + O_2(g) \rightarrow 2CO(g) \quad \Delta H = -110 \text{ kJ}$

     (b) Limestone, $CaCO_3$, is the source of basic oxide for slag formation.

         $CaCO_3(s) \xrightarrow{\Delta} CaO(s) + CO_2(g); \; CaO(l) + SiO_2(l) \rightarrow CaSiO_3(l)$

(c)   Coke is the fuel for the blast furnace, and the source of CO, the major reducing agent in the furnace.

$$2C(s) + O_2(g) \rightarrow 2CO(g); \; 4CO(g) + Fe_3O_4(s) \rightarrow 4CO_2(g) + 3Fe(l)$$

(d)   Water acts as a source of hydrogen, and as a means of controlling temperature. (see Equation [24.9]). $C(s) + H_2O(g) \rightarrow CO(g) + H_2(g) \quad \Delta H = +113 \; kJ$

23.13   The Bayer process takes advantage of the fact that Al^{3+} is amphoteric, but Fe^{3+} is not. Because it is amphoteric, Al^{3+} reacts with excess OH^- to form the soluble complex ion $Al(OH)_4^-$ while the Fe^{3+} solids cannot. This allows separation of the unwanted iron-containing solids by filtration.

23.15   Cobalt could be purified by constructing an electrolysis cell in which the crude metal was the anode and a thin sheet of pure cobalt was the cathode. The electrolysis solution is aqueous with a soluble cobalt salt such as $CoSO_4 \cdot 7H_2O$ serving as the electrolyte. (Other soluble salts with anions that do not participate in the cell reactions could be used.) Anode reaction: $Co(s) \rightarrow Co^{2+}(aq) + 2e^-$; cathode reaction: $Co^{2+}(aq) + 2e^- \rightarrow Co(s)$. Although $E°$ for reduction of $Co^{2+}(aq)$ is slightly negative (-0.277 V), we assume that reduction of water or H^+ does not occur because of a large overvoltage.

Metals and Alloys

23.17   Sodium is metallic; each atom is bonded to many nearest neighbor atoms by metallic bonding involving just one electron per atom, and delocalized over the entire three-dimensional structure. When sodium metal is distorted, each atom continues to have bonding interactions with many nearest neighbors. In NaCl the ionic forces are strong, and the arrangement of ions in the solid is very regular. When subjected to physical stress, the three-dimensional lattice tends to cleave along the very regular lattice planes, rather than undergo the large distortions characteristic of metals.

23.19   In the electron-sea model for metallic bonding, the valence electrons of the silver atoms move about the three-dimensional metallic lattice, while the silver atoms maintain regular lattice positions. Under the influence of an applied potential the electrons can move throughout the structure, giving rise to high electrical conductivity. The mobility of the electrons facilitates the transfer of kinetic energy and leads to high thermal conductivity.

23.21   According to the molecular orbital or band theory of metallic bonding, the 4s and 3d atomic orbitals of the fourth row elements are used to form six molecular orbitals, three bonding and three antibonding. The maximum bond order and highest bond strength occurs in metals with six valence electrons (group 6B). Assuming that hardness is directly related to the bond strength between metal atoms (the stronger the bonds, the more difficult it is to distort the solid without disrupting the three-dimensional structure), Cr, with six valence electrons, should have the highest bond strength and greatest hardness in the fourth row.

23.23   According to band theory, an *insulator* has a completely filled valence band and a large energy gap between the valence band and the nearest empty band; electrons are localized within the lattice. A *conductor* must have a partially filled energy band; a small excitation will promote electrons to previously empty levels within the band and allow them to move freely throughout the lattice, giving rise to the property of conduction. A *semiconductor* has a filled valence band, but the gap between the filled and empty bands is small enough to jump to the empty conduction band. The presence of an impurity may also place an electron in an otherwise empty band (producing an n-type semiconductor), or create a vacancy in an otherwise full band (producing a p-type semiconductor), providing a mechanism for conduction.

23.25   White tin, with a characteristic metallic structure, is expected to be more metallic in character. The electrical conductivity of the white allotropic form is higher because the valence electrons are shared with 12 nearest neighbors rather than being localized in four bonds to nearest neighbors as in gray tin. The Sn-Sn distance should be longer in white tin; there are only four valence electrons from each atom, and 12 nearest neighbors. The **average** tin-tin bond order can, therefore, be only about 1/3, whereas in gray tin the bond order is one. (In gray tin the Sn-Sn distance is 2.81 Å in white tin it is 3.02 Å.)

23.27   An *alloy* contains atoms of more than one element and has the properties of a metal. *Solution alloys* are homogeneous mixtures with different kinds of atoms dispersed randomly and uniformly. In *heterogeneous alloys* the components (elements or compounds) are not evenly dispersed and their properties depend not only on composition but methods of preparation. In an *intermetallic compound* the component elements have interacted to form a compound substance, for example, Cu_3As. As with more familiar compounds, these are homogeneous and have definite composition and properties.

Transition Metals

23.29   Of the properties listed, (b) the first ionization energy and (c) atomic radius are characteristic of isolated atoms. Electrical conductivity (a) and melting point (d) are properties of the bulk metal.

23.31   The *lanthanide contraction* is the name given to the decrease in atomic size due to the build-up in effective nuclear charge as we move through the lanthanides (elements 58-71) and beyond them. This effect offsets the expected increase in atomic size going from the second to the third transition series. The lanthanide contraction affects size-related properties such as ionization energy, electron affinity and density.

23.33   (a) ScF_3 (b) CoF_3 (c) ZnF_2

23.35   Chromium, $[Ar]4s^1 3d^5$, has six valence-shell electrons, some or all of which can be involved in bonding, leading to multiple stable oxidation states. By contrast, aluminum, $[Ne]3s^2 3p^1$, has only three valence electrons which are all lost or shared during bonding, producing the +3 state exclusively.

23.37 (a) Cr^{3+}: $[Ar]3d^3$ (b) Au^{3+}: $[Xe]4f^{14}5d^8$ (c) Ru^{2+}: $[Kr]4d^6$

(d) Cu^+: $[Ar]3d^{10}$ (e) Mn^{4+}: $[Ar]3d^3$ (f) Ir^{3+}: $[Xe]4f^{14}5d^6$

23.39 Ease of oxidation decreases from left to right across a period (owing to increasing effective nuclear charge); Ti^{2+} should be more easily oxidized than Ni^{2+}.

23.41 Chromate ion, CrO_4^{2-}, is bright yellow. Dichromate, $Cr_2O_7^{2-}$, is orange and more stable in acid solution than CrO_4^{2-} because of the equilibrium
$$2CrO_4^{2-}(aq) + 2H^+(aq) \rightleftharpoons Cr_2O_7^{2-}(aq) + H_2O(l)$$

23.43 (a) $Fe(s) + 2HCl(aq) \rightarrow FeCl_2(aq) + H_2(g)$

(b) $Fe(s) + 4HNO_3(aq) \rightarrow Fe(NO_3)_3(aq) + NO(g) + 2H_2O(l)$
(See net ionic equation, Equation 23.28) In concentrated nitric acid, the reaction can produce $NO_2(g)$ according to the reaction:
$Fe(s) + 6HNO_3(aq) \rightarrow Fe(NO_3)_3(aq) + 3NO_2(g) + 3H_2O(l)$

23.45 The unpaired electrons in a *paramagnetic* material cause it to be weakly attracted into a magnetic field. A *diamagnetic* material, where all electrons are paired, is very weakly repelled by a magnetic field.

Additional Exercises

23.47 $PbS(s) + O_2(g) \rightarrow Pb(l) + SO_2(g)$

Regardless of the metal of interest, $SO_2(g)$ is a product of roasting sulfide ores. In an oxygen rich environment, $SO_2(g)$ is oxidized to $SO_3(g)$, which dissolves in $H_2O(l)$ to form sulfuric acid, $H_2SO_4(aq)$. Because of its corrosive nature, $SO_2(g)$ is a dangerous environmental pollutant (Section 18.4) and cannot be freely released into the atmosphere. A sulfuric acid plant near a roasting plant would provide a means for disposing of $SO_2(g)$ that would also generate a profit.

23.49 $HfCl_4(g) + Na(l) \rightarrow 4NaCl(s) + Hf(s)$

23.52 Because selenium and tellurium are both nonmetals, we expect them to be difficult to oxidize. Thus, both Se and Te are likely to accumulate as the free elements in the so-called anode slime, along with noble metals that are not oxidized.

23.55 Assuming that SO_2 and N_2 are the nonmetallic products, two half-reactions can be written:
$$5[MoS_2(s) + 7H_2O(l) \rightarrow MoO_3(s) + 2SO_2(g) + 14H^+(aq) + 14e^-]$$
$$7[12H^+(aq) + 2NO_3^-(aq) + 10e^- \rightarrow N_2(g) + 6H_2O(l)]$$

$$5MoS_2(s) + 14H^+(aq) + 14NO_3^-(aq) \rightarrow 5MoO_3(s) + 10SO_2(g) + 7N_2(g) + 7H_2O(l)$$
$$MoO_3(s) + 2NH_3(aq) + H_2O(l) \rightarrow (NH_4)_2MoO_4(s)$$
$$(NH_4)_2MoO_4(s) \rightarrow 2NH_3(g) + H_2O(g) + MoO_3(s)$$
$$MoO_3(s) + 3H_2(g) \rightarrow Mo(s) + 3H_2O(g)$$

23.57   Silicon has the diamond structure. As with carbon, the four valence electrons of silicon are completely involved in the four localized bonds to its neighbors. There are thus no electrons free to migrate throughout the solid. Titanium exists as a close-packed lattice; each Ti atom has twelve equivalent nearest neighbors. The valence shell electrons cannot be localized between pairs of atoms; rather they are delocalized and mobile throughout the structure. In terms of the band model described in Figure 23.17, Ti has an incompletely occupied allowed energy band. The origin of the different behaviors with regard to structure has to do with the extent of the orbitals in space. Electrons in Si feel a greater attraction to the nucleus than electrons in Ti, so they are more localized.

23.60 Ni^{2+}: $[Ar]3d^8$ ⥮ ⥮ ⥮ ↑ ↑   paramagnetic
                              3d

       Cr^{3+}: $[Ar]3d^3$ ↑ ↑ ↑ ☐ ☐   paramagnetic
                              3d

       Both ions are paramagnetic because they have unpaired electrons in the 3d subshell.

Integrative Exercises

23.65   (a)   malleability

        (b) $\dfrac{1\ \text{troy oz}}{300\ \text{ft}^2} \times \dfrac{31.1\ \text{g}}{\text{troy oz}} \times \dfrac{1\ \text{ft}^2}{(12\ \text{in})^2} \times \dfrac{1\ \text{in}^2}{(2.54\ \text{cm})^2} \times \dfrac{1\ \text{cm}^3}{19.31\ \text{g}} = 5.779 \times 10^{-6}\ \text{cm}$

          $5.779 \times 10^{-6}\ \text{cm} = \mathbf{5.78 \times 10^{-5}\ mm} = 5.78 \times 10^{-8}\ \text{m} = 57.8\ \text{nm}$

23.67 $FeO(s) + SiO_2(l) \rightarrow FeSiO_3(l)$

       $1.2 \times 10^4\ \text{kg ore} \times \dfrac{0.26\ \text{kg FeO}}{1\ \text{kg ore}} \times \dfrac{1000\ \text{g FeO}}{1\ \text{kg FeO}} \times \dfrac{1\ \text{mol FeO}}{71.9\ \text{g FeO}} \times \dfrac{1\ \text{mol SiO}_2}{1\ \text{mol FeO}}$

          $\times \dfrac{60.1\ \text{g SiO}_2}{1\ \text{mol SiO}_2} = 2.6 \times 10^6\ \text{g SiO}_2 = 2.6 \times 10^3\ \text{kg SiO}_2$

23.69   The first equation indicates that one mole Ni^{2+} is formed from passage of two moles of electrons, and the second equation indicates the same thing. Thus, the simple ratio (1 mol Ni^{2+}/2F).

       $66\ \text{A} \times 8.0\ \text{hr} \times \dfrac{3600\ \text{s}}{1\ \text{hr}} \times \dfrac{1\ \text{C}}{1\ \text{A}\cdot\text{s}} \times \dfrac{1\ \text{F}}{96,500\ \text{C}} \times \dfrac{1\ \text{mol Ni}^{2+}}{2\ \text{F}} \times \dfrac{58.7\ \text{g Ni}^{2+}}{1\ \text{mol Ni}^{2+}}$

23.71 $\Delta G° = \Delta H° - T\Delta S°$ (assume ΔH and ΔS are standard)

(a) $PbO(s) + CO(g) \rightarrow Pb(s) + CO_2(g)$

$\Delta H° = \Delta H°_f\ CO_2(g) + \Delta H°_f\ Pb(s) - \Delta H°_f\ CO(g) - \Delta H°_f\ PbO(s)$

   $= -393.5 - (-110.5) - (-217.3) = -65.7\ kJ$

$\Delta S° = S°\ CO_2(g) + S°\ Pb(s) - S°\ CO(g) - S°\ PbO(s)$

   $= 213.6 + 68.85 - 197.9 - 68.70 = 15.8\ J/K$

$\Delta G° = -65.7\ kJ - 1473\ K\ (0.0158\ kJ/K) = -89.0\ kJ$

(b) $Si(s) + 2MnO(s) \rightarrow SiO_2(s) + 2Mn(s)$

$\Delta H° = -910.9 + 0 - 0 - 2(-385.2) = -140.5\ kJ$

$\Delta S° = 41.84 + 2(32.0) - 18.7 - 2(59.7) = -32.3\ J/K$

$\Delta G° = -140.5\ kJ - 1473\ K(-0.0323\ kJ/K) = -92.9\ kJ$

(c) $FeO(s) + H_2(g) \rightarrow Fe(s) + H_2O(g)$

$\Delta H° = -241.82 + 0 - 0 - (-271.9) = +30.1\ kJ$

$\Delta S° = 27.15 + 188.7 - 60.75 - 130.58 = 24.5\ J/K$

$\Delta G° = 30.1\ kJ - 1473\ K\ (0.0245\ J/K) = (30.1 - 36.1)\ kJ = -6.0\ kJ$

23.74 $1\ kg\ ore \times \dfrac{700\ g\ Fe_3O_4}{1\ kg\ ore} \times \dfrac{1\ mol\ Fe_3O_4}{231.6\ g\ Fe_3O_4} \times \dfrac{3\ mol\ Fe}{1\ mol\ Fe_3O_4} \times \dfrac{55.85\ g\ Fe}{1\ mol\ Fe}$

$\times \dfrac{100\ g\ pig\ iron}{97.5\ g\ Fe} = 519 = 5.2 \times 10^2\ g\ pig\ iron$

Substantial shipping costs could be saved by shipping the pig iron rather than the raw ore. The northern refineries went out of business because the hardwood forests were decimated around the mills, and because reduction with coke became a much better procedure.

23.77   Calculate the mass of $Zn(s)$ that will be deposited.

$1.0\ m \times 50\ m \times \dfrac{(100)^2\ cm^2}{1\ m^2} \times 0.30\ mm \times \dfrac{1\ cm}{10\ mm} \times \dfrac{7.1\ g}{cm^3} \times 2\ sides$

$= 2.13 \times 10^5 = 2.1 \times 10^5\ g\ Zn$

$2.13 \times 10^5\ g\ Zn \times \dfrac{1\ mol\ Zn}{65.38\ g\ Zn} \times \dfrac{2\ F}{0.90\ mol\ Zn} \times \dfrac{96,500\ C}{F} = 6.986 \times 10^8 = 7.0 \times 10^8\ C$

(2 F/0.90 mol Zn takes the 90% efficiency into account.)

$6.986 \times 10^8\ C \times 3.5\ V \times \dfrac{1\ J}{C \cdot V} \times \dfrac{1\ kWh}{3.6 \times 10^6\ J} = 679.2 = 6.8 \times 10^2\ kWh$

$679.2\ kWh \times \dfrac{\$0.080}{1\ kWh} = \$54.34 \rightarrow \$54$

24 Chemistry of Coordination Compounds

Structure and Nomenclature

24.1   (a)   Coordination number = 4, oxidation number = +2

   (b)   5, +4   (c) 6, +3   (d) 5, +2   (e) 6, +3   (f) 4, +2

24.3   (a) 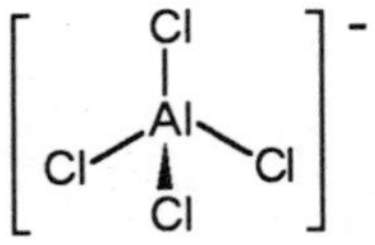
tetrahedral

   (b) 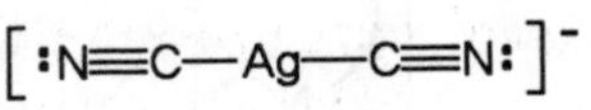 $\left[:N \equiv C - Ag - C \equiv N: \right]^-$
linear

   (c) 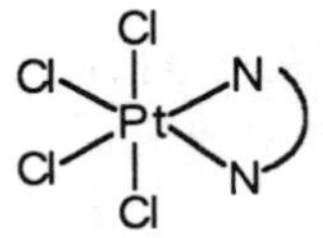
octahedral

   (d) 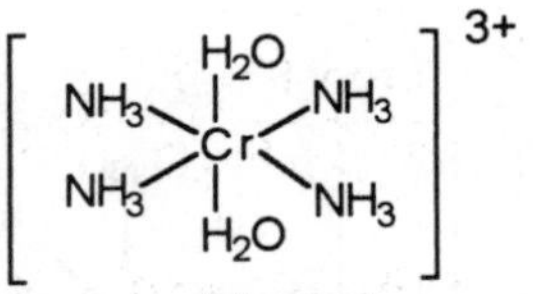
octahedral

24.5   [24.3]   (a)   tetrachloroaluminate(III)
   (b)   dicyanoargentate(I)
   (c)   tetrachloro(ethylenediamine)platinum(IV)
   (d)   *trans*-tetraamminediaquachromium(III)

   [24.4]   (a)   tetraamminezinc(II)
   (b)   aquapentachlororuthenate(III)
   (c)   *cis*-bis(ethylenediamine)dinitrocobalt(III)
   (d)   *trans*-diamminebromohydridoplatinum(II)

24.7   (a) $[Cr(NH_3)_6](NO_3)_3$ (b) $[Co(NH_3)_4CO_3]_2SO_4$ (c) $[Pt(en)_2Cl_2]Br_2$

   (d) $K[V(H_2O)_2Br_4]$ (e) $[Zn(en)_2][HgI_4]$

24.9   (a)   *ortho*-phenanthroline (*o*-phen) is bidentate
   (b)   oxalate (ox), $C_2O_4^{2-}$ is bidentate
   (c)   ethylenediaminetetraacetate, EDTA, is hexadentate
   (d)   ethylenediamine (en) is bidentate

Isomerism

24.11 **(a)**

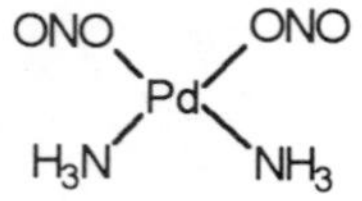

cis 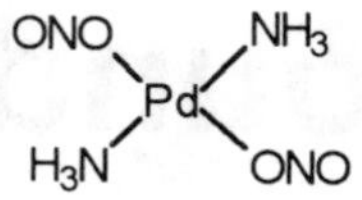   trans

(b) $[Pd(NH_3)_2(ONO)_2]$, $[Pd(NH_3)_2(NO_2)_2]$

(c)

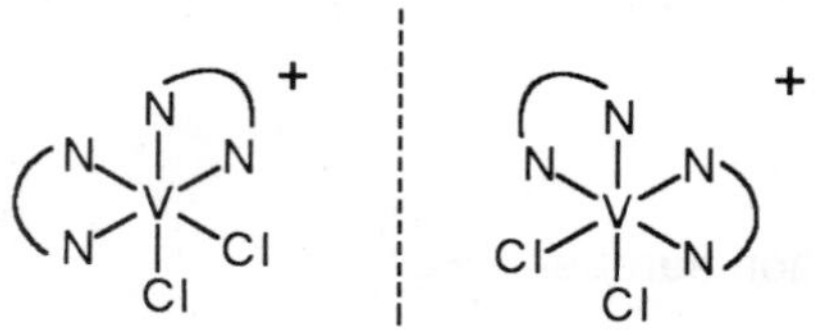

(d) $[Co(NH_3)_4Br_2]Cl$, $[Co(NH_3)_4BrCl]Br$

24.13

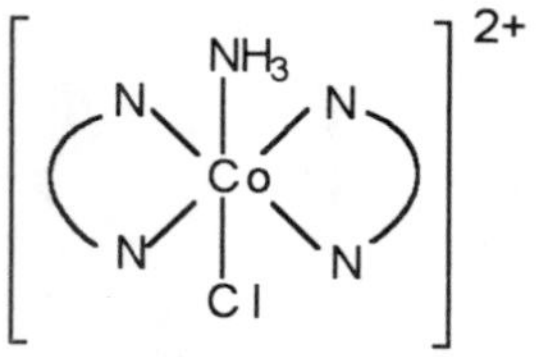

trans 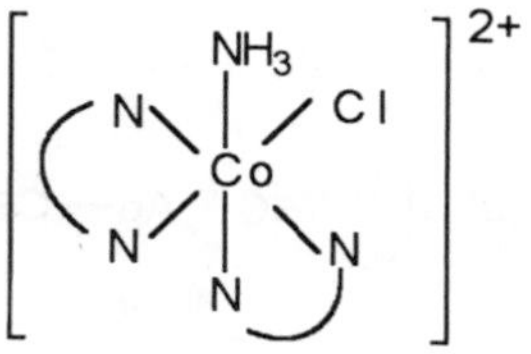   cis

The *cis* isomer is chiral.

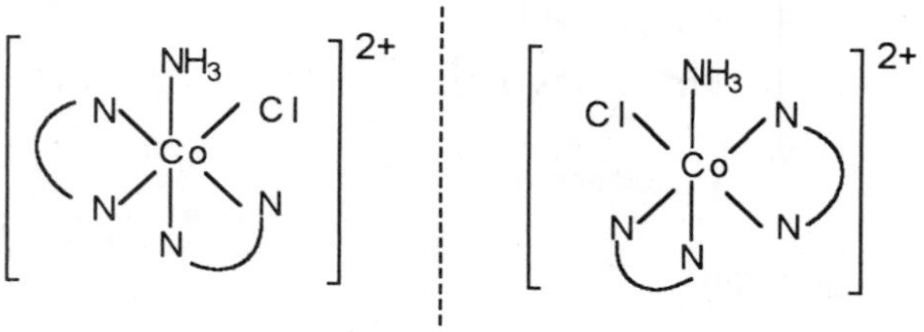

I 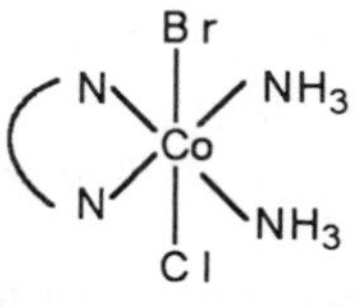   II

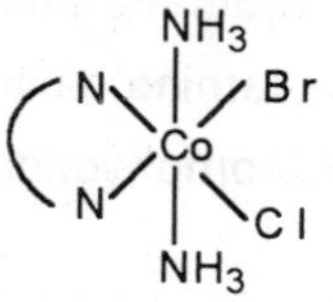

24.15 **(a)** only one:

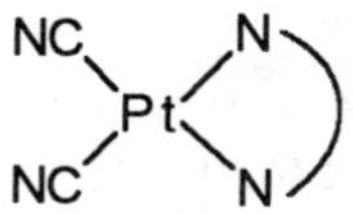

(b)

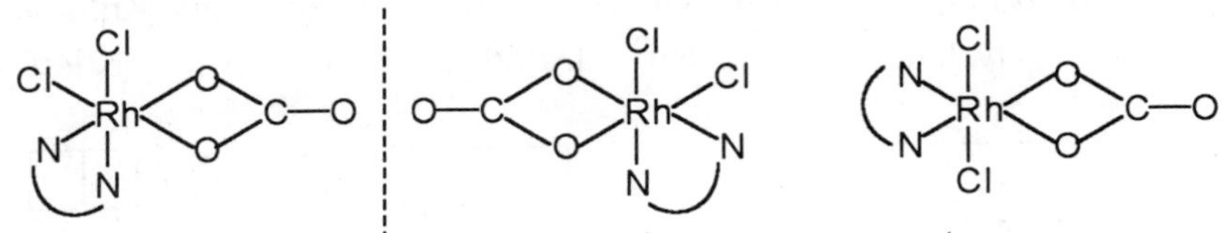

cis   *cis*   *trans*

optical isomers

(The three isomeric compounds in part (b) all have a 1– charge.)

(c)   only one:

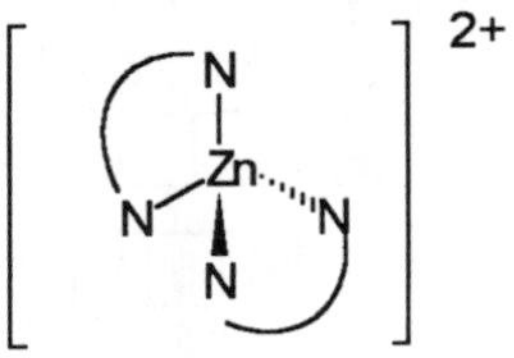

Color, Magnetism; Crystal Field Theory

24.17 Color in transition metal compounds arises from electronic transitions between d-orbital energy levels, or d-d transitions. Compounds with d^0 or d^{10} electron configurations are colorless, because d-d transitions are not possible.

(a) Zn^{2+}, d^{10}, colorless   (b) Cr^{4+}, d^2, colored   (c) Ni^{2+}, d^8, colored

(d) Al^{3+}, d^0, colorless   (e) Cd^{2+}, d^{10}, colorless   (f) Fe^{2+}, d^6, colored

24.19 (a) Ru^{3+}, d^5 (b) Cu^{2+}, d^9 (c) Co^{3+}, d^6 (d) Mo^{5+}, d^1 (e) Re^{3+}, d^4

24.21 Blue to blue-violet (Figure 24.23)

24.23 Six ligands in an octahedral arrangement are oriented along the x, y and z axes of the metal. These negatively charged ligands (or the negative end of ligand dipoles) have greater electrostatic repulsion with valence electrons in metal orbitals that also lie along these axes, the d_{z^2}, and $d_{x^2-y^2}$. The d_{xy}, d_{xz} and d_{yz} metal orbitals point between the x, y, and z axes, and electrons in these orbitals experience less repulsion with ligand electrons. Thus, in the presence of an octahedral ligand field, the d_{xy}, d_{xz} and d_{xy} metal orbitals are lower in energy than the $d_{x^2-y^2}$ and d_{z^2}.

24.25 Cyanide is a strong field ligand. The d-d electronic transitions occur at relatively high energy, because Δ is large. A yellow color corresponds to absorption of a photon in the violet region of the visible spectrum, between 430 and 400 nm. H_2O is a weaker field ligand than CN^-. The blue or green colors cf aqua complexes correspond to absorptions in the region of 620 nm. Clearly, this is a region of lower energy photons than those with characteristic wavelengths in the 430 to 400 nm region. These are very general and imprecise comparisons. Other factors are involved, including whether the complex is high spin or low spin.

24.27 (a) Mn - $[Ar]4s^2 3d^5$ (b) Ru - $[Kr]5s^2 4d^6$ (c) Rh - $[Kr]5s^2 4d^7$
           Mn^{3+} - $[Ar]3d^4$ Ru^{3+} - $[Kr]4d^5$ Rh^{3+} - $[Kr]4d^6$

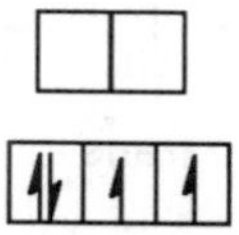

   2 unpaired electrons 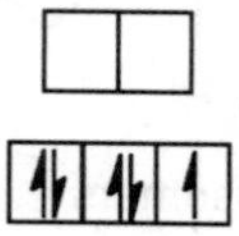   1 unpaired electron 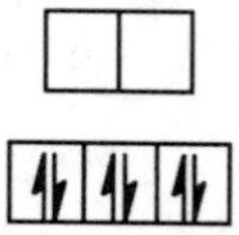   0 unpaired electrons

24.29 All complexes in this exercise are six-coordinate octahedral.

   (a) 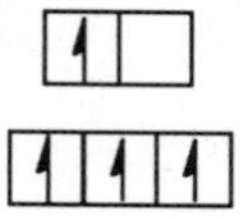 d^4, high spin   (b) 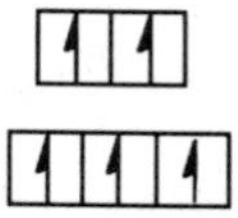 d^5, high spin   (c) 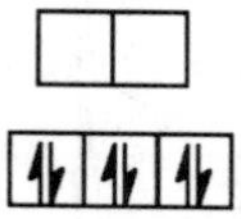 d^6, low spin

   (d) 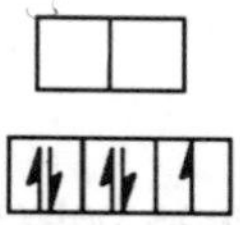 d^5, low spin   (e) 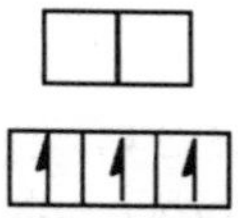 d^3 (f) 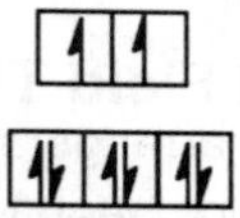 d^8

24.31

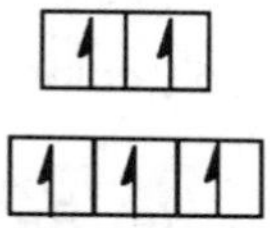

   high spin

Additional Exercises

24.33 (a) $K_2[Ni(en)Cl_4]$; $[Ni(en)(H_2O)_2Cl_2]$
       (b) $K_2[Ni(CN)_4]$; $[Zn(H_2O)_4](NO_3)_2$; $[Cu(NH_3)_4]SO_4$
       (c) $[CoF_6]^{3-}$, high spin; $[Co(NH_3)_6]^{3+}$ or $[Co(CN)_6]^{3-}$, low spin
       (d)   thiocyanate, SCN^- or NCS^-; nitrite, NO_2^- or ONO^-
       (e) $[Co(en)_2Cl_2]Cl$; see Exercise 24.15(b) and 24.16(c) for other examples.
       (f) $[Co(en)_3]Cl_3$, $[Cr(NH_3)_6]Cl_3$

24.36 (a)   linkage isomerism   (b) coordination sphere isomerism

24.38 (a) 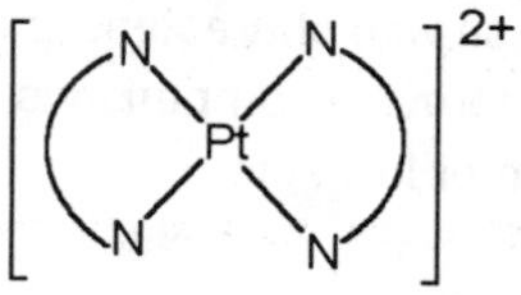   (b) 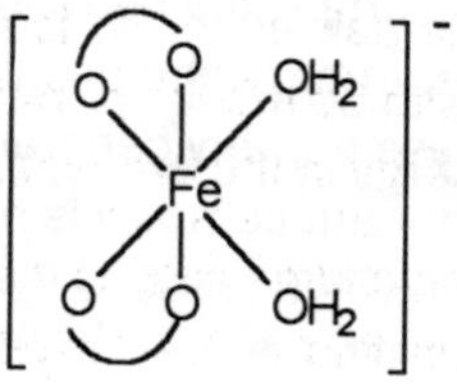   (c)

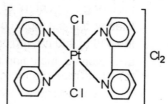

(d) Hg^{2+} is d^{10}, so the complex is probably tetrahedral.

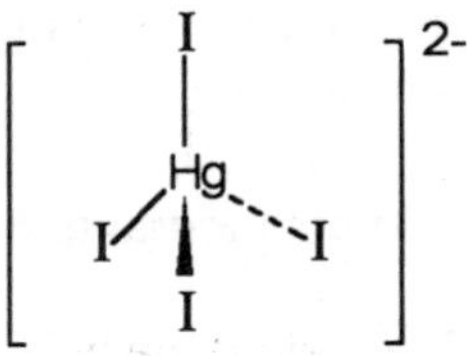

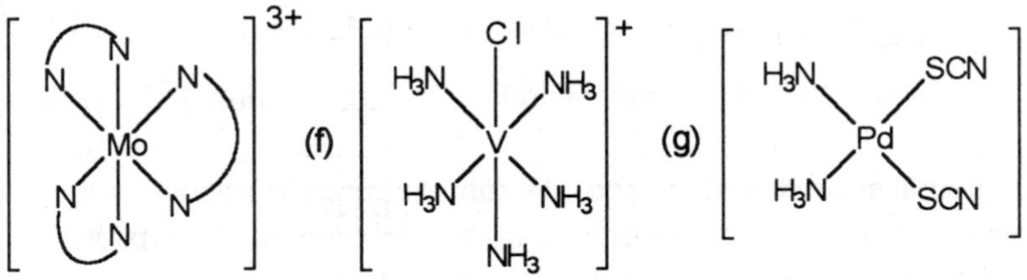

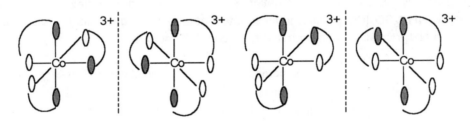

24.40 (a)   Only one

(b)   Two; (*cis* or *trans* arrangement of N and O ends)

(c)   Four; two are geometrical, the other two are steroisomers of each of these; in the figure below, the shaded ovals represent N and the open ovals represent O in the gly ligand.

24.43 According to the spectrochemical series, the order of increasing Δ for the ligands is $Cl^- < H_2O < NH_3$. (The tetrahedral Cl^- complex will have an even smaller Δ than an octahedral one.) The smaller the value of Δ, the longer the wavelength of visible light absorbed. The color of light absorbed is the complement of the observed color. A blue complex absorbs orange light (580-650 nm), a pink complex absorbs green light (490-560 nm) and a yellow complex absorbs violet light (400-430 nm). Since $[CoCl_4]^{2-}$ absorbs the longest wavelength, it appears blue. $[Co(H_2O)_6]^{2+}$ absorbs green and appears pink, and $[Co(NH_3)_6]^{3+}$ absorbs violet and appears yellow.

24.46 (a)   False. The spin pairing energy is **smaller** than Δ in low spin complexes. It is for this reason that electrons pair up in the lower energy orbitals, in spite of the repulsive energy associated with spin pairing, rather than move to the higher energy orbital, which would cost energy in the amount Δ.

(b)   True. Higher metal ion charge causes the ligands to be more strongly attracted, thus producing a larger splitting of the d orbital energies.

(c)   False. Square-planar configurations are associated with a d^8 electron configuration (which Ni^{2+} has) and a   ligand that produces a strong field, leading to large separations between the energy levels. Because cyanide, CN^-, is a much stronger field ligand than Cl^-, $[Ni(CN)_4]^{2-}$ is more likely to be square planar than is $[NiCl_4]^{2-}$.

24.49   The d^3 and d^6 electron configurations in octahedral and tetrahedral fields are shown below:

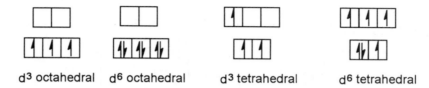

   d3 octahedral   d6 octahedral   d3 tetrahedral   d6 tetrahedral

In an octahedral environment, d^3 and d^6 strong field configurations have electrons only in the d orbitals which are **lower** in energy than the five degenerate orbitals in the isolated metal atom. In a tetrahedral environment (always weak field, see Section 24.6) d^3 has one electron and d^6 has three electrons in the higher energy d orbitals. Thus the octahedral environment is energetically more favorable for metal ions with d^3 or d^6 configurations.

Integrative Exercises

24.51   Cobalt(III) complexes are generally inert; that is, they do not rapidly exchange ligands inside the coordination sphere. Therefore, the ions that form precipitates in these two cases are probably outside the coordination sphere.   The red complex can be formulated as $[Co(NH_3)_5SO_4]Br$, pentaamminesulfatocobalt(III) bromide, and the violet compound as $[Co(NH_3)_5Br]SO_4$, pentaamminebromocobalt(III) sulfate.

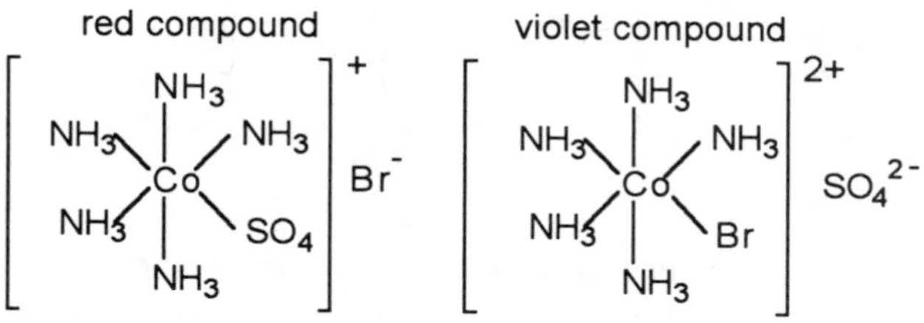

red compound   violet compound

24.53   Determine the empirical formula of the complex, assuming the remaining mass is due to oxygen, and a 100 g sample.

$$10.0\text{ g Mn} \times \frac{1\text{ mol Mn}}{54.94\text{ g Mn}} = 0.1820\text{ mol Mn};\ 0.182\,/\,0.182 = 1$$

$$28.6\text{ g K} \times \frac{1\text{ mol K}}{39.10\text{ g K}} = 0.7315\text{ mol K};\ 0.732\,/\,0.182 = 4$$

$$8.8\text{ g C} \times \frac{1\text{ mol C}}{12.0\text{ g C}} = 0.7327\text{ mol C};\ 0.733\,/\,0.182 = 4$$

$$29.2\text{ g Br} \times \frac{1\text{ mol Br}}{79.904\text{ g Br}} = 0.3654\text{ mol Br};\ 0.365\,/\,0.182 = 2$$

$$23.4\text{ g O} \times \frac{1\text{ mol O}}{16.00\text{ g O}} = 1.463\text{ mol O};\ 1.46\,/\,0.182 = 8$$

There are 2 C and 4 O per oxalate ion, for a total of two oxalate ligands in the complex. To match the conductivity of $K_4[Fe(CN)_6]$, the oxalate and bromide ions must be in the coordination sphere of the complex anion. Thus, the compound is $K_4[Mn(ox)_2Br_2]$, potassium dibromobis(oxalato)manganate(II).

24.56 $\Delta E = hc/\lambda = \dfrac{6.626 \times 10^{-34}\, J \cdot s \times 2.998 \times 10^8\, m/s}{510 \times 10^{-9}\, m} = 3.895 \times 10^{-19} = 3.90 \times 10^{-19}$ J/photon

$\Delta = 3.895 \times 10^{-19}$ J/photon $\times \dfrac{6.022 \times 10^{23}\ \text{photons}}{1\ \text{mol}} \times \dfrac{1\ kJ}{1000\ J} = 234.6 = 235$ kJ/mol

25 The Chemistry of Life: Organic and Biological Chemistry

Hydrocarbon Structures and Nomenclature

25.1 (a) $CH_3CH_2CH_2CH_2CH_3$, C_5H_{12} (b)

, C_5H_{10}

(c) $CH_2=CHCH_2CH_2CH_3$, C_5H_{10} (d) $HC \equiv C-CH_2CH_2CH_3$, C_5H_8

saturated: (a), (b); unsaturated: (c), (d)

25.3 Alkanes have the generic formula C_nH_{2n+2}. C_3H_9 does not have a C : H mole ratio of n : 2n+2; it has too many hydrogen atoms. A molecular formula of C_3H_9 would require at least one carbon atom to form five bonds; carbon never adopts an expanded octet of electrons.

25.5 There are five isomers. Their carbon skeletons are as follows:

25.7 (a) 2,2,4-trimethylpentane (b) 3-ethyl-2-methylpentane
(c) 2,3,4-trimethylpentane (d) 2,3,4-trimethylpentane
(c) and (d) are the same molecule

25.9 (a) 109° (b) 120° (c) 180°

25.11 (a)   2,3-dimethylhexane   (b)   4-ethyl-2,4-dimethylnonane

(c)   3,3,5-trimethylheptane   (d)   3,4,4-trimethylheptane

25.13 (a)

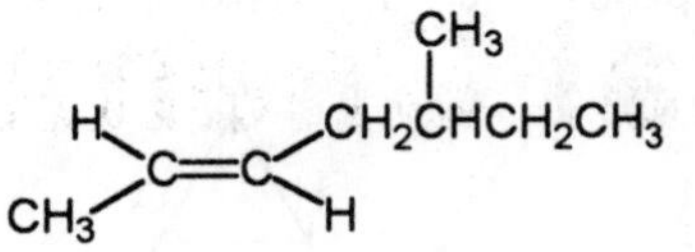

(b) $HC{\equiv}C-CH_2Cl$ (c)

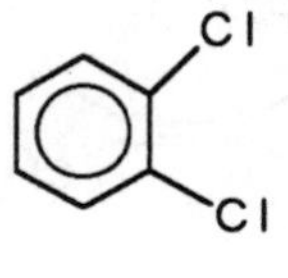

(d)

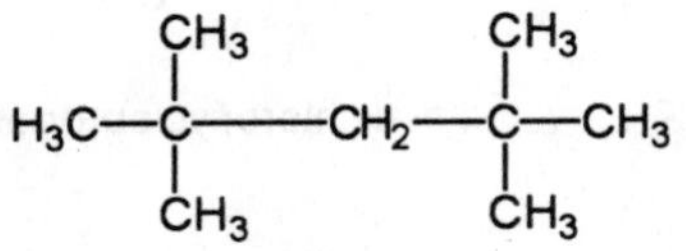

(e) 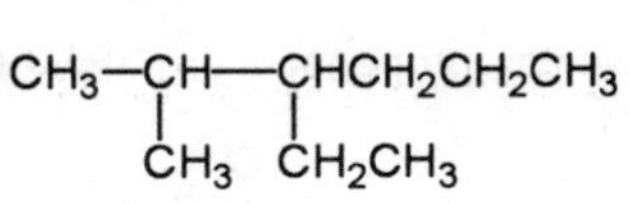 $CH_3-CH-CHCH_2CH_2CH_3$ with substituents CH_3 and CH_2CH_3

(f) 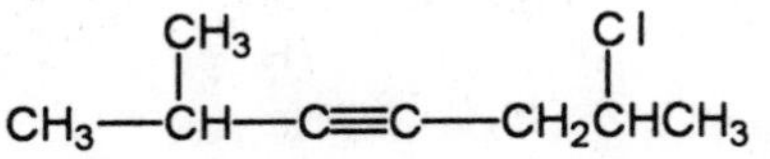 $CH_3-CH-C{\equiv}C-CH_2CHCH_3$ with substituents CH_3 and Cl

(g)

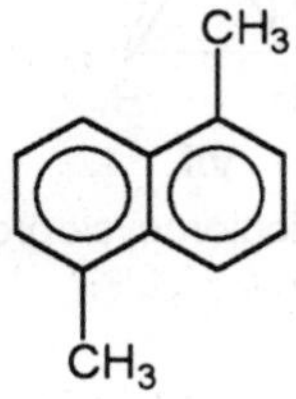

(h) $CH_2{=}CHCH_2CH_2CH_2CH{=}CH_2$

25.15 (a)   2,3-dimethylheptane   (b) *cis*-6-methyl-3-octene   (c) *para*-dibromobenzene

(d)   4,4-dimethyl-1-hexyne   (e) methylcyclobutane

25.17 Butene is an alkene, C_4H_8. There are two possible placements for the double bond:

$$CH_2{=}CHCH_2CH_3 \text{ or } CH_3CH{=}CHCH_3$$
1-butene   2-butene

These two compounds are structural isomers. For 2-butene, there are two different, noninterchangeable ways to construct the carbon skeleton (owing to the absence of free rotation around the double bond). These two compounds are geometric isomers.

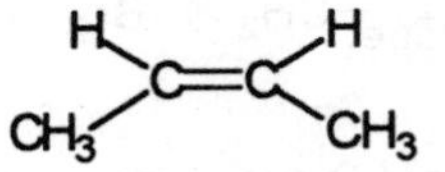

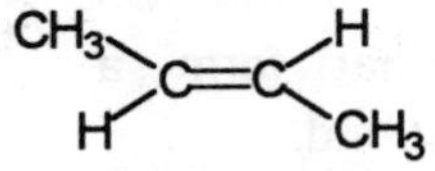

cis-2-butene   *trans*-2-butene

25.19

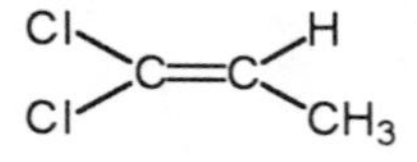

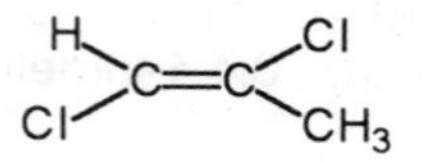

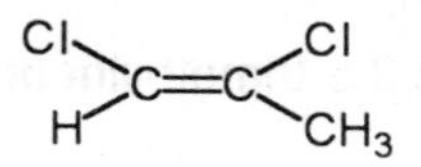

$C_3H_4Cl_2$

I
1,1-dichloropropene

II
trans-1,2-dichloropropene

III
cis-1,2-dichloropropene

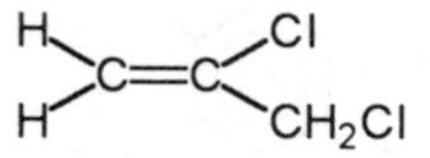

IV
2,3-dichloropropene

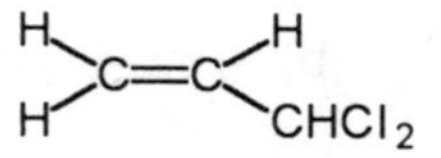

V
3,3-dichloropropene

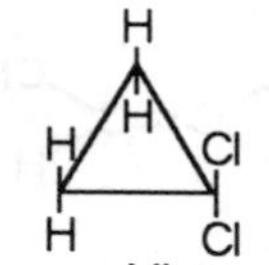

VI
1,1-dichlorocyclopropane

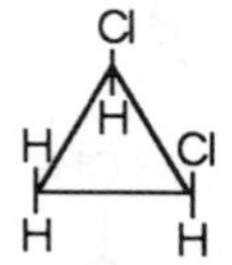

VII
cis-1,2-dichlorocyclopropane

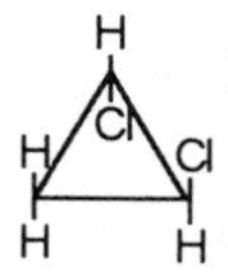

VIII
trans-1,2-dichlorocyclopropane

Compounds II / III and VII / VIII are geometric isomers; the others are structural isomers (different placement of Cl atoms.)

25.21   Assuming that each component retains its effective octane number in the mixture (and this isn't always the case), we obtain: octane number = 0.30(0) + 0.70(100) = 70.

Reactions of Hydrocarbons

25.23 **(a)**   A combustion reaction is the oxidation-reduction reaction of some substance (fuel) with $O_2(g)$.

$$2C_2H_6(g) + 7O_2(g) \rightarrow 4CO_2(g) + 6H_2O(g)$$

(b)   An addition reaction is the addition of some reagent to the two atoms that form a multiple bond.

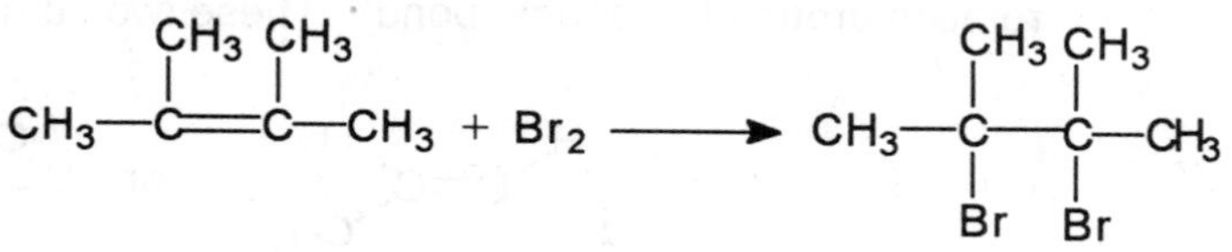

(c)   In a substitution reaction, one atom or group of atoms replaces (substitutes for) another atom or group of atoms.

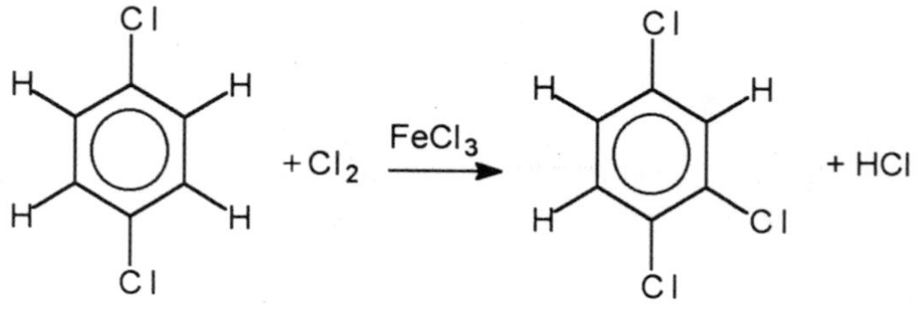

25.25   The small 60° C-C-C angles in the cyclopropane ring cause strain that provides a driving force for reactions that result in ring-opening. There is no comparable strain in the five- or six-membered rings.

25.27   First form an alkyl halide: $C_2H_4(g) + HBr(g) \rightarrow CH_3CH_2Br(l)$; then carry out a Friedel-Crafts reaction:

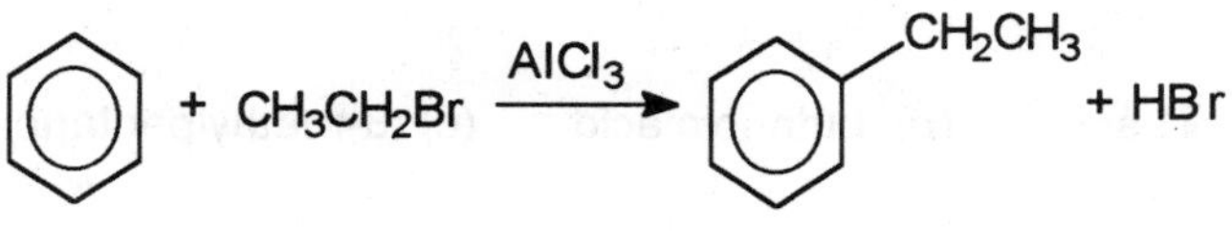

25.29

$$
\begin{array}{ll}
 & \underline{\Delta H} \\
C_{10}H_8(l) + 12O_2(g) \rightarrow 10CO_2(g) + 4H_2O(l) & -5157 \text{ kJ} \\
-[C_{10}H_{18}(l) + 29/2\ O_2(g) \rightarrow 10CO_2(g) + 9H_2O(l)] & -(-6286) \text{ kJ} \\
\hline
C_{10}H_8(l) + 5H_2O(l) \rightarrow C_{10}H_{18}(l) + 5/2\ O_2(g) & +1129 \text{ kJ} \\
5/2\ O_2(g) + 5H_2(g) \rightarrow 5H_2O(l) & 5(-285.8) \text{ kJ} \\
\hline
C_{10}H_8(l) + 5H_2(g) \rightarrow C_{10}H_{18}(l) & -300 \text{ kJ}
\end{array}
$$

Compare this with the heat of hydrogenation of ethylene:
$C_2H_4(g) + H_2(g) \rightarrow C_2H_6(g)$; $\Delta H = -84.7 - (52.3) = -137$ kJ. This value applies to just one double bond. For five double bonds, we would expect about -685 kJ. The fact that hydrogenation of napthalene yields only -300 kJ indicates that the overall energy of the napthalene molecule is lower than expected for five isolated double bonds and that there must be some special stability associated with the aromatic system in this molecule.

Functional Groups

25.31   (a)   ketone   (b) carboxylic acid   (c) alcohol   (d) ester   (e) amide   (f) amine

25.33   propionaldehyde (or propanal):

25.35   About each CH_3 carbon, 109°; about the carbonyl carbon, 120° planar.

25.37   (a) 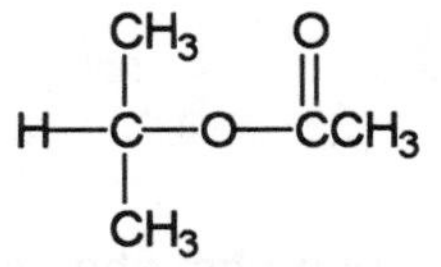   (b) 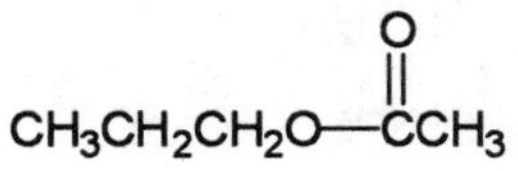 $CH_3CH_2CH_2O-\overset{\overset{\displaystyle O}{\|}}{C}CH_3$ (c) $CH_3CH_2O-\overset{\overset{\displaystyle O}{\|}}{C}-H$

2-propylacetate   1-propylacetate   ethylformate

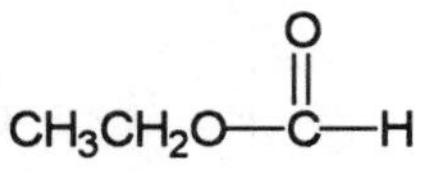

25.39

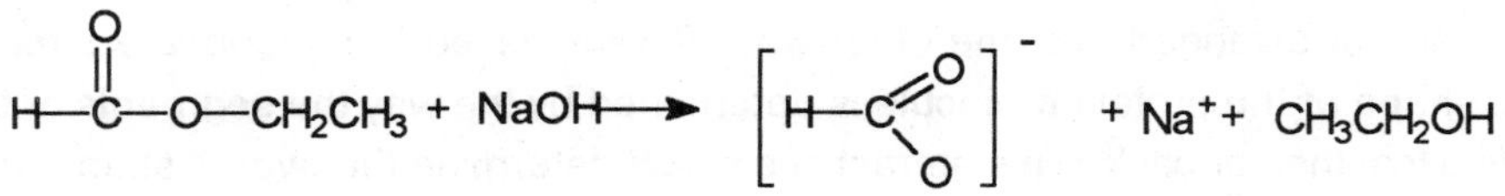

25.41   (a) $CH_3CH_2\underset{\underset{OH}{|}}{C}HCH_3$ (b) $HOCH_2CH_2OH$ (c) $H{-}\underset{\underset{}{\overset{\overset{O}{\|}}{C}}}{-}OCH_3$

(d) $CH_3CH_2\underset{}{\overset{\overset{O}{\|}}{C}}CH_2CH_3$ (e) $CH_3CH_2OCH_2CH_3$

25.43   (a)   methanoic acid   (b)   butanoic acid   (c)   3-methylpentanoic acid

Proteins

25.45   (a)   An α-amino acid contains an NH_2 group attached to the carbon that is bound to the carbon of the carboxylic acid function.

(b)   In forming a protein, amino acids undergo a condensation reaction between the amino group and carboxylic acid:

25.47   Two dipeptides are possible:

glycylvaline   valylglycine

25.49

25.51   Eight: ser-ser-ser; ser-ser-phe; ser-phe-ser; phe-ser-ser; ser-phe-phe; phe-ser-phe; phe-phe-ser; phe-phe-phe

25.53   The *primary structure* of a protein refers to the sequence of amino acids in the chain. Along any particular section of the protein chain the configuration may be helical, or it may be an open chain, or arranged in some other way. This is called the *secondary structure*. The overall shape of the protein molecule is determined by the way the segments of the protein chain fold together, or pack. The interactions which determine the overall shape are referred to as the *tertiary structure*.

Carbohydrates

25.55 Glucose exists in solution as a cyclic structure in which the aldehyde function on carbon 1 reacts with the OH group of carbon 5 to form what is called a hemiacetal, Figure 25.22. Carbon atom 1 carries an OH group in the hemiacetal form; in α-glucose this OH group is on the opposite side of the ring as the CH_2OH group on carbon atom 5. In the β (beta) form the OH group on carbon 1 is on the same side of the ring as the CH_2OH group on carbon 5.

The condensation product looks like this:

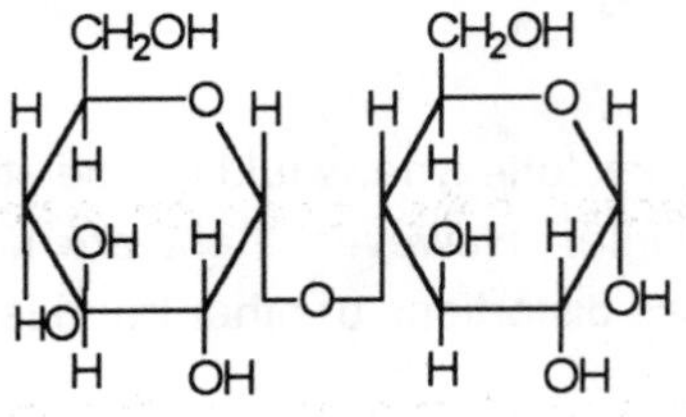

α-linkage 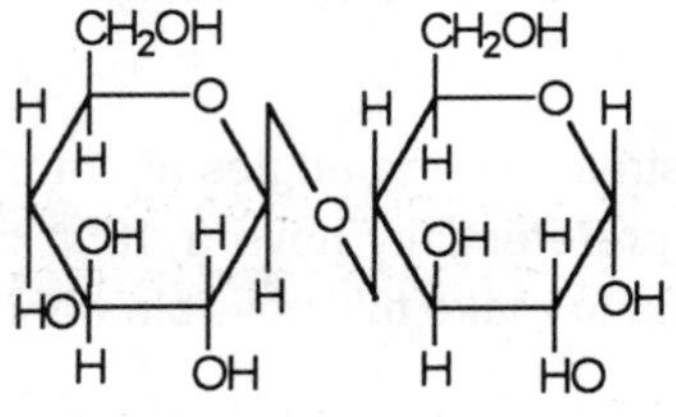   β-linkage

25.57 The structure is best deduced by comparing galactose with glucose, and inverting the configurations at the appropriate carbon atoms. Recall from Exercise 25.55 that both the β-form (shown here) and the α-form (OH on carbon 1 on the opposite side of ring as the CH_2OH on carbon 5) are possible.

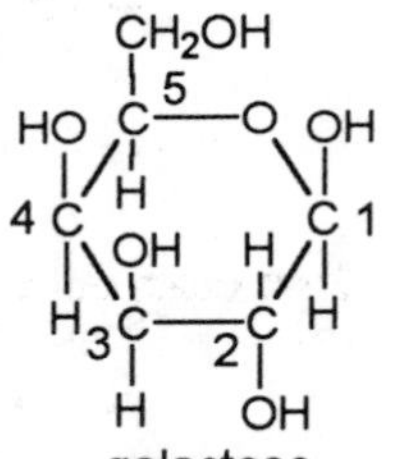

galactose

25.59 In the linear form of galactose shown in Exercise 25.57, carbon atoms 2, 3, 4 and 5 are chiral because they carry four different groups on each. In the ring form (see solution 25.57), carbon atoms 1, 2, 3, 4 and 5 are chiral.

Nucleic Acids

25.61 A *nucleotide* consists of a nitrogen-containing aromatic compound, a sugar in the furanose (5-membered) ring form, and a phosphoric acid group. The structure of deoxycytidine monophosphate is shown at right.

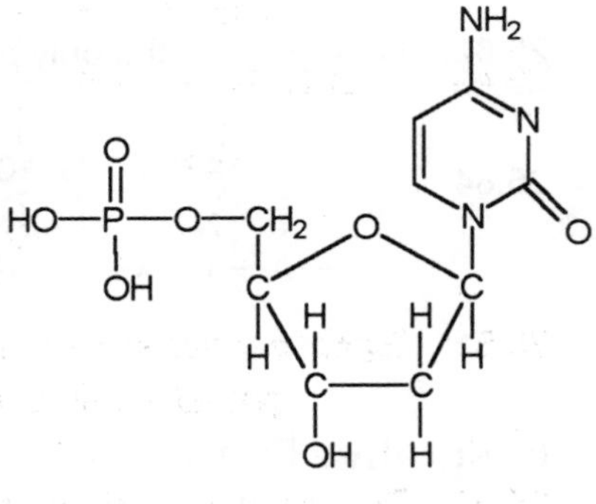

25.63 $C_4H_7O_3CH_2OH + H_3PO_4 \rightarrow C_4H_7O_3CH_2\text{-O-}PO_3H_2 + H_2O$

25.65

complementary strand

Additional Exercises

25.67 $CH_2{=}CH{-}CH{=}CH_2$

          $HC{\equiv}C{-}CH_2CH_3$ $CH_3{-}C{\equiv}C{-}CH_3$ $\begin{matrix} CH_2{-}CH_2 \\ | \quad\quad | \\ HC{=}CH \end{matrix}$

25.70 $\underset{\displaystyle CH_3CH_2CH_2}{\overset{\displaystyle OH}{|}}$ $\underset{\displaystyle CH_3CHCH_3}{\overset{\displaystyle OH}{|}}$ $CH_3{-}O{-}CH_2CH_3$

25.73   Because of the strain in bond angles about the ring, cyclohexyne would not be stable. The alkyne carbons preferentially have a 180° bond angle. However, there are not enough carbons in the ring to make this possible without gross distortions of other bond lengths and angles.

25.76   (a) $CH_3\overset{\displaystyle O}{\overset{\displaystyle ||}{C}}{-}OH$, C_6H_5OH (b) $C_6H_5\overset{\displaystyle O}{\overset{\displaystyle ||}{C}}{-}OH$, CH_3OH

25.79   All growth processes, which require synthesis of new cell materials, require energy. For example, in plants transpiration of water requires energy. In animals, the action of the involuntary muscles, such as the heart, maintenance of a uniform body temperature, synthesis of new cells, motions of limbs, and maintenance of correct body fluid compositions all require energy.

25.82   Glu-cys-gly is the only possible structure.

25.84

```
—T—A—T—G—C—A—
 :   :   :   :   :   :
—A—T—A—C—G—T—   ←— complementary strand
```

Integrative Exercises

25.85 CH_3CH_2OH $CH_3{-}O{-}CH_3$

        ethanol   dimethyl ether

    Ethanol contains -O-H bonds which form strong intermolecular hydrogen bonds, while dimethyl ether experiences only weak dipole-dipole and dispersion forces.

difluoromethane   tetrafluoromethane

CH_2F_2 is a polar molecule, while CF_4 is nonpolar. CH_2F_2 experiences dipole-dipole and dispersion forces, while CF_4 experiences only dispersion forces.

In both cases, stronger intermolecular forces lead to the higher boiling point.

25.87 Determine the empirical formula, molar mass and thus molecular formula of the compound. Confirm with physical data.

$$66.7 \text{ g C} \times \frac{1 \text{ mol C}}{12.01 \text{ g C}} = 5.554 \text{ mol C}; \ 5.554 / 1.388 = 4$$

$$11.2 \text{ g H} \times \frac{1 \text{ mol H}}{1.008 \text{ g H}} = 11.11 \text{ mol H}; \ 11.11 / 1.388 = 8$$

$$22.2 \text{ g O} \times \frac{1 \text{ mol O}}{16.00 \text{ g O}} = 1.388 \text{ mol O}; \ 1.388 / 1.388 = 1$$

The empirical formula is C_4H_8O. Using Equation 10.11 (MM = molar mass):

$$MM = \frac{(2.28 \text{ g/L}) (0.08206 \text{ L} \cdot \text{atm} / \text{mol} \cdot \text{K}) (373 \text{ K})}{0.970 \text{ atm}} = 71.9$$

The formula weight of C_4H_8O is 72, so the molecular formula is also C_4H_8O. Since the compound has a carbonyl group and cannot be oxidized to an acid, the only possibility is 2-butanone.

$$CH_3CCH_2CH_3$$ (with O double-bonded to second C)

The boiling point of 2-butanone is 79.6°C, confirming the identification.

25.89 The reaction is approximately:

$nC_6H_{12}O_6(aq) \rightarrow [C_6H_{11}O_5]_n(aq) + (n-1)H_2O(l)$

$\Delta G° = n(-662.3 \text{ kJ}) + (n-1)(-285.83 \text{ kJ}) - n(-917.2 \text{ kJ})$

$= n(-285.83 \text{ kJ} - 662.3 + 917.2 \text{ kJ}) + 285.83 \text{ kJ}$

$\Delta G° = n(-30.9) \text{ kJ} + 285.83 \text{ kJ}$